Praise for *This Won't Hurt*

'If you get a chance to read this – do! An eye-opening, brilliant work of passion.'

Naga Munchetty

'A vital subject that needs to be discussed.'

Katy Hessel, author of *The Story of Art Without Men*

'A brilliant book… There is so much to unlearn, there is so much that also follows in terms of how medicine could support – rather than fail – half the world's population.'

Helen Pankhurst CBE, author of *Deeds not Words*

'A hugely informative and quietly furious call to arms… Bigg argues convincingly that it is impossible to dissociate medicine and science from culture.'

Irish Times

'Marieke Bigg calls out lingering gender bias… She is balanced in her evidence analysis, forensic in her research.'

Telegraph, ★★★★

'Asking all the right questions about the treatment of women's bodies and more importantly, answering them. Punchy, fascinating and vital.'

Rachel Parris

'A valuable sociological perspective on women's bodies and health and an even more valuable (and optimistic) view of a better future for all.'

Gina Rippon, author of *The Gendered Brain*

Also by Marieke Bigg

Non-Fiction

No Such Thing as Normal: Disorders, Diagnoses and the Limits of Psychiatry

This Won't Hurt: How Medicine Fails Women

Fiction

A Scarab Where the Heart Should Be

Waiting for Ted

A History of Science in 21 Women

MARIEKE BIGG

ONEWORLD

A Oneworld Book

First published by Oneworld Publications Ltd in 2026

A CIP record for this title is available from the British Library

ISBN 978-1-83643-214-2
eISBN 978-1-83643-213-5

Typeset by Geethik Technologies
Printed and bound in Great Britain by Clays Ltd, Elcograf S.p.A.

The authorised representative in the EEA is eucomply OÜ, Pärnu mnt 139b–14, 11317 Tallinn, Estonia (email: hello@eucompliancepartner.com / phone: +33757690241)

Oneworld Publications Ltd
10 Bloomsbury Street
London WC1B 3SR
England

Contents

Introduction

In the Shadows of Giants

In 1675, Isaac Newton wrote to the natural philosopher Robert Hooke: 'If I have seen further than others, it is by standing on the shoulders of giants.' It's an adage I encountered, for the zillionth time, in an introductory seminar as I embarked on a PhD at the University of Cambridge. I was delving into the life and work of an underappreciated in vitro fertilisation (IVF) pioneer – a woman called Anne McLaren (who we'll meet later). And so, I suppose, I was already in a somewhat sceptical frame of mind when it came to accepted truths in science and who they applied to. But scepticism, as any scientist will tell you, is a good place to start.

The lecturers drew on this evocative metaphor to make an important point about academic referencing. We discover the truth, the phrase in this context implies, by building on previous discoveries. To make an *argument*, we have to learn from, and acknowledge, what has come before. Two important consequences follow. The first is that we avoid reinventing the wheel. We will, instead, make *progress*, by moving beyond what is known (the *sine qua non*, we are told, for any research project). The second is that we will give credit where credit is due. That is especially important for research scientists, who rely on the currency of references to build the value that translates into professional status. In short, there is power at stake.

Newton himself wrote in his letter to a man who he had actually fallen out with over a referencing issue (Newton had refused to give Hooke credit for an idea the natural philosophy giant believed to be his): 'what is done before many witnesses is seldom without some further concern than that for truth'.[1]

I was well aware, sitting there, that my professors weren't just introducing me to a system for building and valuing knowledge; they were handing me a key to credibility. This was about more than writing conventions: it was a ritual and a code that separated inside from outside. I was being initiated not just into a method but a social world.

The women in this book were not always so lucky.

In 1870, the American suffragette, activist and sociologist Matilda Joslyn Gage wrote a pamphlet titled, 'Woman as Inventor'. This was a time in America when inventing had burgeoned into a new exciting nexus of science and application. So had the idea that women couldn't do it. Gage, at the dawn of a women's movement, wouldn't let herself be fooled. 'Such statements are made lightly or ignorantly. Tradition, history and experience prove that women possess these abilities to the highest degree,' she wrote.[2]

History does indeed, but often we don't hear about it. The historian Margaret Rossiter came across Gage's pamphlet a century later. In 1993, she published the essay, 'The Matilda Effect in Science.' With a nod to Gage, she described a recurring pattern in the history of science: 'Recent work has brought to light so many historical and contemporary cases of women scientists who have been ignored, denied credit, or otherwise overlooked that a gender-linked phenomenon appears to be at play here.'[3]

In this book, I tell the stories of women scientists in a man-made science. They show us why women have been excluded and, more crucially, how they found ways to work with and against the social codes of their time to carve out space for themselves – in the shadows of giants.

We'll encounter plenty of examples of the Matilda effect at work. Some women were excluded from science's crowning achievements, for example, the Nobel Prize. Take Chien-Shiung Wu, who in 1956 performed an experiment that was even named after her and shattered our fundamental understanding of the physical world – the prize went to her two male colleagues. Or Rosalind Franklin, who in the same decade produced the essential evidence for the helical structure of DNA; the prize famously went to James Watson and Francis Crick and her former boss, Maurice Wilkins, in 1962. Or cases like that of Barbara McClintock, who did eventually get the prize for unearthing the dogma-shattering phenomenon of 'jumping genes' – thirty years after the discovery. Other scientists shared the prize, like Marie Curie, who shared the prize with her husband and the physicist Henri Becquerel in 1903 – but only after her husband advocated for her.

All of this is just the tip of a slippery iceberg. As the journalist Katie Hafner, executive producer of the brilliant *Lost Women of Science* podcast, puts it: 'It's about not being named in a study; to be just an asterisk or a footnote.'[4] It is about that very system for crediting and credibility described to me at Cambridge. There are plenty of women, in this book and many more lost to us, who don't get so much as a footnote. Take the American inventor and scientist of Gage's era, Eunice Foote, who discovered the greenhouse effect but barely managed to have her studies published, let alone recognised. Somehow, even with the pudding to prove their skill and intellect, these women could not find their way into the rituals that ordained great scientists.

But, of course, it has nothing to do with their abilities. The Matilda effect describes a problem in the culture of science.

Referencing is just the symptom. Women were long excluded from the education and opportunities they needed to pursue a career in science. When they did find a role in science, it was often as an assistant, or some other lower-status role, that meant that their work wasn't recognised – Florence Nightingale, who worked below the medical men as a nurse, organising, cleaning and performing the statistical calculations to prevent deaths. Virginia Johnson worked with the doctor William Masters, who had the medical training she didn't; she brought equal scientific skill and also the crucial people skills that made their experiments on couples having sex a reality. Or Katherine Johnson, who was called nothing more than a 'human computer', while she did the manual calculations that sent the first astronauts to the moon. Often, women's informal training or outsider status allowed them to see things their male colleagues didn't: Nightingale saw how sanitation was driving soldiers' deaths in the Crimean War; Jane Goodall, against every belief in the field, knew that chimpanzees created meaning just as humans do. Often, women's way of working in science was, by necessity, more collaborative, more cautious and less driven by the claims to fame that would never be theirs anyway: we have the examples in this book of Anne McLaren, an embryologist who built sprawling networks of collaborations that ushered in an era of IVF and stem cell research, or Frances Kelsey, whose scientific rigour in the face of a pharmaceutical boys' club prevented thalidomide from causing more infant deaths in the US, or Rosalind Franklin, who worked quietly and effectively on the experiments that substantiated the theories of the DNA giants. Their stories didn't fit the narrative of individual giants lifting each other up because no one did lift these women. And yet, they show us a neglected, crucial dimension to how science is, and can be, done.

Today, we are still dealing with this legacy of exclusion. Nobel Prize winners are still predominantly white, male and economically secure. Men are also still cited more than women colleagues. One study highlighting the citation gap put this down to an unconscious bias. Science communication studies have shown that the gender of an author, as well as the gendered stereotypes associated with their area of research, *to this day* impact the perceived scientific quality of their work.[5] Researchers have described this using 'role incongruity theory'; people perceive men, and male-coded skills, as more effective in science because it fits the image that a history of Nobel giants has given us.[6]

The false narrative discourages women from pursuing scientific careers. In 2022, only thirty-one percent of students enrolled for science, technology, engineering and maths (STEM) subjects in the UK were women. The schisms here are the biggest for computer science, where only twenty-three percent are women, and engineering, at twenty-one percent. These disparities are reflected in the STEM workforce, where women make up only twenty-six percent – twelve percent in engineering, and nineteen percent in IT professions. The most contemporary scientist in this book, Timnit Gebru, specialising in artificial intelligence, still works in a field overwhelmingly dominated by men.

We have a self-perpetuating cycle that tells us men are better scientists, discouraging women from pursuing scientific careers, further compounding the false narrative. It is time for a new scientific story. And maybe a new system for crediting them. As Nightingale wrote in her 1852 proto-feminist essay *Cassandra*: 'Why have women passion, intellect, moral activity – these three – and a place in society where no one of the three can be exercised?' The culture of science needs to change.

During my time at Cambridge, I came to prefer my supervisor, the sociologist Sarah Franklin's, metaphor for science.

Research, she said, is a *conversation.* We don't passively inherit knowledge; people negotiate what becomes accepted in the present. A balanced conversation depends on robust relationships. It requires equal airtime; it requires a structure to ensure everyone has their say; and it requires time to speak and to listen. The conversation puts the question of power at the centre. 'Science says', the qualifier for any authoritative statement we read today, becomes 'scientists say'. Who, we ask, is speaking?

I like this metaphor because it gives us a way forward.

The stereotype of a scientist is not necessarily a great conversationalist. The giants we learn about were mavericks and loners or individual adventurers. Their discoveries usually came at the cost of their competitors in the race to be the man on top. The best man, supposedly, always wins. The stories in this book tell us a different story. Some of these women were denied publication or credit because they didn't have the right connections. They didn't get a seat at the table – an equal shot at the race. Social ties, not inherent genius, determine which scientists are heard and remembered. Newton could advance his ideas because he was a valued interlocutor. Men talking to men. Standing on the shoulders of giants. Conversations are shaped by power.

When we envision science as a conversation, we can better understand why women have been excluded from it. Even more importantly, though, when we spend time with these women, and let them speak, we find our priorities in science begin to shift. Progress doesn't look the same: less like a competition and more a collaboration. The great feats that have defined our story of scientific progress, often tied to war, like the atomic bomb, or large-scale pharmaceutical production, are just part of the truth. The violent, often martial metaphors of 'the next frontier', of 'breakthrough', of 'races' and of 'war on disease' don't describe the work many of these women did, or the realities they inhabited.

Science becomes less vertical. No longer the human pyramid of great men but more of a landscape. Not just the space rockets of NASA scientists but the working conditions of data coders; no longer the next breakthrough in genetic engineering, but perhaps the meticulous work of ensuring that healthcare technologies are accessible and sustainably made. A science of giants becomes a science for earthlings. With a new set of tools and a whole new vision, we find there are other truths we can, and other ways we must, come to know. Telling the story of women in science is nothing less than the work of cultivating a new reality.

I

Hypatia (c. 350~370–415 CE)

Mathematician, Astronomer and Philosopher

We begin in the city of Alexandria. Nestled between the Mediterranean and the Nile Delta's western edge, the Byzantine city stands as a shining intellectual beacon and meeting place – quite literally, in fact. A lighthouse, over 100 metres tall, dominates the harbour, its beams visible for twenty-nine miles. This is the birthplace of the Mouseion of Alexandria – a forerunner to the modern university – and home to the immense Library of Alexandria. And yet the city is in turmoil.

The year is 414 CE.

The city has long been a volatile experiment in religious coexistence. Christians are now the dominant group, but they are shot through by sectarian divides. Jews, Gnostics and pagans also form sizeable communities. Violence has long erupted between and among these groups, with the city's intellectual heritage often falling in the crossfire. Many of the great libraries of the Mouseion had been successively burned to the ground.[1]

In 414 CE, any tentative equilibrium has broken once again. Five hundred monks from the monasteries of Nitria have taken to the streets on behalf of the recently acceded bishop, Cyril, to attack the city's governor, Orestes, a Christian deemed too tolerant of other religions. In the midst of the protest, a rock flies at the governor's head and barely misses. Orestes decides it is time for diplomacy.[2]

He needs to build a coalition and he needs someone to lead it. Someone respected, level-headed and adept at this kind of peace brokering.

He chooses a woman to help him. Not a public official, because women didn't take those roles, but a philosopher, astronomer and mathematician. Her name is Hypatia.

*

Who was Hypatia? How did she transcend the limitations put on her gender at the time to become not just a great scientist of her day but, even more unusually, a public figure? And why don't we know more about her?

Hypatia was born in Alexandria around the year 355 CE to an upper-class family.[3] Her father, Theon, was an important philosopher and mathematician. We don't know anything about her mother. What we know about Hypatia's life comes from the little that men have written about her: her adoring student, Synesius, for one, and an occasional mention in the histories of ancient scholars, like the Christian historian Socrates Scholasticus, who wrote about her in his *Ecclesiastical History*, about twenty-five years after her death.

Theon taught Hypatia as if she'd been his son.[4] This meant learning grammar first, then moving on to mathematics, and, finally, the apex of any elite Roman education: philosophy. Most women of Hypatia's class were educated to prepare them for a life as a respectable mother, speaking some literary Greek, but their education was generally aborted unceremoniously after the grammar stage, when they married in their teens. But Theon was unusually committed to his daughter's education,[5] educating her either alongside the boys in the classes at his school or through private tuition. And she advanced quickly.[6] According to historian Philostorgius, writing a decade or so later, Hypatia 'was so well educated in mathematics by her father, that she far surpassed her teacher'.[7]

Hypatia wasn't the only woman to make this ascent. Female mathematicians and philosophers were rare but not unheard of.[8] One seventeenth-century scholar compiled a list of sixty-five women philosophers who were actually referenced in the writings of scholars throughout antiquity.[9] These include Aspasia (c. 470 to c. 400 BCE), whose intellectual salons attracted the likes of Socrates, or the trouser-wearing Hipparchia (b. c. 325 BCE) who had an egalitarian marriage and penned several philosophical works, none of which survived.[10] Hypatia herself had four female contemporaries who were trained philosophers and taught or took up a public role,[11] though none of them held the influence she did. Pandrosion of Alexandria was a mathematician and teacher like Hypatia, though left no historical record and is only remembered through the counterarguments to her maths in the work of male contemporaries. Sosipatra of Pergamum was the daughter of a biographer and benefited from being memorialised, but she only taught her teacher's inner circle from within her home and never stepped into a public role.[12] Asclepigenia of Athens was the daughter of Hypatia's younger rival, the Athenian philosopher Plutarch. Like Hypatia, Asclepigenia was the daughter of an accomplished philosopher; like Sosipatra, she taught only a select subgroup of her teacher's students to supplement the standard curriculum of male teachers. All we know of the final woman, known to us as 'the wife of Maximus of Ephesus' from the ancient Greek historian Eunapius's *Lives of Sophists*, is that her philosophical conduct was superior to her husband's. While Maximus succumbed to fame and status, his wife stayed true to her philosophical virtues and eventually took her life to protect the profession. It was possible for women to rise to the status of philosopher (though not without risk), but Hypatia was the only one to act independently and to do so fully in the public eye.[13]

Hypatia was unusual for her class but even more unusual when we take stock of the rest of her city's population. Most Alexandrian women became mothers – they were expected to deliver at least six children in their lifetime and faced a two percent chance of dying with each birth.[14] Their children rarely survived and, with a population of around half a million crammed into ten square miles (often inside the world's first apartment blocks), people frequently died from the city's atrocious sanitation, accidents and overcrowding. Around three percent of the city's people succumbed every year. Alexandria relied on immigration to maintain its sizeable population. Most were labourers who worked the ports, though they were helped by the enslaved people who made up about twenty percent of the Eastern Roman Empire's population (and that may have been higher in Alexandria, which was a key hub of the slave trade). Besides childbirth, most women did the manual labour deemed appropriate for them, spending their days weaving baskets, in other textile trades or in domestic service. By contrast, Hypatia enjoyed a life of intellectual inquiry. Hypatia was a member of an elite, which gave her access to an education, opportunity and social status. This, as well as her gender, shaped the options available. As is the case for many of the women in this book, however exceptional they may have been as individuals, some aspect of privilege usually helped them along in some way.[15]

*

Hypatia first established herself as a mathematician. At the time, this meant proving your skill by producing commentaries. These were edited versions of mathematical texts that made them easier to understand and that kept them relevant for contemporary scholars: a bit like publishing a new edition.[16] While today we might value revision less than a new discovery, in Hypatia's time editing was a highly valued academic contribution.

Hypatia worked on several texts, including Diophantus' *Arithmetica*, the first-ever text to use symbols for unknown quantities and the same way we use *x* and *y* in algebra today.[17] She also revised Apollonius' *Conics*, which, perhaps less excitingly on the face of it, described the shapes that arise when a plane intersects with a cone.[18] Both these texts were attempts to represent what we would today call higher-order equations, or equations that encompass multiple changes. Diophantus did so using algebra; Apollonius used geometry.[19] By studying both texts, Hypatia joined a new tide of mathematicians who, following Diophantus, now studied algebra as well as geometry. She also absorbed the work of the Alexandrian mathematician Pappus. Pappus was one of the first to use algebraic rather than geometric proofs. Theon had edited Book II of the *Elements* by Euclid (Alexandria's first professor).[20] Hypatia likely used similar methods in her work on *Conics*.

Hypatia joined a lineage of Alexandrian mathematicians who were developing an expanding toolkit for representing complex mathematical problems. Hypatia was, in fact, the last inheritor of their advances; the Alexandrian tradition died with her. And so, her touch is all over the versions of these texts available to us today in her work of refining and communicating ideas that didn't really resurface until the sixteenth century in the work of mathematicians like François Viète, the so-called father of modern algebraic notation, who built on Greek geometry.[21] The mathematician Pierre de Fermat also studied Apollonius and Pappus to develop his analytic geometry.[22] Their iterations of the concepts in these texts in turn gave the astronomer Johannes Kepler the notation he needed to prove that planets travel in elliptical orbits around the sun.[23] Would Hypatia have foreseen any of this?

Probably not. Hypatia's best known commentary, in fact, argued exactly the opposite! She revised her father's work on the mathematician and astronomer Ptolemy's *Almagest*.[24] Book III of *Almagest* outlined the Earth-centric model of the universe and Hypatia likely worked on Books III–XIII, although not all scholars agree on this because – as is a common refrain when women's contributions are in question – the references are missing. However, scholars do generally agree that she worked on Book III to improve on Ptolemy's long division method of calculating how far the sun moved in relation to the Earth each day.[25]

While Hypatia's conclusions may have turned out to be incorrect, they reflected the state of knowledge at the time. Hypatia improved on the maths as it stood and proved an adept translator. Her version of *Almagest* was far more accessible than previous editions, offering a novice a comprehensible introduction to Ptolemaic astronomy without compromising depth. It proved especially useful for teachers trying to communicate the state-of-the-art research on the solar system to their students.[26] Numbers were seen as a higher form of truth: stable entities compared to the ephemeral physical world. Refining numbers meant moving closer to the philosopher's version of divinity.[27] Hypatia was skilled at this work of communicating the science of her day on the page and, as we'll see, in the real world too.

Hypatia's commentaries prepared her for the work that came to define her career: teaching. Theon retired much younger than his contemporaries and, again, most unusually, left his daughter to lead his school. Women became mathematicians in ancient

Greece, but they did not become intellectual leaders. Hypatia didn't just inherit a tradition but was determined to revise it.

Mathematics held a unique place in Hypatia's original epistemology. In antiquity, there was no maths without philosophy; the insights from these two disciplines informed one another.[28] There was much debate among fourth-century intellectual elites about which came first in the hierarchy.[29] Hypatia's father, Theon, placed more emphasis on maths, which reflected the dominant outlook in Alexandria at the time. He believed that numbers and mathematical study offered a higher order of insight into reality than philosophy.[30] Hypatia's education reflected his outlook: maths came first.[31]

Though, as she moved through her education, Hypatia became convinced of the opposite view, that maths was just a gateway to philosophical insight.[32] Once in charge of the school, Hypatia structured the curriculum to reflect her own vision. Hypatia's school still taught maths at the highest level, but philosophy was unapologetically the apex of its training. This wasn't a teenage rebellion but a well-considered choice. Hypatia's commentaries gave her all the credibility she needed. Her demonstrated ability made it plausible that she'd transcended one body of knowledge to access a higher realm.[33]

People believed her. Hypatia was the first woman in the Eastern Roman Empire to really achieve the status of a revered intellectual.[34] Even a generation after her death, Socrates Scholasticus, a well-known ecclesiastical historian, described her as a popular teacher, who attracted 'people from everywhere who wished to philosophise'.[35] With this support, she was able to do more than supplant her father's approach; she influenced the status of philosophy across Alexandria. When Hypatia began her education, Theon's view was and had long been the dominant one in the city.[36] After Hypatia, no Alexandrian teacher would ever posit maths as anything but a foundation for Platonic philosophy.[37] Hypatia defined the intellectual approach in Alexandria of her time and, in doing so, also boosted the status of philosophy throughout the Empire.

Hypatia went against the grain of thinking and was rewarded for it. She was able to do all this by the age of thirty-five and in a political landscape that saw other philosophers censored, attacked or worse. Sectarian strife was mounting in the city. Once again, the Christian government had turned on pagans, a term they used to encompass a whole range of religious beliefs and practices, ranging from those who revered the Olympian deities to various schools of philosophy. Legislation now banned pagan sacrifices and sanctioned the violent squashing of protests. Many Alexandrian philosophers, including one prominent teacher, Olympus, had fled the city. Not just this, but education in the Eastern Roman Empire was increasingly centralised compared to previous centuries. It was more difficult than ever to establish a private school, and this was true for men, let alone for women.[38] Yet Hypatia's school thrived. How was that possible?

Part of the answer, at least, lies in how Hypatia's particular philosophy fitted into the political climate. There were lots of different beliefs floating around, but, outside of Alexandria, tides were generally turning towards forms of Platonic philosophy that used spiritual practices to achieve unification with the divine.[39] Hypatia was a pagan, but she wasn't interested in their rituals. Instead, she promoted the view that philosophers could achieve unity with God through contemplation.[40] This blueprint had wide appeal in an increasingly Christian Alexandria. Christians could follow her teachings and even call themselves philosophers without compromising their religious affiliations; they just had to read their god into the model of unity with the divine.[41] For pagans, it was a purely philosophical system of thought. Whatever their religion, Hypatia instructed her students that philosophy had to be the primary aspect of their identity. Wisdom, justice, temperance; in order to glimpse the divine, students had to organise their lives according to *philosophical*, not religious, virtues.[42] In a city splintered by supposedly irreconcilable views, Hypatia built a unifying middle ground.

Outside Alexandria, Hypatia's approach, still oriented around divinity, was perhaps starting to look old-fashioned; in Alexandria, it worked. The Christian majority not only accepted but welcomed Hypatia's measured approach.[43] Hypatia firmly believed that philosophers needed to play a public role in city life, so coached her students to model peaceful coexistence.[44] Her students remained loyal to her and carried their lessons into their lives. We know this from the few surviving letters written to Hypatia, particularly those by her student Synesius. The content of his letters 'To the Philosopher' reflects an abhorrence for violence and inequity. In one evocative passage he laments the state of his city. He is, he says, 'disgusted with her [Alexandria]' and by the 'men slaughtered like victims on an altar'.[45] Synesius remained devoted to Hypatia as a teacher after leaving her school. He continued to write to her for advice on publication or scientific instruments even after she mysteriously stopped responding to his letters. This is a profound loss to him, worse, he writes, than the many in his life: 'I have lost my children, my friends, and the goodwill of everyone. The greatest loss of all, however, is the absence of your divine spirit.'[46]

With her non-partisan approach, Hypatia secured a loyal following and, as her school continued to model pagan and Christian harmony in the world, increasingly, a *public* status that was practically unheard of for a woman.[47] That visibility came with its challenges. Male philosophers were held to pretty stringent standards. They were supposed to be infallibly honest, encouraged to speak even more freely when advising rulers because, the reasoning went, their commitment to philosophy would always lead them to speak the truth.[48] They were to be impervious to power, wealth or even the threat of punishment if their truth antagonised their leaders. There were, predictably, plenty of recorded cases of philosophers who fell short of these godly criteria and succumbed to mortal temptations, like Maximus of Ephesus, whose wife we briefly met.[49] They (and their wives) experienced cruelly the risk accompanying a public role: criticism, public shame or, at worst, violent punishment when their views fell out of favour.[50]

Philosophers were not just evaluated by their teachings but by how they *lived* their ideals. And so, they often made a display of their virtue. Hypatia gave a distinctly female performance. She had become a renowned philosopher, but she also knew that, in the public eye, she remained first and foremost a woman.[51] She was reminded of this each time she was belittled and mocked while giving public lectures or when walking through the city in her tribon, a Spartan cloak favoured by Cynic philosophers and usually worn by men.[52] She made it known that she was a committed virgin. This was important. Her teaching brought her in close contact with students – predominantly men, of course – and her celibacy helped to dispel any rumours that might undermine her and the school's reputation. Making her sex life (or lack thereof) public was the price Hypatia paid to participate in intellectual life. She was almost as dedicated to maintaining her celibate image as the men who wrote about her after her death.[53] Apparently, she did it with verve. In one account, she carried a rag soaked in menstrual blood to a besotted student, presenting it to him as a reminder of the impurity of his bodily desires towards her compared to the divine intellectual connection pupils and their teachers shared.[54] That's one way to tell a man 'no'.

Hypatia understood that world and navigated it with a pragmatism that, as we'll see, was necessary for many of the women in this book. And her proto-PR campaign worked. During a quieter period in Alexandria's history in the 410s, Hypatia established herself in a new status quo.[55] The city council and imperial governors retained control over political affairs; the archbishop, Theophilus, managed the largest Christian group in the city and Hypatia's circle continued to model inter-religious harmony.[56] Hypatia had crafted an entire philosophy and carefully managed her public image to allow her to play this peace-keeping function in civic life – as a mere philosopher in a Christian world, and a woman in a man's.

By the time Orestes approached her amidst the rising tension with Cyril, Hypatia and her students had exemplified pagan–Christian cooperation for years.[57] She seemed like the ideal choice as the leader of a coalition: a true public philosopher, widely respected for her measured voice and who had, characteristically, not taken sides in the simmering conflict.

Hypatia and Orestes began to meet regularly at his home. Eventually, mixed groups of Christian and pagan elites joined them. They discussed how to manage the city's tensions and mitigate further violence. Simultaneously, Orestes and many other Christian elites stopped attending services that Cyril was likely to attend to protect themselves, very reasonably, from potential attack. But these acts of self-preservation looked suspicious to Cyril and his followers. Cyril's supporters suspected that Orestes and Hypatia were plotting against Cyril.[58] In the end, Hypatia couldn't escape the risks facing her gender. There was a ready script for dispelling women from society.

They accused Hypatia of being a witch.[59]

A group gathered to confront Hypatia in 415 CE. It is unlikely that they set out to kill her. Alexandria had a reputation as a rough city, but mobs – while often used to intimidate – only killed prominent people about once every fifty years between the reigns of Constantine and Justinian (still not great, but not far off today's statistics).[60]

Nonetheless, the mob *did* kill her. The story of her violent murder is well known. They found her somewhere in public, perhaps while she was teaching, or on her way home in her carriage. Either way, they managed to seize her, then drag her through the street, to a nearby church, where she was stripped naked and slashed to death with broken roof tiles. They burned her remains.[61] The style of murder may strike us as a relic of the far-gone past, but women scientists are targeted to this day. The molecular biologist Professor Suzanne Eaton was murdered by a stranger while attending a conference in Crete as recently as 2019. The geneticist Dr Brenda Page was killed in 1978 by her ex-husband, who abused her and accused her of stealing his research.[62] He wasn't convicted for forty-five years. Women, including scientists, are still not safe in public spaces, nor are their murderers always held accountable.

The message was clear. The murder followed the city's script for the purifying the city of pagan spirits.[63] These Alexandrians had classified Hypatia a witch, and so, this was the appropriate course of action. To the rest of the empire, though, it was a horrific assault on their cultural ideals. Hypatia had embodied the role of an unimpeachable, truth-telling philosopher perfectly.[64] She hadn't been appropriately punished; she'd been humiliated. The optics demanded that the emperor order an investigation into her murder. But Cyril only had to bribe the emperor's officials to avoid repercussions.[65]

Besides, the damage to Hypatia's school had already been done. She had left no heir. Without her leadership, her community crumbled and so did the role of philosophers in mediating conflict.[66] She was one of the last philosophers in the Christian Roman Empire to collaborate with government. The stories told about her reflected the change.[67] Hypatia's murder, and the lack of adequate sanctions, fuelled pagan philosophers' distrust of Christian leaders. For the rest of antiquity, collaboration with the clergy was enough to destroy a philosopher's reputation.[68] While Hypatia had dedicated her life to pagan–Christian philosophical consensus, as her former students moved to Athens and recounted her story she morphed into a pagan martyr for philosophy.[69] She became one in a new litany of tales of intellectuals who had resisted Christian prosecution or had attempted assassinations of their leaders.[70] Paganism became the meaning of her life.

Hypatia's death was just the beginning of the life of her misconstrued legacy. The authors in the generation after her who wrote about her death cast her in their histories of Cyril, the Alexandrian church and the onset of a period of social and political upheaval.[71] Hypatia's story was subsumed by a history written by and about men like Cyril and Orestes. Men she'd had nothing much to do with for most of her life. She wasn't an agent in this story; she was a foil for men's stories: a martyr to some, a heretic to others. In all cases, her life's important work was almost totally erased.

The retellings of Hypatia's story since antiquity have cast her in new roles. From the eighteenth century, texts critiquing the religious factionalism of its own day,[72] retellings casting her in the role of the Christian Saint Catherine (underscoring the persecution of Christians, rather than pagans),[73] and condemnations of the Church in the writings of Voltaire to an exemplar of so-called 'manful Christianity' written by the clergyman Charles Kingsley in a Victorian novel,[74] Hypatia's life was repeatedly engulfed by an apparently more important discussion about who sat on the right side of history, often dotted with references to her physical beauty.[75]

Even accounts that could read as somewhat feminist reduce her to a political symbol. In the 2009 Spanish film *Agora*, she spearheads the cause of science, facing multiple gendered barriers. The greatest in the end, though, is the religious extremism of the mob that kills her. The hooded and bearded monks resemble the Taliban.[76] Hypatia finds herself on the right side of history, but still once again appropriated, this time as a symbol of the emancipated white West.

In the end, Hypatia couldn't escape the dangers of being a public woman. She was obliterated according to a specifically gendered script of her time and, again, all too easily in the histories that already reduced women to side characters or whitewashed them entirely.

These histories emphasise her death and consolidate her erasure. None of them show us what was most impressive about Hypatia: that for a while, she established herself as a trusted public scientist and one of the ancient world's great luminaries.

Hypatia took advantage of the access that was available to her as a member of the elite to position herself in the public sphere, despite being a woman. She used her academic work but also her skills of teaching and communication to shape an approach to thinking that left a lasting impact on scholarship. She starts to show us that the necessary skills in science include not only those of reasoning or an ability to construct experiments but also the social skills of building alliances to share ideas. Hypatia is only remembered because she inserted herself into the lineage of male scientists – sometimes through the careful work of revision and in other ways by engaging people across religious divides in a new, shared conversation. This was part of the work of constructing a new scientific worldview, and it is also why her fingerprints weren't entirely erased. Hypatia exemplifies many of the feminised skills of science we'll see written in the margins of men's histories of science throughout this book, excavated by historians through tangential comments and footnotes in men's histories, as we begin to understand the conditions for moving to the centre of the page.

2

Margaret Cavendish (1623–1673)

Natural Philosopher

One spring day in 1667, Margaret Cavendish rattled through the streets around Hyde Park in her black coach. The path leading to the Royal Society, seat of the country's most eminent scientists, was crammed with people, all jostling for a better view. Boys and girls chased after her elegant carriage, just to catch a glimpse of the duchess. 'So many people may come to see her', according to one diarist, 'as if it were the Queen of Sweden.'[1] Emerging in a flowing gown and a man's hat, this was the day she would make history – the first woman ever to be invited to the Royal Society.[2]

*

Cavendish aspired to the status of the ancients. On the cover of her books, she appears part aristocrat, part Greek statue.[3] The ancient roots of Hypatia's culture had found a new and unlikely home in soggy seventeenth-century England. Surrounded by the ancient Roman ruins of her native Colchester, Cavendish was in awe of the portraits of Julius Caesar that hung in her family estate and the autobiographies detailing his military conquests. She was equally impressed by the literary life of Ovid, the philosophy of the Stoics and, indeed, astronomical tools, like telescopes, and even medical theories that had barely changed since Hypatia's day.[4]

The grandiose chimera of classical and modern that burnishes her writing captures her well. Cavendish was an eclectic meeting of parts. Some should have been impossible to reconcile. Was she scientist or poet? Cavendish published twenty volumes of plays, poems, essays and satires exploring topics that ranged from science (called 'natural philosophy' in her day, which encompassed physics, biology and chemistry, as well as mathematics and metaphysics) to animal rights, gender and etiquette – to name but a few.[5] Was she a performer or a serious intellectual? She dressed extravagantly, as if she wanted to be a spectacle, but she was also the only British woman to publish her own serious natural philosophy in the seventeenth century. Throughout her writing she boldly rejects the dominant scientific ideas of her day, choosing intuition and reason over the empiricism and experimentation that would transform the field into its modern form.[6] She wanted greater participation in science and yet also believed that monarchy was the natural social order. She was one of the few women to publish under their own name, the first to be invited to the fellowship of eminent scientists at the Royal Society; she defied conventions and broke gender barriers. And yet, alongside all this, she supported a classist, feudal status quo. All of this she assembled into a performance that left people trembling in her wake and gasping for more.

As a woman in a man's sphere, of course, Cavendish didn't have a choice but to be a contradiction. And maybe it was her strangeness that allowed her to transcend the expectations of her gender. Despite the sexist gossip about her appearance or the state of her mind, people kept looking. Cavendish's dizzying performance distracted her audience, even temporarily, from the limits they imposed on her half of the population, allowing them to glimpse the multitudes that the concept of woman could not contain.

*

Cavendish was born Margaret Lucas in 1623 to Sir Thomas Lucas and Elizabeth Leighton. Her father died when she was just an infant, leaving his land in Essex for her mother to manage. Organised and economising, Elizabeth set a strong example for her daughter as a capable woman in command. As a widow, her social status was complicated. The loss of a husband gave a woman unprecedented legal autonomy to own property but also exposed her to social suspicion and financial precarity. Elizabeth removed the family from Essex society and relocated to their country estate of St John's Abbey in Colchester. The youngest of eight, Cavendish enjoyed roaming the grounds with her older siblings. As the others grew up and married, though, Cavendish became increasingly isolated. This sparked an existential anxiety. She once stood outside one of her sister's bedroom doors for hours, to make sure she was still breathing.[7]

In her loneliness, Cavendish found company in her writing. While her education was uninspiring, centred on transforming her into an eligible aristocratic lady, Cavendish relished the freedom to explore her inner world.

And there was plenty of need for escapism, with England's political climate growing tenser by the day. By the time Cavendish reached her twenties, civil unrest was spreading through the country. Criticism of King Charles I and his policies had been mounting. After eleven years of rule, he was forced to call for parliament to fund the Bishops' War in Scotland, which had erupted after he tried to reform the Scottish Church. His years as a sole ruler came to a definitive end in 1642, after he tried to arrest some especially critical MPs. The clashes escalated into fully fledged civil war. Raging for just under a decade, the English Civil War would claim up to 200,000 lives, or around 4.5% of England's population. Charles was executed in 1649.[8]

Cavendish's aristocratic family supported the king. This became problematic in their hometown of Colchester, deep in the Parliamentarian east, where an increasingly Puritan population saw the civil war as an opportunity to reform the Church. One day in 1642, the local villagers seized the family's land and robbed their house. Cavendish and her mother were marched through the streets and thrown into jail. Six years later, there was a repeat ransacking of the house. This time, the mob cut the hair off the heads of Cavendish's recently deceased mother and sister and wore it as wigs.[9]

Cavendish herself managed to escape the country. She became a lady-in-waiting to Queen Henrietta Maria, the wife of King Charles. Henrietta Maria was living in Oxford with the exiled Royalists after months of riding around the country to supervise the delivery of weapons.[10] Living with the queen may have saved her life, but it did little to alleviate Margaret's isolation. Court was another form of hell to her. She hated the gossip and the small talk. She had near to no time to write. In a play she later wrote of the time, *The Presence*, her character 'Lady Bashful' never speaks.[11]

In the summer of 1644, tensions were rising, and the queen decided to leave England for France. As a lady-in-waiting, Cavendish had to go with her. The move isolated her further in another culture far away from home. Until she met William. Already a marquis and fated to become the first Duke of Newcastle, William was now destitute after defeat by the

Parliamentarians. But he had once been a playwright and a patron to the likes of Ben Jonson and Thomas Hobbes. During his exile, he continued to hobnob with philosophers René Descartes and Pierre Gassendi. Cavendish and William fell in love. He wrote her an impressive seventy poems in the lead-up to their wedding, as well as letters.[12] She was cautious to avoid a 'scandal of reputation', but so unusually for this time, their relationship was built on reciprocal affection.[13] They married in 1645 and later moved to Antwerp, where they lived for ten years while the Protectorate of Oliver Cromwell ruled England.[14] After the Restoration of Charles II in 1660, the couple returned to England, and to William's estates. Order had been restored for the two destitute Royalists.

By marrying William, Margaret had chosen a man with significant intellectual connections. Not just this, but William supported her writing. This might explain how she achieved a feat few women of her time did: publication. This started with poetry philosophical in tone, in the 1650s:

> When *Nature* first this World she did create,
> She cal'd a Counsell how the same might make;
> *Motion* was first, who had a subtle wit,
> And then came *Life*, and *Forme*, and *Matter* fit.[15]

In these playful couplets, Cavendish has personified the elements of what would become a fully fledged natural philosophy: nature, motion, life and matter. What is life made of? From the atomic level to emotions, from love to wind, or the bough of a tree. What makes up this matter? How does it move or change? She was also committed to exploring a question that scientists have veered away from for generations since: do all of these levels have consciousness? A rock? A mountain? A colour? A thought? How does this awareness manifest in the natural world?[16]

Cavendish tackled these ambitious questions in the 300 or so poems she went on to write, and then in a series of philosophical treatises, where she set her ideas out more systematically: *Philosophical and Physical Opinions* (1663), *Philosophical Letters* (1664), *Observations upon Experimental Philosophy* (1666) and *Grounds of Natural Philosophy* (1668). While Cavendish threw her hat in the ring as a serious intellectual, she continued to experiment with genre. She published, for example, what is sometimes regarded as an early prototype of a science fiction novel (although it is so much more than that), *The Description of a New World, Called the Blazing World*, the same year as *Observations*. As an author who also works across genres, I have experienced how different modes of writing don't just help communicate ideas but also develop them; Cavendish was working voraciously in different forms to grasp at a truth she saw. That truth was specific and defined. While the work was wide-ranging, it was all connected by a single developing theory of the nature of reality.[17]

Cavendish was writing during a paradigm shift. Science, as we know it today, was taking shape. The new theory of reality, advocated by philosophers like Thomas Hobbes or Robert Boyle, rejected previous ideas from antiquity that ascribed purpose and intention to nature and instead saw the universe as entirely material, consisting only of parts in motion and the forces between them. They saw the universe as machine. Matter, in this view, had no value or purpose, but obeyed physical laws of motion and collision.[18]

Cavendish had no problem with materialism. In all her work, starting with her first philosophical treatise, she argued that nature is one infinite but entirely material thing. She rejected the idea of supernatural powers or incorporeal beings. 'Nature is material, or corporeal, and so are all her Creatures, and whatsoever is not material is no part of Nature, neither doth it belong any ways to Nature,' she wrote in 1664.[19] Pretty unequivocal.

On the other hand, mechanism – the universe as an inanimate machine – she couldn't get on board with. There were contradictions in the world she observed that it couldn't explain. Looking around, she observed an abundance of different creatures and things, all with unique functions and characteristics – a bird, a tree and a stone are clearly and immediately distinct. At the same time, there was a level of consistency to this multitude; when we study a particular animal, we might, for example, be able to predict its response to a stimulus. A dog, a worm, even a plant will move towards water. If the world was made of uniform matter, transferring motion according to generalisable laws between parts of the 'machine', she reasoned, the universe would be either entirely consistent, or diverse, because it would be connected only by sameness. But it was both.[20]

Cavendish developed a different theory of motion to explain the tension she saw. Different parts move in different ways. Motion can't just transfer between entities, the way the mechanists believed, because motion is a property of these entities and cannot exist without them. Entities interact not through an exchange but in a *dance*. In this dance, each body moves according to its own rhythm: an internal principle distinct from all the others. They never exchange motion, but they *learn* from each other, according to what she calls a 'sympathy' in all matter, whether a body, a liver or piece of furniture, that drives it to mimic other dances according to its own nature; the change comes from within.[21]

We'll return to how off-kilter this theory sounds to modern ears; the idea of inanimate furniture acting like a human body is so far from how we conceive consciousness today. But it remains important to note that Cavendish was making these arguments when the mechanistic worldview was gaining real traction in England, championed by the Royal Society. The organisation, supported by King Charles II, was established by a group of twelve natural philosophers – including the chemist Robert Boyle and architect Christopher Wren – just after the Restoration with the aim of promoting their view of natural philosophy.[22] This authoritative body of scholars, sanctioned by the monarch, were all, to varying degrees, invested in the mechanistic view. They promoted that view of the universe as a machine of uniform parts working according to predictable laws. At the very same time, Cavendish fervently promoted her own theory that addressed the very dimensions this science couldn't or refused to investigate: variety, consciousness, the soul. A theory based on a lively and organic worldview that fundamentally opposed the cold mechanism promoted by the royally sanctioned body of men at the society.[23]

It was not remarkable just that Cavendish was able to disagree with the establishment but that she could publish her theories at all. In the seventeenth century, books by women were a rarity. The few women who did publish tended to publish on religion, maternal instruction or poetry. Certainly not natural philosophy.[24] Between 1651 and 1655 – the years in which Cavendish published four volumes – a dismal 1.3% of total publications were written by women (and even that was a slight improvement on the decade prior).[25] Cavendish had work published on the most cutting-edge debates in the science of her time. And all this, remember, without a formal education in natural philosophy.

Of course, this was made possible by her privileged class. Even at the beginning of her career, while banished in France, she was able to publish in Cromwell's London. Her husband's connections gave her a clear in. Back home, Cavendish's boldness was rewarded. She became a celebrity. And yet, as her reputation grew, so did the gossip. The famous diarist Samuel Pepys – a fellow of the Royal Society himself – kept a record of the worst admonitions. In May 1667, he wrote: 'The Duchesse hath been a good, comely woman; but her dress so antick, and her deportment so ordinary, that I do not like her at all, nor did I hear her say any thing that was worth hearing.' He called her 'a mad, conceited, ridiculous woman'.[26] Accusations of madness became a recurring theme. Dorothy Osborne, one of her first readers, claimed she had seen 'soberer people in Bedlam'. Cavendish soon acquired the nickname 'Mad Madge'.[27]

Cavendish had strayed too far outside the confines of feminine propriety. Such women were generally accused of being promiscuous, mad or both. Like the label of witch in Hypatia's day, these scripts were designed to keep women in their place.[28] But alongside the power of its male titans, Cavendish coveted the whorish posturing of the female emblems of the ancient world.[29] A Restoration Cleopatra, she claimed her public status along with her sexuality. By 1667, now a duchess, she'd ramped up her performance, flashing her 'scarlet-trimmed nipples' at the theatre.[30]

Cavendish's antics outraged society, but she was never totally ostracised. She was the Duchess of Newcastle after all, and her behaviour was usually treated as the high-society performance it was. When it came to her science, too, her eccentricity worked both for and against her. It was sometimes used as leverage to discredit her – her appearance a sign that her arguments were irrational, even insane. At the very same time, it sometimes opened doors.

The same year as the theatre incident, Cavendish made her debut at the Royal Society. As we saw, her arrival sparked quite the commotion. The enthusiasm had nothing to do with her science. Cavendish was the object of curiosity. And that didn't change once she entered the building to meet the society's fellows.

There was, by this point, controversy surrounding not only Cavendish's personality but her science too. Cavendish had supposedly attacked the Royal Society in her 1666 text, *Observations upon Experimental Philosophy*,[31] and the society's members had joined the crusade against her. They argued that Cavendish was an 'anti-experimentalist' who failed to understand the true significance of the methods employed by Boyle, Hooke and the organisation's other luminaries.[32]

This was harsh criticism indeed. Experimentalism was the method at the core of the mechanistic worldview. The universe was a machine that could be objectively observed and definitively understood. The society itself had been explicitly set up to promote 'experimental learning'.[33] To be anti-experimentalist was to be anti-Royal Society.

Cavendish's view threatened the certainty of this method with the uncertainty it exposed. Her stance undermined not just their method, but their status and power too. In her egalitarian view of a reality of interacting parts, the whole was elusive. Cavendish argued that this kind of pure objectivity is unattainable through any method of inquiry and definitely not through the method of observation they were using. Individual parts, human or not, she wrote, 'cannot know the truth of those Infinite parts, being but a finite part itself'.[34]

Cavendish saw how this argument was not just a critique of a scientific method but an attack on the authority of experimentalists. This, not a commitment to truth, was the true source of their outrage. She diagnosed them with a god complex accordingly:

> But I perceive man has a great spleen against self-moving corporeal nature, although himself is part of

> her, and the reason is his ambition; for he would fain be supreme, and above all other creatures, as more towards a divine nature: he would be a God, if arguments could make him such.[35]

Cavendish understood that the backlash against her ideas, and mechanistic philosophy itself, were tied up with claims to power. She understood that she had triggered the defences of those fortifying their authority in the field; she knew that she was criticised not on the basis of what she was arguing, but for arguing at all.

The Royal Society took fire where it would hurt the most. What better place to discredit Cavendish than by attacking her views on a new, publicly admired invention? The microscope was a useful tool to experimentalists interested in observation. The technology had been around for a while, but scientists had started turning them to biology with eye-opening success. In England, Robert Hooke in particular held audiences at the Royal Society where he demonstrated the power of the lens on life. Hooke drew on earlier designs of compound microscopes that used successive lenses to magnify the object under investigation. He used two: the smaller lens closer to the specimen and the longer lens near to the observer's eye. He also designed a nifty method for concentrating light on specimens, diffusing the light from an oil lamp through a water-filled glass flask. His technical skills and artistic rendering together illuminated the samples in extraordinary detail.[36] He published his studies in *Micrographia* in 1665. This thick folio volume requires two arms to hold and contains 246 pages of Hooke's descriptions – given accessibly in English rather than Latin – of five classes of objects, from minerals and feathers to insects, at the microscopic level. Most striking, though, were the thirty-eight, full-page, fold-out plates that revealed life as never seen before. The most famous of the images, the flea, stunned readers with bulging thorax, spindly legs and face; as the pages unfolded, so did a creature that they'd previously only known as an unremarkable speck. The public were in awe of the microscopic worlds science revealed. Samuel Pepys stayed up into the early hours reading it, proclaiming it 'the most ingenious book that ever I read in my life'.[37] The book introduced the world to the fundamental building block of life, the cell, for the first time, and its success helped solidify the Royal Society's status as a leading scientific institution.[38]

Cavendish had made her views on the microscope clear. The society members pointed to her claims that Hooke's invention was 'deluding Glasses' rather than 'true Informers' about the nature of objects.[39] This, of course, sounded ridiculous to the awe-struck audience who had seen the microscopic images with their own eyes. It was easy for the Royal Society to ridicule her; her argument already went against the grain. The fellows, now more authoritative than ever, hammered the nail in the coffin and denounced her arguments as misguided, uninformed and 'child-like'.[40]

Let's take a look at some of their accusations.

Cavendish's views of the value of microscopes were not based on ignorance. She was, in fact, particularly well read on the subject and even had practical experience with them in the laboratory. Her well-connected husband gave her access to many cutting-edge instruments. During their exile, the couple acquired a whole collection of microscopes and telescopes.[41] Here, she noticed that early prototype microscopes often performed poorly, the imperfect lenses creating distorted, blurry images. Cavendish pointed out that 'a Glass that is flaw'd, crack'd, or broke, or cut into the figure of Lozanges, Triangles, Squares, or the like, will present numerous pictures of one object'.[42] The lighting of specimens could also distort the objects under inspection. Hooke himself had confronted this particular obstacle when observing the eyes of a fly.[43] Cavendish realised how difficult it was to draw accurate conclusions when you didn't really know what you were looking at. More than this, she added, the flaws of inventions are always revealed through their new iterations. The next model of microscope showed us aspects of reality we couldn't have imagined with the previous one; the microscopic worlds contained within Hooke's *Micrographia*, for example. The flip side to this was that there would be fatal flaws to this equipment that might emerge in time, limiting our view today.[44]

But, more interestingly, and more radically, perhaps, Cavendish held a more philosophical objection to this mode of scientific endeavour. In her view, observation alone would never be sufficient.[45] Instead, we had to think about *how* we were using these tools to properly interpret what they were showing us. A microscope might show us the details of a flea – the hairs on its legs, its hooked feet, distended abdomen – but it does not necessarily tell us what host it feeds on, the diseases it carries or how it reproduces. The view is, by definition, microscopic, and scientists must synthesise these observations into a working theory, speculating, perhaps, on how those hooked feet might help a flea cling to its host or leap between them. In fact, Hooke agreed, saying that his methodological intention was to alternate between 'Artificial Improvements' and 'Observations of Nature'. Microscopes were *tools* to be used to get to know the natural world, not arbiters of reality.[46] And so, again, if Madge was Mad, then so was Hooke, and so were Thomas Sydenham and John Locke, who collaborated on an essay echoing Cavendish two years after *Observations* was published.[47]

Cavendish, of course, went further. She was sceptical of the very idea that 'sense', or observation, could ever get to the nature of things. 'The truth is, our exterior senses can go no further than the exterior figures of Creatures, and their exterior actions, but our reason may pierce deeper, and consider their inherent natures and interior actions.'[48] She believed that experimentation should be guided, ultimately, by reason. In this, she did diverge from the experimental enthusiasts in the Royal Society at the time. But again, she wasn't alone in her criticism: most famously, the English philosopher Thomas Hobbes concurred that the world could not be understood through observation alone and must always be mediated by reason.[49]

Though she differed from Hobbes too. When it came to the question of motion, for example, Hobbes divided it into internal and external components. Cavendish believed that the important distinction was between animate and inanimate matter. The animate parts were active and moving, the inanimate passive and static. As we saw, Cavendish's model of reality was made of equal and self-moving parts. They existed in everything, and their interaction gave rise to new forms. And so it made no sense to distinguish between inside and outside to understand them. It also didn't make sense to think in atomic terms, as so many scientists did, of bodies or entities as made up of smaller parts.[50]

The idea that all matter is part of an interconnected, self-moving process might sound quasi-spiritual to modern scientific ears, but it's honestly not far from the emergence theory that has crystalised as a challenge to scientific reductionism across the sciences through the twentieth century. Complex systems exhibit properties that are greater than the sum of their parts. We can't explain water's wetness only by understanding individual water molecules; we can't explain how an ant colony functions based on the physiology of a single ant. Cavendish didn't foresee that microscopy would develop into an ever-more refined source of evidence in modern science. But she did offer a sound counterpoint to her era's dominant theories. Conflicting hypotheses *are* the engine that drives scientific discovery; in that sense, both contain truth. The idea that we can understand the world in terms of processes and systems, rather than parts, is now integral to the science of life, from mycelial networks to human consciousness.

Cavendish attended the Royal Society despite her ongoing disagreement with them and continued to make her arguments despite their hostility because she stood, above all, for scientific discussion. Philosophers couldn't know anything definitively and so ideas could and should be disputed to advance the knowledge base. She invited criticism, as long as it was convincing.[51] She didn't shy away from conflict but leaned bravely into it. And always with flair. Nowhere is this more apparent than in her novel *The Blazing World*. Tacked unassumingly onto the end of her 1666 philosophical text, here she is, once again, blurring the lines of acceptable science. Not just in content, but in form.[52] In a scientific climate that emphasised plain observation and objective statements, Cavendish writes an unapologetically fictional, entirely fabricated and subjective piece that *shows* people her science.

Science fiction is often thought to be a contemporary genre born out of industrialisation. *The Blazing World* came earlier, but it carries many of the hallmarks we associate with classic sci-fi.[53] The story itself is another vessel for Cavendish's natural philosophy – and an opportunity for a more pointed jab at the Royal Society. After marrying an emperor, an empress of a new world tries to explain and investigate her land full of animal–human hybrids. She gives all her subjects a role: 'bear-men' become experimental philosophers, 'bird-men' astronomers, 'lice-men' mathematicians.

Needless to say, they all work beneath her. In a world where she is ridiculed by the men in charge, she writes a utopia in which learned men value women's intelligence enough to accept her authority over them.[54] Philosophical experiments follow, featuring the same microscopes and scientific debates of *Observations*, only with some lesbian love affairs thrown in the mix. You can only imagine how radical this was at the time.[55] Many of the characters in Cavendish's other creative work are in fact lesbian. As one of the characters in her 1668 play, *The Convent of Pleasure*, asks: 'But why may not I love a Woman with the same affection I could a Man?'[56] We can't know Cavendish's own sexual orientation definitively, but we do know that her works are true genre, and gender, benders.

While some of Cavendish's writing is startlingly modern, at other times it reflects her conservative roots. In one section, Cavendish discusses her views on experimentalism and claims that these scientists are like servants who should defer to the superior rationalist thinkers.[57] The analogy announces Cavendish's views of monarchy as another form of natural order. With the civil war still a recent memory, this was a pretty provocative statement to be making, particularly for Fellows of the Royal Society, who continuously stressed the 'nobility' of their programme to attract patronage and funds.[58] Many of these Fellows were 'commoners' by birth; notably Robert Hooke and Isaac Newton. Cavendish challenged the status of a boys' club of science by invoking a class-based social order. That Cavendish didn't see her monarchism as oppressive makes sense; her aristocratic status benefited her. She was brazen and unapologetic, but she was always defending the chimera of social status that she was.

Contradictory, genre-defying, unabashed – Cavendish was fundamentally contrarian. She wasn't interested in being liked. Unlike Hypatia, who made an intellectual home in her politically divisive climate, Cavendish sat boldly outside of it. A collection of parts, all moving to their own rhythm. She let them dance. Her philosophy reminds us that, for all the advances experimentalism has brought us, it reveals only part of the picture. She reminds us that, as people, we participate in a reality and don't hover omnisciently over it. That while new perspectives offer further parts of the whole, the whole will likely remain elusive.

And while Cavendish's cultivated outlandishness was part of her approach, we shouldn't let it distract from fair appraisal of her thought. Seen in the context of her contemporaries, Cavendish's views were not eccentric, but well-developed contributions to some of the most challenging academic debates of her time.[59] She was a self-taught natural philosopher participating in the debates that gave shape to the field we now call science as it incorporated the new disciplines of astronomy, geography, mathematics and physics.[60] Her coherent natural philosophy was encoded not just in the content but in the form of her work and embodied in the way she carried herself. Her work was innovative and thorough and continues to challenge us today, with the partial truth of its collected parts. Margaret died in 1673 and was buried at the epicentre of British monarchy, Westminster Abbey. And yet, the inscription on her tomb has less of the grandeur and more of the understated levity she appreciated. The ode, written by her husband, reads: 'This Duchess was a wise witty and learned lady, which her many books do well testify.'[61]

3

Ada Lovelace (1815–1852)

Mathematician

1843. Britain booms with the fruits of a new industrial age. Machines – steam engines, mechanised looms, telegraphs – transform how people work, live, the air they breathe. But what if they could do more? What if a machine could mechanise complex calculations, completing in seconds sums that otherwise might take days, months? Moreover, what if those numbers were only symbolic? Then they could stand for other forms of information: music, colour, words…*anything*. The machine could then manipulate information according to a rule, drawing pictures or composing music. Would such a machine be intelligent or a mere automaton?

These questions consumed Ada Lovelace. Twenty-eight years old and without a formal education, she had just conceptualised the modern computer, over a century before the computational revolution.

*

Today, we hold science in opposition to artistic expression. From our schooldays, we are asked to identify as either scientific or artistic, left-brained or right-brained, then sifted into opposing camps – STEM or humanities. These are the 'two cultures', as C. P. Snow dubbed them in 1959, immiscible as oil and water.

But to Lovelace, imagination, creativity, *poetry* was the driving force behind scientific endeavour. 'Imagination', she wrote, '*is* the Discovering Faculty...which penetrates into the unseen worlds around us, the worlds of Science.'[1] The faculties of art and science were part of the same drive to discover. Maths was also how Lovelace herself articulated her wonder. She called it 'poetical science'.

Lovelace was the child of poetry and science. Born Augusta Ada Byron, her father was none other than the Romantic poet Lord George Gordon Byron – 'mad, bad and dangerous to know', as his lover, later stalker, Lady Caroline Lamb declared. Fresh from said affair and an incestuous liaison with his half-sister, Augusta Leigh, Byron married Annabella Milbanke somewhat reluctantly in 1815. Annabella was a social reformer from an aristocratic family with a deep interest in maths and education. They were both young (he was twenty-seven; she was twenty-two) when they married, but Annabella, of course, had lived a protected life until then. She had always been clear about what she wanted in a husband; she wanted someone steady and attentive. In a tale as old as time, Byron convinced her that she could save him from himself. He presented himself as a 'fallen spirit'. At first, he called her his 'princess of parallelograms'; later, a 'mathematical Medea'.

Maybe part of him wanted to change. Born disabled with a clubfoot that caused him pain and shame throughout his life, Byron had been sexually abused as a child and was deeply conflicted by his attraction to men, which his Calvinist upbringing deemed immoral.[2] Today, much of his temperamental behaviour and intense mood swings might have been connected to these compounding traumas.[3] Nonetheless, their story was doomed. Opposites attract and then repel.

The marriage lasted only a year. Annabella left Byron when Ada was just five weeks old, afraid his increasingly erratic moods and newly revealed homosexual tendencies ('kindly' brought to her attention by Lady Caroline) would impact their daughter.[4] Annabella began to devise a scheme to have him declared insane. It never came to that. When Ada was four months old, he left England, never to return. When Lovelace was eight, Byron died leading a campaign in the Greek War of Independence against the Ottoman Empire in Missolonghi from a fever that some attribute to malaria relapse. Lovelace didn't see so much as a picture of him until she was twenty.[5]

Having secured safety for her child, Annabella became determined to create a mind that was as reasonable, logical and un-Romantic as her own. As a social reformer, Annabella took education very seriously. She was dedicated to imparting her own passion for science and maths to her daughter while steering her from the literature that had 'corrupted' her father.[6]

It wasn't entirely unusual for girls of Lovelace's class to get an education, but hers was particularly stringent. Annabella demanded dedication and discipline from her daughter. She employed numerous tutors, who taught Lovelace from rented country estates, and one of whom she fired for giving inadequate attention to the all-important mathematics. One of her tutors was the renowned mathematician and astronomer Mary Somerville (her work itself so interdisciplinary that the reviewer of her book, William Whewell, used the word 'scientist', analogous to 'artist', in print for the first time, to connote the students of the material world collectively)[7]. The family was well connected, part of a social elite, and this environment allowed Lovelace's talent to grow even in an unconventionally matriarchal family.

And she was talented. From the age of four she showed an aptitude for mathematics: not just following patterns but questioning the underlying rules. By age twelve, Lovelace had fallen in love with mechanical engineering. She decided to apply her thinking to the task of flying. She studied the anatomy of birds to deduce the proportion between wings and body. She investigated which materials would be suitable for the wings. Paper? Silk? Feathers? She considered her equipment methodically and even incorporated the Industrial Revolution's great technological innovation: steam power. She collated her findings in a book titled *Flyology* which presented her research alongside drawings of her very own flying machine.[8]

Her mother's mathematical education seemed to be bearing fruit; Lovelace's interest and ability in this area were growing (albeit with persistent creative flair). But, nonetheless, she hadn't quite managed to dispel her poetic proclivities. Despite the strict educational regime, Lovelace seemed to have retained not just her creativity, but also her rebelliousness (some see this as evidence of her father's temperament; I see it more as a very normal, teenage response to being controlled). In a moment of defiance, she wrote a letter to her mother: 'You will not concede me philosophical poetry. Invert the order! Will you give me poetical philosophy, poetical science?'[9]

Her mother generally left Lovelace to a group of no-nonsense spinster friends, who Lovelace tellingly christened 'the Furies'. A bout of illness in her mid-teens left her bedridden with what her Furies claimed was a form of hysterical paralysis in the legs: another Byronic form of protest against order and good sense. At eighteen, after she'd recovered, Lovelace turned to flight again and tried to elope with one of her tutors.[10]

Lovelace's relationship with her mother would grow only more strained in adulthood. When it came to maths, though, Annabella shared her daughter's enthusiasm and encouraged her to follow her passions. This meant socialising with the great scientific luminaries of the day. At seventeen, Lovelace attended a soiree hosted by the renowned mathematician Charles Babbage. Babbage was known for his eccentric inventions. The son of Benjamin Babbage, a wealthy banker, he had been left a comfortable fortune (around £10 million in today's terms) that sustained him in his polymathic proclivities. He studied maths at Cambridge and was elected Lucasian Professor of Mathematics there in 1828, a position held by Sir Isaac Newton before him (and Stephen Hawking in our time). But he neglected his teaching responsibilities and fell out with his colleagues as he tried to push for more inclusive education and a broader syllabus. He was interested in social reform and tried to run for parliament twice in his life. He was close friends with Charles Dickens and shared his progressive outlook, though some of their campaigns were less than idealistic, not least their war on street musicians. He wasn't shy of conflict, though still found his way to the centre of influence.[11] His house at 1 Dorset Street, Marylebone, London, was also the meeting place for London's intellectuals, scientists and socialites.

For Babbage, the evening was an opportunity to garner support and funding for a new research programme. He was showcasing an invention that he called the 'Difference Engine'. This was a two-foot tower of wheels, made of brass, that could make reliable calculations by turning a handle.[12] Lovelace, at seventeen, was more fascinated by the calculator than any new romantic prospects at the soiree (it happened to be her official society debut). Mrs Sophia De Morgan, who had tutored Lovelace in childhood, remembered other guests gawping 'with the sort of expression… that some savages are said to have shown on first seeing a looking-glass'. Meanwhile, 'Miss Byron, young as she was, understood its working, and saw the great beauty of the invention.'[13]

Lovelace wanted to inspect the blueprints of the machine more closely. Her mother took her, a few days later, to visit him.

He demonstrated the device in his drawing room. Lovelace followed up with letters on how she saw the prototype's potential and her mathematical aspirations, and he replied with details about his plans.[14] A touching relationship blossomed; their intellectual relationship was also a proxy for their emotional needs. Lovelace was fascinated by the work of the man who was roughly the same age as her father, and she filled the void the death of his sixteen-year-old daughter, Georgiana, had left. Many of the letters between Lovelace and Babbage have, amazingly, been preserved. They wrote to each other often, pre-empting the computer age in more ways than one with their email-like correspondence, sometimes composing several letters a day. Together, the letters offer a record of an unusually reciprocal, often witty and touching working friendship as well as an intimate view into the conversations that shape scientific ideas. 'My dear Babbage', Lovelace repeatedly calls him, while to him, she is the 'Enchantress of Number'.[15]

Meanwhile, Britain was caught in a storm of scientific and industrial innovation. The 1830s saw the creation of the first railway from Manchester to Liverpool, the first sewing machine, telegraph line and daguerreotype photography. In 1837, Queen Victoria, just a few years younger than Lovelace, came to the throne, overseeing Britain's transformation into the world's most powerful empire. Charles Darwin set sail on the HMS *Beagle* two years before Babbage and Lovelace's fateful meeting, while Michael Faraday and Joseph Henry discovered electromagnetic induction the same year.

Lovelace found herself at the centre of a scientific renaissance. However, it was a slightly older invention that caught Lovelace's eye. In 1804, the French weaver and merchant Joseph Marie Jacquard invented the eponymous Jacquard loom, which could effectively be 'programmed' to weave pictures. The loom was operated by punched cards to create a sequence that automated a loom, with an on or off switch. The cards told the loom to raise or not raise the thread. This was the kind of digital code that would become the basis, we will see, for the first computer. Lovelace was captivated by the looms (all this while her father, in 1812, spoke in the House of Lords in defence of the followers of Ned Ludd, who were vandalising the mechanical looms out of fear that they would put weavers out of work). Annabelle took Lovelace to the Midlands, not far from that dark core of industry in Britain; the 'Black Country', named for the smoke and soot from the thousands of ironworking foundries and the coal-rich soil fuelling them.[16] They visited the surrounding areas, where around 8,000 Jacquard looms now supercharged its textile factories, transforming the skilled artisan 'cottage industry' into industrial-scale, mass production.[17] Lovelace was lucky enough to be able to travel to study these inventions. Her privilege, talent and enthusiasm coalesced to open a new world of mathematical possibilities.

In a sense, it was the perfect time for a person with Lovelace's mathematical passion to thrive. The industrial age saw a society-wide enthusiasm for the ways in which science could transform society. Babbage's aims, too, were ever expanding. Babbage had first landed on the idea for the Difference Engine out of frustration with the manual calculations he and the mathematician and astronomer John Herschel were working on. At this time of technological advancement, not just mathematicians, but navigators, engineers, surveyors and bankers relied on printed mathematical tables to perform calculations that required more than a few figures of accuracy. These mathematical tables were essential to military and commerce. And yet they were, as Babbage was experiencing, not just laborious but also, crucially, prone to error. Babbage made a point of the money the government lost through these errors, due to the infrastructural and navigational errors it caused. In an 1821 meeting with Herschel, Babbage famously exclaimed, 'I wish to God these calculations had been executed by steam.'[18] His Difference Engine was one step in that direction, but it was just a calculator; it worked only with the numbers you put in.[19] Babbage wanted to mechanise more complex calculations, to move beyond a calculator that relied on humans to input numbers, and extend the model of industrial production from things to ideas.

By the early 1840s – and after a decade of correspondence with Lovelace – Babbage had moved on to the Analytical Engine: a machine with thousands of cogwheels that could perform more functions with greater accuracy. It was just a sketch; Babbage didn't have the funding to actually build it, but it embodied almost every logical feature of the modern digital computer. Babbage produced thousands of technical drawings that conceptualised the computer. The Analytical Engine was programmable, general purpose and had the same internal architecture of later computers. The machine, then, took the leap from mechanising arithmetic to digitising computation.[20]

The design for the machine was probably more remarkable than Babbage realised. He couldn't fully grasp the implications of creating a multi-purpose machine that automated complex operations. It took Lovelace's imagination and understanding to extrapolate the impact this Analytical Engine could have on society.

Lovelace continued to correspond with Babbage as she began to take up more conventional roles as a mother and wife. She had submitted to her mother's regime after she was dragged home from her escapade with the tutor and married her husband, William King, a friend of Mary Somerville's and by all accounts a kind but boring man, in 1835 at nineteen.

They had three children and her only daughter, Anne Blunt, inherited Lovelace's polymathic streak, becoming an accomplished writer, violinist, artist and horse breeder, in addition to speaking six languages. William was made an earl under Queen Victoria for his government work in 1838, and Ada Augusta King became Countess of Lovelace.[21] While these aspects of her life set her up as conventional and respectable in the ways that were expected of a lady of her status, Lovelace continued to prioritise maths. Her mother remained her most vigilant critic in all this, often policing her child rearing and household management. Nonetheless, by 1841, Lovelace writes that she found herself becoming less 'secretive & suspicious' towards her mother. The moment coincides with a growing confidence in her mathematical abilities. 'I believe myself to possess a most singular combination of qualities exactly fitted to make me pre-eminently a discoverer of the hidden realities of nature,' she writes. Lovelace was gaining confidence in what she was seeing; on the border of the dry mathematical world that even her mother herself seemed to commit to only in theory, she was discovering the wonders of what her mathematical imagination could show her. Suddenly, her mother's opinion mattered less.[22]

Lovelace saw the potential in Babbage's machine before anyone else. Babbage had appealed to the British government for financial backing, but they'd refused, nor did they enjoy dealing with Babbage's undiplomatic personality. Babbage had already been given a sizeable sum of government money to build the Difference Engine (around £17,000, or the cost of two naval warships at time). All this to no avail. There was uproar about the waste of funds in parliament. Babbage didn't try to appease them. He only pushed for more. After one row with Babbage, the prime minister, Robert Peel, pled, 'What shall we do to get rid of

Mr. Babbage and his Calculating Machine? It would be worthless as far as science is concerned.'[23]

Babbage saw he wasn't getting anywhere with his government and travelled to Turin, Italy, to try his luck there. After multiple rejections from various prominent mathematicians, he finally convinced one Luigi Federico Menabrea to write a paper on the machine. The paper was published in a Swiss academic journal in October 1842.

Lovelace had been looking for an opportunity to get involved. In February 1843, she decided to translate the paper into English and try to get it published in an English journal.[24] Babbage welcomed the help; his chaotic nature made writing as difficult for him as diplomacy. This is how, at age twenty-seven, Lovelace came to write her most lasting contribution to what would become computer science.[25]

Lovelace did much more than translate Babbage's paper. In the nine months she spent working furiously on the paper, she added her own mathematical corrections and, most importantly, drew out the design's significance. Lovelace's version was almost three times longer than the original. At the end of her intellectual pregnancy, she wrote to Babbage, 'I am very much satisfied with this first child of mine. He is an enormously fine baby and will grow to be a man of the first magnitude and power.'[26]

It was a wonder-child indeed. In the *Notes*, she first describes clearly and accessibly how Babbage's device would work. She drew on her knowledge of the Jacquard loom to explain how Babbage's machine weaved algebraic patterns the same way the weaving machine produced images using a chain of punched cards.[27] She also *showed* how the machine would perform a calculation, writing out a plan for what many have dubbed the first computer program. And then, she went even further.

Perhaps her poetic disposition made her see that the form of computation Babbage had described represented a new language. Numbers, in this machine, weren't just expressions of sums but potentially of something other than quantity. Numbers could represent, for example, musical notes or symbols. If numbers could represent concepts, then computers could be used to produce not just calculations but whole other systems of ideas. This insight has become a key concept in the digital age: the idea that music, text, pictures, numbers, symbols and even sounds can be expressed in digital form and processed by machines.[28] Lovelace described how the Analytical Engine could manipulate information according to rules. In doing so, she devised the concept of the computer. She saw the potential of the machine; she *imagined* its future. That insight, imaginative as it was, relied equally on an intimate understanding of what the machine could do. It was the perfect fusion of two influences that had split her from birth. With one parent on either side of the divide, Lovelace's *Notes* reconciled these ways of knowing in a way that she personally never did.

Lovelace's creative logic led her to some incredibly prophetic questions about the potential social implications of Babbage's design. Given the possible shift of thought from the minds of humans to computers, she asks, would machines themselves one day be able to think on their own? Perhaps she'd been inspired by the movement that, since her father's time, had always prompted people to look beyond science's mechanistic universe. Now, literary horror stories like Mary Shelley's *Frankenstein* primed her on the possibility of human inventions taking on a life of their own. In the case of her hypothetical creation, Lovelace addressed the matter of what we might today call 'artificial intelligence' and quickly put the notion to bed. 'The Analytical Engine', she wrote, 'has no pretensions whatever to originate anything. It can do whatever we know how to order it to perform.'[29] She went on with the definitive statement that 'Only when computers originate things should they be believed to have minds.' This is what Alan Turing, the father of modern computer science, would later refer to as 'Lady Lovelace's Objection'.[30] Today, computer scientists refer to Lovelace's test as the condition for 'intelligence'; Turing's own test, which assesses whether a machine can convince an operator if it is human in conversation, is too easy.[31]

Lovelace's *Notes* remain the most comprehensive account of computing at the time. They are astoundingly modern; the birth of modern computer science would not truly begin until the 1930s. Indeed, the lack of continuity between Lovelace's notes and this later science have caused some to challenge her impact on the field. And it is true that nobody studied Babbage's notes until the 1970s. Historians of science have only retrospectively revealed the connections between the thinking at these disparate times. The principles of computing, so beautifully laid out by Lovelace, had to be reinvented. The scientists developing computing for the digital age did know of Babbage's work but not in detail. Turing himself, while he referred to Lovelace's objection, said that he had come across these ideas in the writing of the mathematician Douglas Hartree and so probably only read a summary, and didn't lay eyes on her sketches and calculations.[32] It seems that Turing arrived at the same thinking as Lovelace independently when he devised his theoretical machines that would manipulate symbols on a strip of tape according to a table of rules. In doing so, though, he made the same leap to the insight that machines could be generalisable; that they could be used to do much more than simple arithmetic. He went on to demonstrate the principle when he designed the Automatic Computing Engine (ACE) in 1945, after his time at Bletchley Park during the war. Indirectly, the computers we have today are testament to the veracity of Lovelace's insight. In 1952, as he was chemically castrated by the British government for the 'crime' of being gay, Turing applied his thinking to biology with his paper, 'The Chemical Basis of Morphogenesis'. Here, he explains mathematically how uniform cells can differentiate into complex patterns – based on the same principle of emergence that Cavendish had gestured to. It is an apt metaphor for how scientists' ideas emerge, as seemingly spontaneous insights, forming unexpected but logical patterns, replicable, at various points in history, as they spiral through time.

Babbage immediately recognised the importance of her papers: 'Their author has entered fully into almost all the very difficult and abstract questions connected with the subject.'[33] Not that Lovelace needed his approval. She had always been decisive in the face of the man known for his bullishness. But the *Notes* became a real turning point in a relationship still defined by Babbage's superior expertise. Now, Babbage clearly needed Lovelace's help; they were equals. And Lovelace leaned happily into her new role.

In the five letters a day the pair exchanged as Lovelace was compiling her *Notes*, she confidently fought to keep complete control. In the run-up to publication, for instance, she writes: 'I am much annoyed at your having altered my note. You know I am always willing to make any required alterations myself, but that I cannot endure another person to meddle with my sentences.'[34]

She continually challenged Babbage when she disagreed with him. When he suggested adding an anonymous preface to the translated paper stating that the British government had failed to fund the project, Lovelace thought this was a bad idea and told Babbage so. Babbage insisted and even threatened to withdraw the paper. In the end, though, the paper was published…without a preface.

When it came to the issue of funding, of course, Lovelace knew of Babbage's fraught history with the British government. Again, she wasn't afraid to set him straight. One of my favourite interactions between the pair is their correspondence around Lovelace's request to manage the business operations for the engine.[35] In a sixteen-page letter from 14 August 1843, she asks him to stick to the behind-the-scenes making of the machine (where he wouldn't offend anyone) and to let her do the public-facing work required to raise funds. Her tone is, as ever, witty but firm. 'Your affairs have been, & are, deeply occupying both myself and Lord Lovelace… And the result is I have plans for you.' She proceeds to make her request. She offers her rationale: 'My own uncompromising principle is to endeavour to love *truth & God before fame & glory*', while 'Yours is to love truth & God…but to love *fame, glory, honours, yet more.*' She ends her letter charmingly, 'I wonder if you will choose to retain the lady-fairy in your service or not.' Babbage resists for a while. Lovelace writes to her mother the following day: 'I am uncertain as yet how the Babbage business will end… I have written to him…very explicitly; stating my own *conditions*… He has so strong an idea of the *advantage* of having *my* pen as his servant, that he will probably yield.' By 10 September, Lovelace writes to Babbage: 'You are a brave man to give yourself wholly up to Fairy-Guidance!' Babbage signs his next letter to her with 'Your faithful Slave'.[36] Lovelace held her own in a very authentic (and amusing) relationship between equals.

Lovelace and Babbage never got the funding they needed. Lovelace increasingly lost herself to the Byron-like tendencies her mother had tried so hard to suppress. Some historians argue that Lovelace turned to gambling to raise the money needed to build the Analytical Engine,[37] but this adventure also coincided with her growing interest in a whole range of rebellious activities – everything from mesmerism to sexual exploration, drugs and phrenology.[38] The turn had come after she'd discovered more about her parents' tumultuous relationship. These activities, she said, were part of a new direction of her career, what she now called 'poetical science', that would reconcile the divide in her, between maths and feeling. Whatever her precise motivations, it seems that Lovelace was expressing a pain she still hadn't been able to salve when she turned to horse racing.

Lovelace surrounded herself with a circle of men who lent her money, while her mother controlled her access to the family fortune. This group was a mixture of well-respected men and unsavoury characters. One was possibly Florence Nightingale's father, another was John Crosse, son of Andrew Crosse, an electrical scientist who may have been the inspiration for the character of Dr Frankenstein, who also became her lover. Lovelace overestimated her abilities. In the spring season of 1851, she placed a series of bets that left her £3,200 in debt: around half a million today. She now faced extortion from some menacing individuals and was forced to admit defeat. Though she didn't admit her gambling debts to her mother until after she fell ill with uterine cancer a few years later, at the age of thirty-six.[39]

Lovelace, in the end, was *over*confident. Though some accounts of Lovelace's life go further and point to this phase to claim she'd become mentally ill. There is no evidence of this. Around the same time that she was hedging her bets, Babbage was developing his own plan for a fundraiser. He'd designed an engine that played tic-tac-toe. He hoped to tour it around the country to raise money for his projects. His misguided attempts never earned him a diagnosis.[40]

Lovelace asked to be buried alongside her father, who had died at exactly the same age – thirty-six. Her turning towards him at the end of her life also begs us to compare their legacy. Compared to Cavendish, Lovelace has been treated relatively kindly, not just by Babbage but by the critics of her day. On her death, she was given the highest of compliments for a woman in science at the time: masculinity. 'With an understanding thoroughly masculine in solidity, grasp and firmness, Lady Lovelace had all the delicacies of the most refined female character,' one obituary read.[41] According to this author, at least, she'd struck the socially acceptable balance of masculine and feminine. At the same time, assessments of her life often depict it as a disappointment: her descent into gambling and drugs, or the failure to have the Analytical Engine built, as the sad conclusion to her short life. Her early death, of course, is tragic. But the choices Lovelace made in her personal life should not discredit her important intellectual contributions. They didn't for Babbage, nor for Byron. If anything, her father's escapades consolidated his status as a creative genius. By asking to be buried alongside him, Lovelace challenges us to reconsider her legacy – again, with an eye to posterity.

I can't help but feel for Lovelace's mother, outliving her only child. She touchingly had one of Lovelace's sonnets read at her memorial, embracing her daughter's poetic side, allowing her to rest alongside the man whose influence she had tried to shield her from during her life. She had tried to protect her, however clumsily. Now, she did what she couldn't while her daughter was alive, for her to rest in peace.

Today, Lovelace is one of the best-known women scientists of the period. In STEM, she has become a feminist icon. And yet, my hope is that we remember Lovelace for the full and messy person she was. That we can remember her personal struggles without letting them define her or her work. She was an unusually gifted and confident thinker who made leaps of thinking that far surpassed her contemporaries and that connected maths to the real world in the same way that recognised founders of computer science have done. She was also fraught, conflicted and thrill-seeking – in part, because she continued to wrestle with the unresolved divide imposed on her, partly by her parents, but ultimately by a social climate that pushed women into narrow boxes. Now, her messiness really was part of her ingenuity and her legacy. Her presence confidently on the page with Babbage today continues to challenge us to reconsider the neat borders between creativity and science. 'My kingdom…is not to be a temporal one,' Lovelace once wrote.[42] She was talking to us.

4

Eunice Foote (1819–1888)

Physicist and Inventor

The effects of climate change are impossible to ignore. From catastrophic flooding to unprecedented droughts, from rapidly intensifying hurricanes to mass coral bleaching. And yet we do. Even as we sweat through record summers and tropical winters, even after teachers instil the science, despite ongoing international movements to incite action…we fail to act. We prefer to be dazzled and distracted by the outputs of a system established back in the nineteenth century.

As Ada Lovelace fell in love with the possibilities of the Jacquard loom, a new generation across America and Europe was witnessing the possibilities of industrialisation. And, like Lovelace, they were enamoured. They were witnessing how steam engines amplified rates of production, how railways expanded territories. Thanks to these inventions, they also saw how agriculture ramped up production to meet the demand of burgeoning city populations. Or how machines relieved the burden on labourers and horses as they took over the brunt of their gruelling work. Or how the alternative to manual labour created the possibility of legislating against exploitation and abuse (at the same time that it created the conditions for a new form of industrial workers' exploitation). In 1877, Anna Sewell, in her novel *Black Beauty*, took the opportunity to challenge people who 'always seemed to think that a horse was something like a steam engine, only smaller'. There were seductive benefits to industrialised production. And so, people gave little attention to the effects of the carbon-based fuel blazing like never before to power their revolution. They didn't pay much notice to the palpable impact it was having on their health or the planet.

And yet, a few individuals did. Remarkably, one of the first scientists to describe what became known as the greenhouse effect was an amateur scientist and inventor, working from a makeshift laboratory in her home on the outskirts of New York. This was Eunice Foote, hoping to leave her mark on a science that, for all its progress, still refused to listen to her.

*

Foote was born Eunice Newton on 17 July 1819 in Goshen, Connecticut. Her father was a cattle runner. When she was three years old, Foote and her ten brothers and sisters abandoned their rural beginnings to travel to New York in a covered wagon, in search of a better life.[1] New York was where it was at. It was a hub of scientific and social innovation. The city attracted a whole range of people: from aspiring middle-class professionals to entrepreneurs to abolitionists. It was also a major stop in the underground railroad, a secret network of routes and safe houses for enslaved African Americans to escape to free states and into Canada. Interracial marriages were allowed here, as well as utopian religions: revived evangelical sects, millenialists and new religious groups, like the Mormonists, Seventh Day Adventists or the 'Shaking Quakers', so-called for their ecstatic services, who practised egalitarianism, pacifism and sexual equality.[2] The place vibrated with the sense of possibility.

Middle-class families in New York exposed women to science as never before, and in ways that neither the lower classes, who were forced to work from a young age, nor upper classes, who were primarily prepared for marriage, were not.[3] By contrast, there were now academies for middle-class women; some of their textbooks were even written by them. There were museums and open lectures women could attend. Plenty of opportunities, as the founder of Foote's school, the educator and activist Emma Willard, wrote in 1819, for women to be educated as 'the companions, not the satellites of men'.[4]

Foote's parents were at the forefront of this wave of progress. They sent her to Troy Female Seminary, the first school in the country offering a scientific curriculum on a par with men's colleges. The school had its own laboratory, but it was also adjacent to Rensselaer Polytechnic, which was the top science institute in the country; Troy students could drop in on its classes.[5] It is hard to grasp now just how unusual this was at the time – a school where women studied science and even conducted experiments. Where women were taught how to be scientists.[6] And yet, this utopian reality had an expiration date.[7] Some things hadn't changed. The women at Troy knew that, once they married, they'd enter into the same old historic bargain – their husbands would protect them in exchange for the considerable sum of their civil rights.

The man Foote married was different. Elisha Foote was an eligible prospect – ten years Eunice's senior, he was an attorney with a bright future ahead. Foote's sister had hired him to tackle the debt their father had left after his death in 1835,[8] and Elisha successfully represented the sisters in court in the spring of 1840. But this wasn't a white knight situation. Foote was far more intrigued by his work as an inventor.[9] The two shared a passion for science. Sparks flew, literally. They married in 1841 and started inventing together. First, they created a whole range of little gadgets, from rubber shoe inserts to paper-making machines to thermostats for stoves (quite revolutionary – it allowed people to control the heat of their stoves for the first time).[10]

This collaboration was only possible because the couple also shared the same politics. Elisha had attended Albany Academy, a similarly progressive school. Both these schools instilled the idea that men and women should be partners. Elisha may have been ten years older at a time when men were very much patriarchs but, in their relationship, they were equals.

The couple's unconventional marriage was supported by their community. They had moved to Seneca Falls the year they got married, and the town sat at the heart of the American women's rights movement. They lived next door to pioneering feminist and abolitionist Elizabeth Cady Stanton, just as she was emerging as a leader in women's rights and suffrage. Stanton was a philosopher by training but, like Eunice, she was educated at Troy Female Seminary and so shared her thorough scientific education.

Women's right to participate in science was a key item on the agenda of the first ever women's rights convention, held in Seneca Falls itself. News of the event had been circulated in the press, and the event attracted a much bigger crowd than expected. At 10 a.m. on Wednesday 19 July 1848, Eunice and Elisha rode their wagon pulled by stout farm horses down a street full of family carriages, farm wagons, people on horseback and on foot. The organisers themselves barely made it to the door of the Wesleyan chapel, where the convention would be held.[11] Once there, they realised they had forgotten the keys. They thought on their feet and hoisted a professor from Yale through the window to open the door. The crowd poured in.[12] There was no time to waste on etiquette.

The leading minds in the women's rights and abolitionist movements had gathered, alongside cautiously curious members of the public, to imagine a daring and provocatively emancipated future. Journalists hovered around, pens at the ready to report the controversy of the event across the less progressive states.[13] On the second day of the convention, the organisers presented their 'Declaration of Sentiments'. This manifesto for the equality of men and women would eventually win women the right to vote. There were initially ten resolutions proposed at the convention, including the most fundamental, 'That woman is man's equal... and the highest good of the [human] race demands that she should be recognized as such.' And yet one, added last minute, gained more attention from women across America than any of the other resolutions approved at the convention: women's right to work across 'the various trades, professions and commerce' became the first to be enacted.[14] Work rights were perhaps easier for the majority to get on board with than political rights at the time. As Elizabeth Cady Stanton speculated, 'Though few women responded to the demand for political rights, many at once saw the importance of equality in the world of work.'[15] Foote's career would come to show that the right to work as a scientist depended on a society that was willing to hear them.

Once the resolutions had been voted in, the audience stood up to sign, according to the news reports, in the same way as the Founding Fathers had signed the 'Declaration of Independence' seventy-two years earlier.[16] The women rose first. Eunice joined the others on the podium and signed her name fourth on the list, under Stanton's. Elisha signed his just above that of the renowned abolitionist Frederick Douglass.[17] It was a historic moment, and yet, still only a third of the attendees dared to add their names. The rest – mainly men – feared the derision they knew would come once the document was publicised, and, for some, the damage it might do to their political careers.[18] But Eunice and Elisha stood firm.

Foote's scientific education, her supportive marriage and her political awareness all supported her career. The historian of science, Professor Sally Kohlstedt (one of the earliest historians to recover Foote's contributions to climate science) agreed that it wasn't enough for Foote to be 'pretty brilliant and pretty smart and pretty persistent to do that kind of work'. She was 'immersed in a world' that 'took her seriously'.[19] And so she took herself and her work seriously.

Foote's luck kept improving. In 1846, Elisha's mentor, Joseph Henry, had become the first director of the Smithsonian.[20] Henry had already been one of the most important scientists in America at this time; he discovered the property of inductance in electrical circuits, where a conductor stores energy as a magnetic field when an electrical current flows through it (coincidentally at around the same time as Michael Faraday in England, who we'll meet shortly, and who was the first to publish the results). The discovery powered many inventions, including the electrical motor. Henry was well connected, even to European scientists, who were very much America's bigger brother when it came to science.[21] Now at the head of the country's leading research organisation, though, he was in a position to really wield influence.

Elisha was Henry's protégé and they were collaborating on meteorological research, Henry's main scientific interest. America's primary exports at the time were agricultural and understanding the weather was essential to the industry. Henry wanted to collect long-term data to better understand the climate across the US to help farmers manage their crops.[22]

Henry saw the sun as the primary mover of weather.[23] The cornerstone of meteorology, he said, was 'that nearly all the changes [in climate] which now take place at the surface of the Earth are due to the action of the sun'.[24] The Footes decided, very sensibly for two emerging scientists, to tackle a scientific question that would interest Henry. They each designed an experiment to test various aspects of the sun's effect on the planet.

Foote was well placed to design an experiment in this area. Troy had moulded her into an exceptional chemist.[25] Nineteenth-century scientists in this field were interested in three principal molecules: oxygen, hydrogen and carbon dioxide.[26] Aware that gases like carbon dioxide, nitrogen, hydrogen and oxygen were present in the Earth's atmosphere, Foote decided to test the effect of the sun's heat on them. What she discovered outstripped anything her textbooks had taught her. It was nothing less than astonishing.

*

Eunice wasn't alone in asking this question. Over the century, several scientists had wondered whether the temperature on Earth was influenced by its atmosphere. Henry himself had suggested an answer. To do so, he'd referred to the work of the eighteenth-century Swiss meteorologist, geologist and physicist Horace Bénédict de Saussure (a name that set him up to make an impression!). Saussure, like many others, had noticed that the more widespread use of glass windowpanes meant rooms were considerably hotter. To test how well glass trapped the sun's heat, Saussure built a series of miniature greenhouses and exposed them to the sun evenly over several hours, then measured the temperatures inside. 'Fruits exposed to this heat were cooked and became juicy,' he observed.[27] The culinary applications of the technology thrilled Henry; he was captivated by the astronomer John Herschel's variation on Saussure's experiment, wherein the Englishman had used 'the heat accumulated in the wooden box in cooking the viands of a festive dinner'.[28]

But the implications extended well beyond cooking. Henry was certain that these experiments demonstrated 'the mode in which the atmosphere retains the heat of the sun and increases the temperature of the Earth's surface'.[29] He also speculated that water vapour in the atmosphere and the density of air moderated this heating effect.[30]

One of Henry's predecessors, the early nineteenth-century mathematician and physicist Joseph Fourier, had come to a similar conclusion. He argued that, if the Earth's atmosphere became solid like glass, it would trap the sun's heat in the same way. 'The resultant heat would lose all at once its power of passing through the transparent solids and would accumulate in the lower strata of the atmosphere, which would thus acquire very high temperatures.'[31] He was describing what became known as the greenhouse effect. However, none of these scientists conducted the experiments that would determine the effect of the atmosphere's various components on said heating effect.[32]

This is where Foote diverged[33]. To test the heat-trapping effect of a range of gases in different states, Foote filled glass cylinders with various compositions of gas, before placing them in either the sun or shade. The first experiment compared dense air with a vacuum, and the next examined air at standard, sea-level pressure. She then turned to dry and damp air, before finally looking to carbon dioxide.[34] Separating the different gases into these different states allowed her to compare how these individual variations – barometric pressure or humidity – each mediated the heating effect. Foote noticed, for example, that dense air exposed to sunlight grew hotter than thin air, and damp air hotter than dry air. No one, man or woman, had done this before.

Now, Foote turned her efforts to comparing different gases present in the Earth's atmosphere. The atmosphere is a mix of mainly nitrogen (around seventy-eight percent) and oxygen (twenty-one percent), alongside a sprinkling of noble gases, largely argon. Carbon dioxide makes up a tiny fraction.

At school, Foote had learned that burning coal will combine with oxygen to form 'carbonic acid gas' (carbon dioxide),[35] which then remained in the atmosphere.[36] But they hadn't covered the thermal effect of the sun on carbonic acid. So Foote repeated her experiment, this time isolating carbon dioxide in one cylinder, and air in another, and exposed them to sun. The cylinder of carbon dioxide grew significantly hotter than the air. Even after she'd moved it out of the sun, she wrote, 'it was many times as long in cooling'.[37] The carbon dioxide inside the cylinders absorbed and stored the sun's heat. Foote had identified a greenhouse gas.[38]

This was itself a huge discovery. But Foote didn't stop there. She wrote up her findings in a paper, 'Circumstances affecting the heat of the sun's rays', which was published in the November 1856 issue of the *American Journal of Science*. Here, she drew conclusions about what her findings indicated about the Earth's temperature. Foote reasoned that, given its heat-absorbing properties, more carbon dioxide would mean a warmer atmosphere: 'an atmosphere of this gas would give to our Earth a high temperature.'[39]

Foote's experiment demonstrated a principle, but the connection she drew between levels of carbon dioxide and the Earth's climate changed the way scientists would make sense of the Earth's history. In the mid-nineteenth century, the scientific world was discovering that the Earth was much older than previously thought, estimates shifting from tens of thousands of years to hundreds of millions (though we now know, it was more like billions). New research into climate, meteorology and the effects of the sun and planetary motion had led geologists to posit that the climate had been very different in our planet's infancy.[40] The Earth had once been much hotter. Scientists wanted to understand why.

A book by the Scottish geologist Charles Lyell, published in 1830, titled *Principles of Geology*, had suggested that volcanoes, glaciers and erosion accounted for much of these dramatic changes. Lyell was still up against scientists who explained the geology of boulders and other coarse sediments like sand or gravel in biblical terms, as the relics of God's catastrophic floods. But there had been other scientific attempts to explain the changes in temperature. Joseph Fourier, for example, had begun to describe the physics of atmospheric heating in 1824; he suspected that changes in how humans used the land were the cause of the prehistoric temperature shifts.[41]

Foote's insights overturned all the previous theories and the implications were immediately obvious. An 1856 article in *Scientific American* summarised the impact of Foote's work. Before Foote, they wrote, the leading scientific minds believed the 'great atmospheric heat to the [once] elevated temperature of the earth' was due to the law of entropy.[42] The law stated, in the words of one well-regarded proponent of the theory, the German physicist Hermann von Helmholtz (known, among other things, for his contributions to 'field theory'), 'the inexorable laws of mechanics [that] indicate that this store of force [heat energy], which can only suffer loss and not gain'.[43]

Foote's experiment suggested a different explanation. Scientists studying the Earth's geological history had also found that there had been periods when coal deposits had formed in swampy seas, which correlated with lower atmospheric carbon dioxide; perhaps coal trapped carbon dioxide from the atmosphere.[44] Foote's theory drew connections between this newly unearthed history and her observations, attributing the cooling of the Earth to a decrease in carbon dioxide in the atmosphere. Foote's paper, as *Scientific American* concludes, 'leaves but a small foundation for [the former] theory to stand upon'.[45] Foote had carved out a new way of understanding a change in the Earth's climate over geological time.[46] But more than that, she opened up what was possible for other women in science, as *Scientific American* declared: 'Some have not only entertained, but expressed the mean idea, that women do not possess the strength of mind necessary for scientific investigation…the experiments of Mrs. Foot [sic] afford abundant evidence of the ability of woman to investigate any subject with originality and precision.'

Foote believed in her own findings and thought she'd earned the right to showcase them. Besides publishing, she also submitted it to the American Association for the Advancement of Science (AAAS). The AAAS was a travelling showcase of the most cutting-edge scientific advancements and the country's first national science association. If you presented there, newspapers would report on your ideas, and you could also get covered in the write-up of the *Proceedings,* a published record of the papers presented at the meeting. It was a good opportunity to build a reputation.

Unfortunately, the AAAS was also a boys' club.[47] It wasn't that women weren't *allowed* to participate…it was that they simply didn't. Women sat in the audience, quiet, demure – seen but not heard. Speaking publicly as a woman was generally considered a transgression into a man's domain, and it could mark you as an outsider to elite social circles: unfeminine, brash – a bluestocking.

Foote, though, had the connections to get her foot in the door, namely her closeness to Joseph Henry. During his travels, Henry had met Mary Somerville in England, *the* leading female scientist of the time (and as we saw, tutor to our friend Lovelace).

He was surprised to discover that she was attractive – he had expected a beard, a 'manly-looking' woman to be avoided at all costs, of course... But meeting this beautiful scientist entirely changed his attitude to women in science. Since female scientists remained attractive, he had nothing to fear (that, of course, was the main concern here). He decided he would become the initiator of women entering organised science.[48]

Of course, Foote's research was very much in his ballpark and Henry recognised that Foote had the skills of an excellent scientist, and that her work deserved recognition. She was ripe to become the American Somerville as far as he was concerned. As long as she maintained her evening grooming ritual.[49]

Henry clearly wasn't a revolutionary. He was a powerful but moderate man and, aware of the reaction Foote would get in front of a crowd, he decided on a less provocative approach to share Foote's science.[50] He would present her paper himself.

This is how Foote came to attend the presentation of her findings at the tenth annual meeting of the AAAS in New York on 23 August 1856 – as an audience member. With her husband, she travelled to Albany, where their scientific educations had begun. There, Elisha got to read his own paper, while Foote got the privilege of listening to the esteemed professor, Henry, read her paper from the senate chamber of the State Capitol. Henry prefaced his presentation with an attempt to reconcile his idea of 'womanhood' with science. 'Science was of no country and of no sex. The sphere of woman embraces not only the beautiful and the useful, but the true,'[51] he announced. (As long as she was beautiful, 'ey?)

We will never know how Foote felt about Henry presenting her findings. Perhaps she was happy to be getting a hearing at all – she was the first woman to ever have a paper presented at the AAAS, after all. Moreover, when she wrote another paper the following year, she presented it herself. Her first paper was also published in the *American Journal of Science and Arts*.[52] And, most importantly, she and Elisha were the only two from the meeting to get a write-up in *The Annual of Scientific Discovery* – this, we'll see, became a key piece in her recent reappearance in the history of science.[53] Perhaps she was consciously tiptoeing her way into the room. Or perhaps she felt the same, subtle disappointment many women feel when a colleague takes credit for their idea in another all-male boardroom. Desperate for a seat – any seat – at the table.

In the end, though, the AAAS did not provide the springboard Foote hoped for. It was, after all, just one institution in a world full of institutions determined to keep women out of science. Her paper wasn't included in the *Proceedings*.[54] And, while Henry made a point of telling newspapers how important the research was, here too the reception was mixed and, with feminism still in its infancy, centred mainly on discussions about women's ability to think scientifically at all.[55]

The content of Foote's work barely registered in this fraught political terrain. The AAAS didn't open any meaningful doors to her. Discussions of her work in America and Canada quickly faded. And her findings never really reached the scientific centres of Europe. A few German and Scottish journals covered her work, but, like Henry, didn't understand the broader implications. There was also no mention of her in any correspondence between physicists of the time – neither in Europe nor America. As the historian of science Sir Roland Jackson said, 'her work sort of blazed like a meteor and then disappeared'.[56]

And yet, her insights made a curious reappearance just a few years later. An Irish scientist called John Tyndall had reached very similar conclusions. He'd arrived at insights by very different means. Tyndall was not working from a home laboratory. Far from it. Tyndall was a professor of natural philosophy at the Royal Institution in London. Attending the University of Marburg in Germany, the intellectual centre of European physics, he'd studied under Robert Bunsen (yes, of the Bunsen burner) and been noticed by Michael Faraday, the renowned English chemist, physicist and father of the electric motor and generator among other inventions (and a supporter of Lovelace). After Tyndall joined the Royal Institution, he gained access not just to the latest research, but to its high-tech equipment. He began to investigate the heat-absorbing properties of gases. He followed a very similar method to Foote, using the same basic components of gases, heat and thermometers. But his experiment was elevated. Rather than relying on sunlight, he used boiling water in a copper tube, which emits long-wave infrared radiation. He used a spectrometer and had assistants to help him measure his results. Tyndall drew similar conclusions to Foote about the heat-absorbing effect of carbon dioxide and went on to show that it wasn't actually the direct sun radiation that caused the atmospheric heating but infrared energy that was re-radiated from the Earth's surface, due to absorption and emission of radiation by oscillating molecules.[57] With contacts all over Europe and a strong position in England, Tyndall published his findings to great acclaim and long-lasting status as the father of climate science.[58]

Sir Roland Jackson, mentioned previously, also researched and wrote extensively on Tyndall. He argued, along with other historians, that Tyndall added a rigorous explanation of the phenomenon Foote had demonstrated.[59] Perhaps that justifies his status as a founding father. He effectively drew together a succession of discoveries, completing the chain. Foote built on the findings of Saussure, Fourier and Henry by demonstrating the heat-trapping effect experimentally. Tyndall explained the mechanism. Given this lineage, it is difficult to determine who coined the idea. But one insight is clear: that, while Tyndall credited many of the scientists who came before him, there was no mention anywhere in his paper of Eunice Foote.

Historians are divided as to why Tyndall failed to mention her. Jackson, for example, argues that it wasn't intentional; he simply hadn't seen her work. The worlds of American and European science were quite separate. Scientific communication was slow, and it is possible that Tyndall wasn't so tuned into the American scene. After all, as we have seen, she wasn't widely publicised in Europe, so other scientists of the time also didn't know about her work. The sequence of Tyndall's experiment supports this theory. In her experiment, Foote had demonstrated that carbon dioxide and water vapour were good heat absorbers, while the other gases in mixed air weren't. But Tyndall started his experiments by testing hydrogen, oxygen and nitrogen, and only later moved on to methane, coal gas and, finally, carbon dioxide. If he'd known about Foote's work, wouldn't he have jumped straight to carbon dioxide?[60] It is also possible, of course, that Tyndall wanted to confirm Foote's findings before moving on.

Or maybe he wanted to stage a convincing 'first'. Tyndall had failed to credit the work of Joseph Henry in a previous study on sound waves, which became an international scandal.[61] Perhaps plagiarism was a bit of a habit for Tyndall...[62] Tyndall could have encountered Foote's paper at so many junctures that it's hard to believe he wasn't at least aware of it. The historian John Perlin is writing a book on Foote and has researched her extensively. He points out that the Royal Institution received the November 1856 issue of the *American Journal of Science and Arts* that included Foote's paper, as well as an article on colour blindness by Tyndall. Surely, Tyndall would have had a flick through to look at his own article, at least glancing over Foote's piece as he looked for his page? Another major scientific journal, *Philosophical Magazine*, also published Foote's article – Tyndall was its *editor*. How could he not have seen her piece?[63]

There is the other, no less disconcerting, possibility that Tyndall wilfully skipped over Foote's paper even if he did come across it. Tyndall didn't think that women could be scientifically creative. He conceded that they could probably *understand* the experiments men had invented, but he didn't believe they were capable of devising meaningful experiments themselves. This leads to the possibility that, on seeing the name 'Eunice' under a paper, he simply decided not to read it.[64]

Tyndall may equally have ignored the paper due to the author's nationality, not her gender. Many European scientists had a superior attitude towards the work happening in America; that the science there, which was more focused on inventing, was not 'proper science'. The field of physics was emerging in Europe with some paradigm-shifting insights, like the conservation of energy, light waves, heat as the motion of particles and the functioning of electricity, all using new, expensive tools and expressed in advanced mathematics.[65] America, they believed, didn't have much to show for itself in comparison.

Of course, Tyndall wasn't the only one to ignore Foote. As we saw, her work disappeared in the ether not long after her presentation at the AAAS. Jackson, in fact, argues, that Tyndall's scientific rivals would have jumped on the opportunity to expose him for failing to credit Foote, had they known of her work. It is possible that she was simply collectively overlooked amidst the swell of other findings by more 'credible' scientists.[66]

All these explanations are frustrating. Despite Foote's access to a first-rate scientific education, equipment and time, it still wasn't enough to get her work credited. She never entered into the annals of science 'proper'. Instead, her career progressed in the way that would have confirmed Tyndall's pejorative attitudes towards her 'American' branch of science. Foote wrote one more scientific paper, on static electricity, but quickly moved on to a life of inventing. One of her inventions, an oven thermostat, made her rich. She continued inventing well into her forties, also filing patents for a rubber shoe insert to 'prevent the squeaking of boots and shoes'. A far fall from ground-breaking discovery. She was now better known as the wife of a leading government figure (and the mother-in-law of another; her daughter married Senator John Henderson, an intimate of Abraham Lincoln).[67] She eventually became a full-time housewife. In a letter to her daughter, Foote wrote, 'I look after all things in the house now.'[68] The AAAS meeting had sealed her fate as a largely unknown name in climate science.

Until today.

There has been more talk of Eunice Foote recently, with a flurry of articles and podcasts recounting the forgotten history of her contributions to climate science. Most of them start with the unexpected discovery of her work by the retired petroleum geologist Ray Sorenson in 2011. It makes for a good story: the amateur historian, with a quirky interest in old scientific journals, landed on Foote's article while flipping through that 1857 *Annual of Scientific Discovery*, which, as we saw, was one of the few places Eunice's work was documented. Sorenson was surprised to see this early paper describing the greenhouse effect that he thought Tyndall had discovered. Tyndall's paper was published three years after this one. Sorenson posted a blurb online and wrote a paper about his discovery,[69] which sparked a wave of articles: everywhere from the *New York Times* to *Scientific American*.[70] This is Foote's second chance to get the recognition she deserves. A chance to become more than a flash in the history of science. This time round, will we remember her?

Only if we are prepared to rewrite history.

In Foote's time, science fuelled a toxic yet intoxicating machine. Drawn to the vision of a more productive, technologically advanced society, scientists wanted to consider the damage their inventions – the combustion engine, the electric generator, the machine gun – were wreaking on the planet, as little as they wanted to hear from women. While Foote faded into a life of domesticity and inventing until her death at age sixty-nine (just a few years after Elisha), her work lives on. It persists with an alternative history of the world, containing warnings about the cost of unfettered pollution, as well as an early suggestion in the history contained in this book: that the man-made structures used to credit science might be as fallible as the glittering narrative of scientific progress. Since Foote published her finding, the concentration of carbon dioxide in the atmosphere has increased by fifty percent. Around the world we are seeing the collapse of ancient ecosystems. It is not enough to recognise Foote; we must *listen* to her.

5

Florence Nightingale (1820–1910)

Nurse and Statistician

Florence Nightingale is not just the most iconic nurse of all time, Nightingale *is* nursing. All over the world she has become a symbol for a profession, embodying the qualities we expect from nurses today. She earned the name 'Lady with the Lamp' serving during the Crimean War, where her capacity for caring took her to the soldiers' bedsides in the night, lantern in hand to light the way. The ideal of Victorian gentle femininity, and yet she was so much more. She was also a ground-breaking statistician who revolutionised the sanitation practices in hospitals across the world. She was a devout Christian mystic called to her mission by God. A social campaigner for sex workers' rights. A woman who injected her vision unrelentingly into the male-dominated hospitals in which she also worked. The ideal of a retiring (in fact, even bedridden) Victorian woman on paper, and yet, in practice, outspoken, unmarried and unflappable.

Nightingale has become synonymous with a gentle, caring and white femininity that travels into the present and inflects our image of nursing. May is National Nurses' Month in the US because Nightingale's birthday was on the twelfth. The World Health Organization made 2020 the Year of the Nurse and Midwife to mark Nightingale's 200th birthday.[1] But Nightingale as symbol of nursing misses both some of her most essential contributions to science, as well as the more troubling parts of her legacy. Perhaps more than any of the women in this book, when we scratch just a little beneath the surface of Nightingale's caricature, we find a complex, breathing human impossible to reconcile it with.

*

Nightingale was named, a little obviously, after her birthplace in Florence, Italy, by her wealthy British parents, William Edward Nightingale and Frances Nightingale Smith, though the pair raised her and her elder sister, Frances Parthenope, in England. Her father's family wealth came mainly from the Derbyshire mines, but some of it also from Smedley's Mill, a cotton mill which, like all the others at the time, used cotton grown by enslaved people in the Caribbean and American South.[2] This past sits uncomfortably alongside the legacy of her mother, who came from a family of liberal progressives. Her grandfather and uncle were both Liberal MPs and her grandfather, William Smith, an anti-slavery campaigner.[3]

Nightingale, we will see, took from both sides.

These progressive, if aristocratic, reformers shaped Nightingale's early education and outlook. Her father, a former MP, dedicated much of his time to home-schooling his daughters. He defied the conventions for upper-class girls' educations at the time that were designed to prepare women for courtship.[4] He passed his own University of Cambridge education onto them. Nightingale took to the learning. She was particularly interested in maths and statistics, an emerging field at the time.[5]

But the most valuable aspect of Nightingale's upbringing and education, however, was that it prepared her to hold her own against men in power. This skill was crucial to everything Nightingale went on to do. Class divides in British society at this time were stark and unquestionable.[6] At every juncture of Nightingale's career, her class was relevant: sometimes it compensated for the ways in which she transgressed gender lines, granting her access to male spaces, lending her the authority. When dealing with MPs or military officials, she matched them academically with a confidence learned from childhood.[7]

While Nightingale's education empowered her outside the home, it also taught her about the world's problems. Her family upheld the upper-class tradition of visiting the homes of poor people on their estates, bringing food and even providing healthcare. But Nightingale grew frustrated at her limited impact.[8] By age sixteen, she saw that she had academic capabilities to be of greater use, but that she couldn't use because, as per the conventions for upper-class girls, she was confined to the home. Nightingale did not enjoy the life of a housebound, corseted Victorian woman. In her 1852 proto-feminist essay 'Cassandra', she laments 'suffering, sad, female humanity!' and describes her existence as a kind of death in life. 'Why', she urges, 'have women passion, intellect, moral activity – these three – and a place in society where no one of the three can be exercised?'[9]

In the end, a serious influenza epidemic set her free.[10] Nightingale, still a teenager, stepped up and cared for the entire household. In 1837, towards the end of the outbreak, she believed she received a visit from God, imploring her to do his work.[11] Nightingale's religious calling was sincere, but it was inextricably connected with an alternative vision of her own life. Her new-found purpose offered a way out of the stifling confines of Victorian domesticity. She never returned to it; she never married. Her work during the epidemic gave Nightingale the sense of autonomy she craved. Reflecting years later, after she had started her formal training, she recalled it almost fondly; it had allowed her, she wrote, to feel useful in an otherwise 'wholly unpractical' life.[12]

It wasn't just the epidemic, but a changing world, that made a career in nursing possible. As Dr Richard Bates, lecturer in history at the University of Nottingham, explained to me in an interview, this was a time where new medical developments were creating a demand for skilled nursing.[13] This included new, more daring operations – tying off aneurysms, operating on fused joints, excising diseased bones (this was considered 'conservative' surgery since it didn't amputate the entire limb). The risk of haemorrhaging after an operation was high, and it required the constant support of trained nurses.[14] In the 1840s, first dentists and then doctors took up diethyl ether, nitrous oxide and chloroform (gases that had caught the public interest mainly as party drugs) to temporarily sedate patients for surgery. General anaesthesia transformed excruciating surgeries into relatively painless processes where patients were kept in an unconscious state. In 1846, Boston dentist William Morton publicly demonstrated the use of ether as general anaesthetic for surgery at Massachusetts General Hospital. In Edinburgh, professor of midwifery James Young Simpson advocated for chloroform to be offered to women during childbirth. Anaesthesia was definitively proven safe when none other than Queen Victoria inhaled it during the birth of two of her children. Doctors needed nurses to assist in dosing the gases during surgery.[15] General pain relief was also becoming a vital part of care that, while usually managed by doctors, was delegated to nurses in hospitals.[16]

Then there were new ideas about hygiene. Slowly, physicians were turning from the miasma theory that held that bad smells spread disease to the first insights into contagion and sanitation in spreading disease. The findings of the Hungarian physician Ignaz Semmelweis showed that handwashing reduced mortality rates on his maternity ward.[17] Social reformers, like Sir Edwin Chadwick in Britain, were starting to expose the link between the filth of industrialisation (not least its open sewage systems) and cities and disease. As hospitals began to prioritise keeping equipment and surgical rooms clean, the work, again, fell to nurses. The most time-consuming duty of all, though, was feeding the patients nourishing diets and frequent meals. The brunt of the burden of the new, much improved medical care fell on nurses. They were responsible for administering everything from beef teas (bone broths) to enemas, opium to chloroform.[18]

There was a higher demand for nursing than ever before. Throughout society, there was a movement of wealthier women taking up jobs outside of the home.[19] Social norms were slowly changing. But not soon enough. Nightingale struggled to gain her family's support. She had spent her early adulthood visiting hospitals across Britain, Europe and even Africa, observing nurses and doctors in action. This was all fine – a gentlewoman's curiosity – but now she wanted to train professionally she faced opposition. She had found a school that taught the basics of nursing, record-keeping and hospital organisation at a school in continental Europe. But her parents were against it. No wonder. Women may have been rattling the bars of their domestic cages, but nursing was still considered an unsuitable profession for young women of Nightingale's background. One of the best-known nurses before Nightingale was Dickens' Mrs Gamp in his *Martin Chuzzlewit* – an incompetent drunkard who embodies the very deviation from respectable femininity. According to Dickens: 'Like most persons who have attained to great eminence in their profession, she took to hers very kindly; insomuch that, setting aside her natural predilections as a woman, she went to a lying-in or a laying-out with equal zest and relish.'[20]

The low status of nursing also made it a dangerous profession. Working-class women were more often sexually assaulted, and Nightingale's parents worried that, working with male doctors and tending to male patients in hospitals, she would be exposed to the same dangers. Not at all an unreasonable concern. As Dr Bates explained to me: 'this was not the modern NHS, and hospitals did not have good employee safeguarding policies, to put it mildly'.[21] It is hard to grasp now just how disreputable nursing was in England at the time; it was considered dirty work, lower than that of a servant. All this helps explain why Nightingale's parents reacted, in her own words, 'as if I had said I wanted to be a kitchen-maid.'[22]

Eventually they conceded. In July 1850, Nightingale travelled to continental Europe, to experience nursing with the Protestant Deaconesses in Kaiserswerth, now in modern-day Germany. Nightingale couldn't have chosen a more respectable institution. The institute was established in 1836 by Pastor Theodor Fliedner and his wife, Friederike, to train young women as 'Protestant attendants' for the 'deserving' poor, orphaned children and the sick.[23] They called their graduates 'deaconesses' to evoke a Christian tradition of solidarity with the poor. There was an urgent need for it. The Industrial Revolution had drawn swarms of rural families to work in urban centres that brought more problems than possibility. Cities were overcrowded and unsanitary, wages were low and insufficient, the working conditions exploitative and dangerous. Poverty and disease were on the rise.

Besides this need, deaconesses offered unmarried women a socially acceptable independent lifestyle. The women were celibate and lived in a 'motherhouse,' a convent-like community. (Though in a letter Nightingale suggests otherwise. Writing to her friend, the feminist writer Mary Mohl, Nightingale boasts about having excited the 'passions' of the 'Prussian Bäuerinnen', and other women, she shared a bed with… Nightingale may not have been the chaste virgin we imagine; perhaps – shock, horror! – not even heterosexual.)[24] The deaconesses didn't dress like nuns but wore the dress and bonnet of married women (this bonnet later became the nurse's cap). The community modelled the correct social hierarchy, putting an aristocratic spinster in the position of supervisor, compensating for their subversiveness by adhering to the social order in all other ways (the rest, they did behind closed doors).[25]

These elements of religious decorum, gender compliance and class structure helped to establish the school's reputation and distinguish it from the 'immoral' nursing of the past. The education was rigorous and standardised, with the prototype nurses taught medicine by a licensed physician.[26] The curriculum included ethical standards and medical theory, and gave supervised practice for hospital and home nursing.[27]

After a three-month course, Nightingale left the motherhouse, but she would always remember the place fondly as her 'spiritual home'.[28] In 1851, she wrote a booklet, published anonymously, promoting the school's approach, especially its emphasis on practical training. This would become key to her nursing philosophy. Hands-on experience was generally lacking in women's education, she observed, 'while the intellectual foot has made a step in advance, the practical foot has remained behind. Woman stands askew'.[29]

Returned to London in 1853, Nightingale became 'Lady Superintendent' at the Institution for the Care of Sick Gentlewomen (or, educated women with moderate means).[30] The original nursing home had existed in Chandos Street, just off Cavendish Square, since 1848. But it had only six beds. When Nightingale arrived, the hospital moved to a new three-storey building with twenty beds on Harley Street. Now it needed serious reorganising.[31] The position proved that Nightingale had absorbed more than just the medical training at Kaiserswerth. The emphasis on propriety at the school had been a masterclass in how to make her presence acceptable, pragmatically, to get things done. She had at the time the most inconspicuous assistant: a tiny owl named Athena she'd rescued in Greece. Like the owl in her pocket, Nightingale slotted inoffensively into the hospital hierarchy. She subordinated herself and her staff to the doctors on the wards, deferring to them for decisions on treatment. When it came to policies and procedures, though, she channelled her aristocratic confidence to challenge leadership and even went over their heads when she felt it served the patients. She insisted that hot water was piped to each floor, demanded a dumbwaiter to carry meals from the basement kitchen, and installed the equivalent of a buzzer; a bell by the patient's bed that would 'ring in the passage outside the nurse's door' and a 'valve which flies open when its bell rings, and remains open in order that the nurse may see who has rung.'[32] Nightingale didn't stop at practical innovations. She also challenged the committee on fundamental values, pushing them to open the hospital to all faiths, and not just the Church of England as they had proposed. Nightingale soon figured out how to bend the committee to her will. She wrote to her father at the time:

> Now I perceive that I do all my business by intrigue. I propose in private to A., B. or C. the resolution I think A., B. or C. most capable of carrying in committee and then leave it to them. And I always win. I am now in the heyday of my power.[33]

Nightingale was carving out a place for herself in the hospital system. She drew on her understanding of social hierarchy, as well as her own position in that system, to implement changes where she could get away with them. The job was a good training ground for the work ahead of her, but the hospital was small and soon bored her. Nightingale resigned after just a year.[34]

*

The Crimean War brought a challenge Nightingale was unlikely to outgrow. In 1853, the Ottoman Empire was in decline. The power vacuum presented a threat to some and an opportunity to others. Russia saw the chance to expand its influence over the Balkans and gain access to the Mediterranean, while Britain and France were invested in the current balance of power, along with their trade routes. Religious tensions finally tipped the balance, when, in 1853, the Ottoman Empire's government denied Russian Tsar Nicholas I protection of the Orthodox population of Palestine. Russian troops mobilised. They occupied the Ottoman region of the Danubian Principalities in July 1853. The Ottomans declared war on Russia in October. With Russia spreading its fingers through Europe, England and France moved to join the Ottoman Empire against Russia in 1854. Nightingale took her post just as the alliance was about to declare war against Russia.

Nightingale's connections got her the job. Her close friend, Sidney Herbert, the secretary of war, authorised her to bring a team of thirty-eight volunteer nurses to the military hospital in Scutari, in modern-day Turkey. Herbert would have been sceptical of sending any other woman to issue orders in the male hierarchy of the military world. But Nightingale possessed some unique qualities. As he explained to her: 'your knowledge and your power of administration, and among greater things your rank and position in Society give you advantages in such a work which no other person possesses.'[35] Above any experience or ability she might have, Herbert saw Nightingale's social status as a key qualification. The presence of a woman was a clear transgression in the male military world. Nightingale, though, was all-but aristocratic, from a political family with government connections, and with a man's education. She was the most suitable a woman could be.[36]

Nightingale was determined to prove that female nursing could make a difference in the army.[37] This meant she needed the strongest possible team. She only recruited nurses with ample experience. Some were middle- and upper-class women, but most of the group were nuns (twenty-four of the thirty-eight).[38] With her team of 'respectable' women in place, she would use the same approach as she had at the gentlewoman's hospital. She instructed the women to defer to the authority of the sceptical medical officers on the wards. If there were no medical orders to give a patient fluids, Nightingale would not allow it – even if this proved fatal. She was rigid about the hierarchy. Some of her nurses resented her for it.[39]

Nightingale also demanded respectability from her nurses. She was aware of how much there was to prove, and of the damage a single misstep could do to women's future involvement in military nursing. She banned her nurses from drinking or fraternising with soldiers.[40] She was especially suspicious of her working-class nurses, and she never let them go for a walk unaccompanied by a lady nurse, just in case a soldier scandalised them. This was how Nightingale won respectability for the profession – less by challenging attitudes, than by managing and replicating them.[41] As always, walking the tightrope of social expectations: risking ostracisation and destitution with every step in this 'unfeminine' domain.

It quickly became clear where there was much work to be done. Army hospitals at the time were dire places. Officers and high-ranking officials avoided them. They were sent to places like the Şehzade Kiosk in Haydarpaşa, or the Sultan Kiosk in Tarabya – bright, breezy rehabilitation centres. Hospitals were for the regular troops. For all the medical advances of the previous century, hospitals were still largely seen as 'gateways to death'.[42]

Scutari, the main British military hospital in Crimea, was no exception. When Nightingale and her nurses arrived, the walls, floors and ceilings were coated in dirt; rats scurried under the beds; and the soldiers' clothes were infested with bugs, lice and fleas.[43] There were just fourteen baths for the 2,000 soldiers at the hospital and an undersupply of towels, basins and soap.[44] Infections were rife, and the death toll was higher than any other hospital in the region.[45] Nightingale and her team had little time to adjust. In their first week in Scutari, the already straining hospital received a new influx of soldiers from the battles of Balaklava and Inkerman, which injured thousands of Allied soldiers.[46] Nightingale described the place in a letter to her friend, Dr William Bowman, in 1854:

> But oh! you gentlemen of England, who sit at home in all the well-earned satisfaction of your successful cases, can have little idea from reading the newspapers, of the horror & misery (in a military Hospl.) of operating upon these dying and exhausted men…I have no doubt that Providence is quite right and that the Kingdom of Hell is the best beginning for the Kingdom of Heaven, but that this is the Kingdom of Hell no one can doubt.[47]

Nightingale quickly diagnosed the problem in Scutari. Like most medical professionals at the time, she believed the miasma theory of disease. According to the theory, diseases were caused and spread by a poisonous and foul-smelling vapour carrying suspended particles of decaying matter.[48] Nightingale's fundamental assumptions about how diseases spread proved inaccurate – scientists began to prove the relationship between germs and disease in the 1860s, and by the 1880s they'd established germ theory.[49] Nonetheless, the emphasis on hygiene was key. Nightingale saw that diseases could be transferred among sick patients. Her approach was not to cure but to prevent these diseases by tackling what she thought was the mechanism by which they spread.

Nightingale put the death toll in Scutari down to three key issues: diet, dirt and drains.[50] In her first month, she established a functioning hospital laundry system. As the army's purveying department was incredibly inefficient,[51] Nightingale sometimes used her own money to buy supplies: 200 Turkish towels for the soldiers, clean shirts, soap and kitchenware. She imported food from England, cleaned up the kitchens and set her nurses to cleaning up the hospital wards.[52] She also made sure to express her disappointment to her friend Herbert for sending her so unprepared: 'You have sacrificed the cause, so near my heart. You have sacrificed me, a matter of small importance now.'[53]

Nightingale wasn't the only one to take issue. Crimea, unlike previous wars, was well reported. Public criticism of the government's management in Crimea was growing. In January 1855, the House of Commons finally passed a motion for a committee to investigate and improve on the conditions in the Crimea. The Sanitary Commission did some of the more heavy-duty cleaning in the hospital: just everyday chores like detoxifying toilets, clearing out sewers and removing a dead horse from the water supply.[54] It seemed to help. Within just a few months, the mortality rate at the camp dropped from 42.7% to 2.2%.[55]

The major cause of death in Scutari, it turned out, wasn't war. It was poor hygiene. Nightingale had witnessed a real-life experiment in the power of sanitation. But this connection between living conditions and health, a cornerstone to medical care today, wasn't obvious at the time. Nightingale believed now, more than ever, that sanitation standards should be standard in all hospitals across the board. But the policymakers back home would need convincing.

*

I first became interested in Nightingale after my mentor at the British Library revealed her diagrams to me. In a small room, I and a few other students huddled around the unfolded parchment in the dimmed light. They were captivating; an elegant rose of pastel blues and pinks was my map connecting Nightingale to the military frontline. The intricate wisps of her hand's lettering gave me a view of her story as she'd seen it. 'Diagram of the causes of mortality in the army in the east.' Bold, ambitious, at the centre of power. This was a side to Nightingale I'd never seen.

Working with Dr William Farr, the leading medical statistician of the day, she would present evidence to the Royal Commission set up to investigate healthcare during the Crimean war. They collated data from military hospitals in Constantinople that confirmed Nightingale's suspicions: far more British soldiers had died of disease than in combat. During Nightingale's first winter at Scutari, ten times as many soldiers died from diseases associated with poor hygiene than from their wounds. Moreover, deaths dropped dramatically once hospitals at the front were properly cleaned. To this, they added data from military hospitals in Great Britain, where, bereft of Nightingale's stringent sanitation measures, mortality rates were even higher than those at Scutari.[56]

While on paper the numbers spoke for themselves, Nightingale knew they needed to go further. Statistics was an emerging field and poorly understood among the general public. She would need to *show* the difference she had made. The diagrams were, in her words, there 'to affect thro' the Eyes what we may fail to convey to the brains of the public through their word-proof ears'.[57] Nightingale was inspired by other statisticians in the burgeoning field. In 1801, William Playfair had published a visual argument for lowering taxes; the first published pie chart showed that Brits paid more tax compared to the citizens of other countries. Besides this, a revolution in the use of statistics in social reform was underway. Led by Farr himself, the General Registry Office at Somerset House had begun to systematically record births, deaths and marriages in the UK in 1837. The registration statistics were a rich resource for new studies into the causes of social problems.[58]

In their pie chart, Farr and Nightingale split a circle into wedges the size of which corresponded to the mortality rate per month.[59] The colours of the wedges represented the cause of death: blue for preventable deaths, red for wounds and black for other causes. The charts covered two years: the first year, 1854–5, following her arrival in the region, and the second, the year after the arrival of the Sanitary Commission. Together, these strikingly visualised the direct relationship between improved sanitation and decreased mortality rates.[60] Her work helped bring the issue of overcrowded barracks to the Houses of Parliament, which moved to reform the sanitary conditions of the army.[61] This remains Nightingale's best-known diagram and was widely reproduced in the press.

Nightingale's statistics became her toolkit for a whole range of reforms. In the early 1860s, the British government planned to enact legislation to combat syphilis in the armed forces. Their solution? Police-enforced fortnightly vaginal examinations for suspected prostitutes in garrison towns. In 1862, Nightingale published a report critiquing the government's Contagious Diseases Acts (CDAs). The horror of using women's bodies as a proxy STD test for men was not lost on Nightingale. She saw the social, usually economic, conditions that drove women to sex work and did not consider these women immoral, as others did, but fully deserving of their freedom. Ever the pragmatist, Nightingale devised a scientific, rather than a political, case against the CDAs. The title of her report is unambiguous: 'Note on the supposed protection afforded against venereal disease by recognising prostitution and putting it under police regulation.' Nightingale used comparative statistics from jurisdictions with and without compulsory examinations, which proved that the measures made no difference to STD rates.[62] 'The assumed protection of the troops at Malta, Gibraltar and Corfu [which had police regulation] is simply a myth,' she wrote.[63] Clear as day. The government ignored her advice and went ahead anyway. Nightingale persevered and collaborated with leading feminists, including the theorist Harriet Martineau and activist Josephine Butler, to publish her findings in newspapers and campaign for repeal. In 1886, they were successful. A true feminist triumph, also made possible by Nightingale's unconventional, arguably queer, lifestyle. Whatever her sexuality, she dedicated her life to women: 'my experience of women is almost as large as Europe,' she wrote to Mohl.[64] She prioritised her female friendships above all others (Ada Lovelace was another!). She corresponded with them frequently, rejecting any marriage proposals that tried to distract her from them and their cause.

Over the next decade, Nightingale continued to apply her statistical thinking to a range of public health questions. She continued to collaborate with sanitation reformers like Farr and Chadwick to campaign for proper drainage and sewage systems for public buildings, hospitals, barracks and working-class houses in London, calling for these to be built according to the sanitary principles she had learned in Crimea.[65] Her diagrams represent a side to Nightingale's skillset lost in the image of her as a military field nurse. She spent the largest part of her career not as a nurse but as a statistician and reformer.[66]

Nightingale knew how to run a persuasive campaign. But she was also riding a wave of publicity that had already made her the centre of public attention. The British press had become enamoured with Nightingale and her saintly dedication to their countrymen's recovery. The reality was that she spent more time organising wards than she did tending to patients. Ward management continues to be the bulk of a head of nursing's work today. This, in addition to preventative healthcare, like screenings and immunisation, and sanitation, like cleaning, education and managing outbreaks, continues to be devalued compared to the work of doctors and medicine 'proper'. Still, the image of the lady nurse persists, feminising the practical work in order to denigrate it. The practical work was always more crucial and, at the same time, more intellectually demanding than the image of the bedside nurse did justice to. And it continues to be.

The Times war correspondent Sir William Howard Russell did much to cast Nightingale in her most feminine light. After a visit to the hospital in February 1855, he described her for the public:

> She is a 'ministering angel' without any exaggeration in these hospitals, and as her slender form glides quietly along each corridor, every poor fellow's face softens

> with gratitude at the sight of her. When all the medical officers have retired for the night and silence and darkness have settled down upon those miles of prostrate sick, she may be observed alone, with a little lamp in her hand, making her solitary rounds.[67]

Nightingale also received a front-page spread in the *Illustrated London News*, which included the famous image of her carrying a lamp. The piece cemented her status as a national icon.[68] Nightingale fever took hold of the nation, complete with memorabilia and also plays, songs, poetry, ships, cafes and even racehorses with her name.[69] In all this, she was presented as angelic and maternal, very much the gentlewoman. She played her part as the domestic mother – so well, in fact, that Queen Victoria herself began to feel threatened in her status as 'mother' supreme.[70] Nightingale had once again made a place for herself in the existing social order, all the while claiming unprecedented authority for a woman, and especially a nurse.

Nightingale fever seemed to override any reservations the public may have had about women becoming nurses. In November 1855, a group of Nightingale fans held a public meeting in London to raise funds for the world's first professional nursing school. Nightingale opened her school at St Thomas' Hospital in London in 1860. The training reflected her general preference for practical training – hands-on, in wards and under the supervision of a ward sister.[71] Nurses would be trained in regular civic hospitals, regardless of their own class or educational background. This practical training was supplemented with classes taught by medical doctors, just as at Kaiserswerth, and it also included the same emphasis on ethical standards for patient care (although they wouldn't have used the term 'ethics').[72] The training was rigorous and set the profession on a par with the intellectual work of establishing hospital policies. 'There should be a professorship of hospital administration, hospital construction and hospital nursing,' she argued.[73]

Nightingale's training school established nursing as a secular, respectable profession for women – gone were the Mrs Gamps. It was undeniably a leap from the profession Nightingale had fought so hard to enter into just a decade earlier. And yet, that old world stayed with her. Ideologically, it was sometimes hard for her to shed.

A few years after she had established her school, Nightingale introduced a two-tiered system that distinguished between middle- and upper-class 'ladies' and working-class 'nurses'.[74] Ladies were required to pay for their training but were also instructed more closely and were exempt from some of the more menial tasks. The nurses, on the other hand, were paid a stipend but did the more laborious work.[75] While this sounds classist and unfair, there Nightingale had her reasons.[76] There was no state-funded education at this time, meaning that many of the nurses too poor to afford schooling could not do the reading or case reporting the job required.[77] Nightingale's training didn't aim to rectify these educational disparities; it worked pragmatically with them.

Nightingale's aspiration, though, was that her nursing hierarchy would transcend class. A 'lady,' in Nightingale's definition, was a well-educated woman who was primarily defined by her character, not her social status. She had self-control, good manners, confidence and the work experience required to take a leadership role. For all her good intentions, Nightingale's ideal here reflects her privileged worldview. She didn't see how the values she prized in her nurses, like confidence and manners, had been instilled largely through upbringing and education. Nightingale employed both poor and rich women who qualified as ladies, and there were wealthy women who would not qualify – but most did. By posing ideals without opportunities for women to rebalance the scales, Nightingale's hierarchy reinforced the idea that social constructs, like upper-class etiquette, were innate personal qualities. In doing so, she perpetuated the kinds of ideas that reinforced class divides within her nursing unit – despite her attempts to include working-class women in the profession.[78]

Nightingale had to be assertive, inventive and sometimes subversive to claim space for her profession and reforms within male-governed institutions. But in trying to establish her profession's authority, she sometimes also re-enacted the violence, arrogance and absurdity of the power structures that had trapped her.

Nowhere is this clearer than in her support of British colonialism.

After Crimea, Nightingale's next best job opportunity came from another warfront of the British Empire. In India, Britain was struggling to retain its imperial hold.[79] She would have travelled there, as she had to Crimea, but Nightingale's health had rapidly declined since the war. A range of inexplicable symptoms, including spinal pain, muscular weakness, depression and fatigue, plagued her for most of the rest of her life. She was retrospectively diagnosed with brucellosis, which she probably caught drinking infected milk in Crimea.[80] Being bedridden could have returned Nightingale to the domestic quarantine imposed on most Victorian ladies, but, ever determined, she worked from bed.[81]

From her bed, Nightingale's views on world politics also evolved. While paving the way for Victorian women at home, she started her career as an unabashed imperialist. From her home office, she wrote a letter to the newly appointed viceroy,

Sir John Lawrence, in 1864, urging him to conquer 'India anew by civilization…by knowledge instead of the sword'.[82] She presented her sanitation reforms as a tool in this 'civilising' mission. And while her instructions were based on statistical research, they were at first also informed by ignorance. Everything she knew was based on the sanitation conditions she had witnessed in the British army. With what can only be described as a colonial arrogance, she simply expanded her findings to that of rural Indians without ever having visited the country.[83] She recommended, for example, light and free circulation of fresh air for barracks and hospital designs, despite the country's entirely different climate and specific diseases.[84] Indian newspapers of the time appropriately took issue, pointing to the fact that Nightingale had never visited their country as a major hindrance to her being effective in her (self-appointed) role as sanitary advisor and social reformer. 'Indian villages are not the "happy villages" of England', one editorial read.[85]

Over the next three decades, Nightingale changed her position. She became increasingly concerned with the living conditions of Indian farmers. She began to see British rule as part of the problem. She was especially troubled by Britain's role in causing famine in India.[86] In her 1878 article, 'The people of India', published in the liberal journal *The Nineteenth Century*, Nightingale comes out with a bold criticism of government and public inaction in the face of the mass numbers of dead and starving in the famine (in modern estimates, over 8 million) that was then ongoing in Madras. 'We do not care for the people of India,' the article begins. The ministering angel of the British public had turned on her doting audience and implored them to act. She condemned the government's economic and social policies in India: aimed at extracting cash crops with little care for people's welfare.[87] She started to see that structural reforms were needed to meet people's needs. These reforms included land tenancy, agricultural practices and education – none of which would directly benefit the colonisers.[88] Her recommendations became ever more radical. She befriended members of the Indian National Congress in the 1880s and became an advocate for the country's independence.[89] She began to see the cruelty of disrupting local customs to impose foreign practices on colonised people.[90] She argued that local cultures and norms should be respected, that nurses should aim to empower the locals themselves to do the work. She even taught that British nurses could learn from some of the practices in India.[91] Increasingly, she moved towards a vision of reciprocity: what she called friendship. In 1894, she wrote about rural health education: 'Let not England lag behind – especially not in the conviction that nothing *can* be done without personal friendship with the women to be taught. It is a truism to say that the women who teach in India must know the languages, the religions, superstitions, and customs of the women to be taught in India.'[92] Compare this to her earlier approach and we can see that Nightingale underwent a true education in cultural humility.

Nightingale challenged herself and her country to face the violence of its imperial rule. Her imagination stretched farther than most but never quite beyond the dominant worldview. She stopped short of challenging the system itself. She saw the Indian people, in her words, as 'fellow subjects', and so never questioned whether they should have been subjects at all.[93] She also assumed that, ultimately, they wanted the bourgeois domesticity and 'cleanliness' she offered.[94] Even as her views progressed, she continued to default to the assumption that the Indian people needed Western education and guidance. When she designed a nursing programme for India, she recommended that English-trained nurses were sent over to set it up. Even after Indian nurses had been trained (according to her standards), she held onto the idea that the matrons should be European.[95] She underwent a dramatic transformation from unapologetic imperialism to supporter of self-governance, but she never fully transcended her context.[96]

'Could Florence Nightingale get cancelled?' a group of historians asked in 2022.[97] Some believe that championing Nightingale as a symbol of nursing condones her imperialist views. This is part of the reason, for example, why the New Zealand Nurses Organisation (NZNO) refused to celebrate Nightingale's birthday in 2020, after they uncovered her correspondence with colonial Governor of New Zealand George Grey. For the nurses of the NZNO, glorifying Nightingale condones the violence of his rule.[98] Whatever we make of this relationship (a debate in its own right), the new wave of attention for Nightingale's colonial involvement may well knock her from her pedestal as the 'mother supreme'. And maybe that's for the best. Nightingale was never the paragon the Victorian press made her out to be. She was a person working from within her historical and personal limitations. For all the work she did, she undoubtably reinforced some of the most harmful views of her time.

I don't think we should stop talking about Nightingale, though. For one, this would deprive us of a history that can show us how colonial thinking shapes our healthcare systems today. Besides that, though, so much of her work contains a vision of equal healthcare. Nightingale gave us a set of observations, an analysis and a profession built on the insight that poor and rich, colonised and coloniser are connected by an environment that can make us sick or healthy. She contributed to a profession that assumes that health is not intrinsic to any kind of person but requires the right infrastructure and resources to support it.

Nightingale's story challenges us to learn from her insights *and* to improve on her vision of health as a universal right.

Nightingale was many things, and she was also a nurse. She raised nursing into a modern era, but even today, nursing isn't the 'respectable' profession it should be. Nurses still don't hold the status doctors do, and most nurses continue to be women. The two facts are related. Nursing continues to be overworked and underpaid; nurses still do disproportionate amounts of emotional labour and unpaid work. During COVID-19, they put their lives on the line without much meaningful recognition. Their work continues to be crucial, both scientific *and* highly practical, and yet, undervalued. If there is anything Nightingale's multifaceted career proves, it is how much she saw *because* she was a nurse. Working from hospitals as an organiser and in a supporting role, she saw the patterns and people that other medical professionals missed. She took that perspective and turned it into statistical evidence and campaigns on issues that affected the people those systems rarely served. She produced data, warm with the care and compassion that should be central to all scientific research.

Despite her lifelong illness, Nightingale lived to ninety. She died from heart failure at her home in Mayfair. Never married, independently minded, determinedly autonomous. Her relatives declined the offer to have her buried in Westminster Abbey. Nightingale had requested a simple burial: understated and undeniable.

6

Marie Curie (1867–1934)

Physicist and Chemist

In May 1921, Marie Curie, at the height of her fame, visited the US. The 53-year-old had discovered the elements radium and polonium, the phenomenon of radioactivity and had won the Nobel Prize – twice. But on the tour, US President Warren Harding, while gesturing to her 'great attainments in the realms of science and intellect', focused his praise on her womanhood. 'We lay at your feet the testimony of that love which all the generations of men have been wont to bestow upon the noble woman, the unselfish wife, the devoted mother.'[1]

Marie Curie is one of history's most famous scientists, let alone female scientists. She broke new ground for women – before her no one, man or woman, had won multiple Nobel Prizes – and she became a symbol of what was possible. And while these milestones reflect a truly remarkable scientific career, there is, as ever, a danger in telling these stories without examining the conditions that helped and hindered her. Harding's speech begins to reveal the contours of the world Curie was working in. He praises Curie as a woman, not a scientist. Her significance seems to flow from her adherence to a feminine ideal, rather than her participation in the carefully guarded male world of science. His speech does more to reaffirm gender divisions than to illuminate the real stories of complex humans navigating systemic obstacles and occasional opportunities. In

Curie's case in particular, the stereotype is not just unhelpful but blatantly ridiculous. Let's try again.

*

Marie was born with the name Maria Skłodowska in November 1867, in Warsaw, Poland. From the beginning, her world was under siege. Her parents, Bronisława and Władysław, both came from families of the minor Polish nobility, who had been stripped of their land and status under the Russian occupation. Poland, since 1795, didn't officially exist. The country had been split between Russia, Austria and Prussia progressively throughout the 1800s, until it was entirely subsumed. Russia was the biggest occupying power. In 1863, four years before Curie was born, the Polish people had rebelled against their country's occupation. The Russians responded by annihilating all signs of Polishness. Thousands of Poles were imprisoned, exiled or sent to Siberian labour camps. The Russian viceroy stationed himself in the former home of the Polish monarch. Polish history and language were banned from schools. Even Polish street signs were replaced.[2] Conditions for Polish women, at this time, were especially precarious. Many had lost their husbands in the 1863 January Uprising. Widowed women became the sole breadwinners. That is if they could find work at all, which was difficult without an education.

The world around her was unequivocally hostile, but Curie was born into a vestige of a more hospitable Poland. Her parents rebelled against the Russian occupation by championing their people's rights to education. Both came from intellectual families – Curie's grandfather had risked his job as a school principal by insisting that 'talented peasants' should attend the same schools as the children of nobility – and became teachers.

Władysław taught physics and maths at a government-run school under the constant surveillance of Russian supervisors ready to squash rebellious teachings. He presented himself as the 'perfect government official' so as not to arouse suspicion.[3] Bronisława taught at a private school for girls, which was less policed, but still subject to unannounced drop-ins from Russian inspectors.[4]

For Curie's parents, there was hope in education.[5] They were determined to pass this potential onto their daughter. Sometimes, this demanded resilience. Curie's parents expected their children to commit themselves to their studies, their careers and, eventually, to disseminating that knowledge – regardless of the personal risk involved.[6] Curie worked well under pressure. Other children froze when the Russian inspectors dropped in on her class at primary school, but the teacher could always rely on Curie to list the Tsars perfectly in fluent Russian.[7]

Curie's childhood wasn't always easy – she lost her older sister, Zofia, at nine, and her mother at eleven. But she was always emboldened by her parents' values. On their way home from school, Curie and her best friend, Kazia Pryborowska, stopped, every day, in Saxony Square in Warsaw to spit on the pompous obelisk dedicated 'To the Poles Faithful to the Sovereign'.[8]

Curie graduated first from her class when she was fifteen (just as her older brother Józef and sister Bronisława had – this was a high-achieving family).[9] And yet, it was unclear how any of her talents would come to fruition. Curie wanted to continue her studies, but women were banned from Warsaw University. The Sorbonne, a university in Paris, did accept women, but it was unaffordable.

Curie took her only option: evening classes at a clandestine university for women. The 'Flying University' moved around to avoid detection by the Russians. Scholars and scientists met secretly with small groups – in apartments, homes and shops – to offer them the higher education outlawed by the occupation. Here, Curie read in French, German, Russian, and, of course, in Polish, devouring the works of Jean de La Fontaine, Heinrich Heine and Dostoevsky.[10] The members of the Flying University saw education as the foundation to incremental political change. As the Polish positivist Józef Kraszewski put it in 1868: 'We believe in slow and gradual progress…through reforming individuals, increasing enlightenment, encouraging work, order and moderation.' Their aim was not revolution, but 'evolution in the social system'.[11]

Curie took pains to maintain that measure and steadiness; a stance of 'disinterestedness' towards her science and her life. And yet, just like the Polish positivism, her attitude simmered with the anger of the girl who had spat on a Russian monument to assert her agency against an oppressive regime.[12]

Curie was more determined than ever to continue her studies.[13] Together with her sister, Bronisława, she finally devised a plan to fund them both. They would pool their resources.[14] Curie would work as a governess and support her sister first, and Bronisława, once she became a physician, would support Curie's studies in return. It was a long-term plan that demanded patience on Curie's part, but there was no other option for these two women bound doubly by their gender in an oppressive regime.[15]

While waiting for her time in Paris, Curie found more clandestine schemes. Her cousin, Jozef Boguski, had returned from Warsaw after studying in Russia with chemist Dmitri Mendeleev, who in 1869 devised a revolutionary system for categorising elements: the periodic table. He'd set up a modest laboratory, another undercover operation for Poles. Under his tutelage, Curie discovered the highs and lows of experimentation.[16] 'Even though

I learned to my cost that progress is neither rapid nor easy', she recounted, 'this first trial confirmed in me the taste for experimental research in the fields of physics and chemistry.'[17] Living a stifled existence under an oppressive regime, Curie found her academic home in the lab.

At twenty-four, Bronisława came good on her side of their bargain, and Curie enrolled for physics, chemistry and maths at the Sorbonne in Paris.[18] In 1891, Curie joined a vanguard of twenty-three women in a cohort of 1,800 students enrolled in the faculty of sciences. In Paris, she joined a community of Poles, Americans and Eastern Europeans who had all been drawn to the French Third Republic for the opportunities and freedom their own countries denied them. In fact, most doctoral science degrees awarded to women at the Sorbonne during this time went to women with surnames that weren't French.[19] While the state supported women's right to education, the families who could afford to send their daughters there still usually opted for a more traditional life for their daughters.

Paris had lit up; the *Belle Époque* was defined by nostalgia for a pre-war France, but it sparked a whirlpool of artistic movements, political resistance, urban development and scientific invention. The Eiffel Tower attracted two million visitors to its reveal at the Paris exhibition in 1889. The first motion picture was projected at the Salon Indien du Grand Café in 1895. On 1 May 1890, the city's socialists organised May Day, the first international celebration of labour, while anarchists planted bombs in restaurants, hotels and the homes of the city's elite. Artists, writers and radicals dialogued alongside lesbians and sex workers at the cafes, cabarets and dance halls in the bohemian Montmartre. Curie herself was studying under some of the most eminent scientists: Gabriel Lippmann, for example, who went on to win the Nobel Prize in 1908 for his work on colour photography, and the mathematician Henri Poincaré, who coined new theories in celestial dynamics.[20]

Curie revelled in her new-found freedom. She found her own garret apartment in the Latin Quarter, up six flights of stairs, so cold that the water froze in the washbasin, but it was close to the university. Here, she was free to work in the mode that became her preference: 'solitary years exclusively devoted to the studies, finally within my reach, for which I had waited so long'.[21] She often became so absorbed that she forgot to eat. On one occasion she fainted, and, when her brother-in-law questioned her, he discovered that she had only eaten radishes and cherries all day.[22]

Curie's dedication to her intellectual freedom sometimes became a cage, alienating her from the world around her. She abandoned her Polish community of friendships to focus on her work. She was a brilliant student and took first place in the exams at the end of her master's in physical science (called a *license* in France). With the help of a small scholarship, she was able to extend her time in Paris to pursue a degree in mathematics. As Curie discovered her calling, she distanced herself from her family too. She changed her name from the Polish 'Maria' to the French 'Marie' to express her commitment to integrating in France. She visited Poland every summer, but she found herself increasingly distant from her family. 'I hardly need to say that I am delighted to be back in Paris,' she wrote to her brother, having returned to complete her maths degree. 'It is my whole life that is at stake. It seemed to me, therefore, that I could stay on here without having remorse on my conscience.'[23]

In 1894, the 26-year-old Curie met Pierre, a 35-year-old physicist who had been studying crystals and magnetism at a French technical college. Curie was attracted to Pierre's detached attitude. He was a kindred spirit, a 'dreamer absorbed in his reflection'.[24]

Pierre was definitely a freethinker. He'd been tutored by his revolutionary father, who had tailored his education to his son's more practical learning style. The approach seemed to have worked. Pierre, like Curie, advanced quickly and matriculated to university at age sixteen, completed his master's at eighteen, and then immediately began to conduct and publish original research.[25] He quickly became internationally recognised for his incisive research and thinking. But he also remained an outsider in the French scientific establishment. Pierre had no interest in academic rituals or conventions. He'd published more than enough work to earn him a doctorate, but he never bothered to write a thesis and get a PhD. When he met Curie, in 1894, he had been in the same position at the very bottom of the academic ladder at a prestigious trade school for eleven years, and it didn't seem to bother him.[26]

Curie and Pierre shared an orientation in life, a set of values that prioritised education, social justice and a shared disregard for social mores. For both of them, all this was subsumed by the quest for scientific discovery. Their relationship was never going to be traditional. Pierre spelled out the only life he could offer Curie in a letter, convincing her to stay in Paris with him after she finished her second degree. 'It would be a fine thing, in which I hardly dare believe', he wrote, 'to pass our lives near each other, hypnotized in our dreams: *your* patriotic dream, *our* humanitarian dream, and *our* scientific dream.'[27] Curie decided to stay with him, choosing to keep Poland in her heart but to build the life she wanted.

The couple centred their relationship and their lives around the lab. They relinquished all unnecessary luxuries. Curie wore a practical, dark blue suit to their wedding in 1895, that allowed her to check in at the laboratory after the ceremony and served as a lab outfit for many years after. Their honeymoon was a hiking trip. They lived in a sparsely decorated apartment on the rue de la Glacière within walking distance of their labs.[28]

Curie, by this time, had started working on a paid commission from the Society for the Encouragement of National Industry to study the magnetic properties of various metals.[29] She supplemented her income with a job at a girls' secondary school, as women did not get paid for research positions.[30] She also took on the conventional chores of a housewife. Pierre was emancipated enough to share lab work with his wife but not quite enough to split the domestic labour. Meanwhile, Curie did the bare minimum to keep the house running: simple stews and bare furnishings. By 1897, she also had the duties of motherhood to add to her list, when she gave birth to her first daughter, Irène.[31] She enlisted the help of a nurse, and Pierre's father stepped in to offer child support. With this support in place, she turned her attention to her doctoral thesis.

Pursuing a doctorate in science was rare for a woman, let alone a new mother. She was eager to leave her mark. She found her subject shortly after the German physicist Wilhelm Conrad Röntgen released his paper announcing the discovery of X-rays in 1895. Röntgen had discovered X-rays while playing with a cathode ray tube. These tubes, used in older TVs and monitors, work by tunnelling electrical currents through a vacuum. The currents generated glowing, phosphorescent light, as well as electrical and magnetic disturbances that needed to be explained. Röntgen had encased a tube in black cardboard inside a dark room before firing the current. He'd noticed a glow, at a distance from the tube itself. It wasn't light – but what was it?[32]

The discovery captured the public imagination. The X-ray ('X' standing for the unknown, invisible light) was as mysterious as it was illuminating. Röntgen recruited his wife, Anna Bertha, to X-ray her hand. It materialised: the hand of a corpse bride, bony fingers adorned by a ring. On seeing it, Bertha supposedly gasped, 'I have seen my death,' and never set foot in his lab again.[33] The first published image of Mrs Röntgen's hand was followed by countless photos of bullets in shins, coins in stomachs, toes inside shoes. Besides the general absurdity, the practical uses of X-rays inspired too. They could be put to work to diagnose broken bones without the former invasive surgeries, improve complexions and remove hair.[34]

For scientists, the rays held their own kind of magic. Professor of physics Antoine Becquerel at the École Polytechnique in Paris wanted to find natural sources of the phosphorescence that the cathode ray tubes produced. In the next few weeks, he reported that, of the many phosphorescent materials he tried, only uranium salts could fog a photographic plate.[35] The effect was weaker than that of X-rays but lasted longer. What were these substances emitting? Becquerel had recorded his findings, but they didn't stir the same public curiosity. He moved on.[36] The British physicist Lord Kelvin turned the Curies' attention to the questions Becquerel had raised. Kelvin had published the results of an experiment using an electrometer that Pierre had made for him, showing how these 'uranium rays', like Röntgen's X-rays, 'electrified' the air.[37] Curie had a subject for her doctorate and, as it turns out, a ground-breaking career.

Curie wanted to measure a particular feature of the rays that Becquerel had discovered. Kelvin had developed a way of measuring how much electricity the air around uranium rays conducted. Curie believed measuring electrification more precisely would reveal more about the rays' properties.

Initially, Curie studied emissions from various minerals.[38] She wasn't fussy, testing whatever samples of metals, metallic compounds and minerals her colleagues gave her.[39] It was meticulous work, as one mentee remembered:

> She sat before the apparatus, making measurements in the halfdarkness of an unheated room to avoid variations in temperature. The series of operations involved in opening the apparatus, pushing down the chronometer, lifting the weight, etc. as the piezoelectric method requires, is accomplished by Madame Curie with a discipline and a perfect harmony of movements. No pianist could accomplish with greater virtuosity what the hands of Madame Curie accomplish in this special kind of work.[40]

Within a few weeks, Curie had discovered that thorium emitted the same kind of rays as uranium. The phenomenon she was studying existed in nature, and it was not tied to a single element. She continued to test materials, now focusing on the intensity of the rays. She discovered that the mineral pitchblende, which contained uranium alongside many other elements, had a much stronger electrifying effect than any other substance.[41]

Pierre realised at this point that Curie's work was worth prioritising. He abandoned his own project on magnetism and helped her in preparing and testing materials.[42] They worked efficiently. Before long, they had achieved the pinnacle of a chemist's career: they discovered an element.[43]

On 20 July 1898, Curie announced her discovery of a new substance more active than uranium. She sent the results to her cousin at the same time as they were presented to the Academy of Sciences by Becquerel on their behalf (neither of the Curies were members), so that the news was published in Warsaw in that month's photographic review, *Swiatlo*. 'If the existence of this new metal is confirmed', the Curies declared, 'we propose to call it polonium, from the name of the original country of one of us.'[44] Science had brought Curie home once again.

The paper, 'On a novel radioactive substance contained in pitchblende', took the scientific world by storm. The academy granted Curie a prize within a month.[45] Although the academy broke with tradition by awarding the prize to a woman, they wouldn't stoop so low as to tell her directly. Instead, Becquerel and the French chemist Marcellin Berthelot both wrote letters to Pierre, informing him that his wife had won and he should probably let her know.[46]

A few months later, the Curies, together with their colleague and laboratory chief Gustave Bémont, officially announced the possible existence of another previously unknown element in pitchblende.[47] Radium was easier to identify than polonium – we now understand that it loses particles from its nucleus at a slower rate, making it easier to study.[48] In their papers announcing the discovery, they finally coined a term to describe the effect they had been studying across the different elements: radioactivity.[49]

Curie's most ground-breaking insight was just a small step from here. Her measurements had shown her that the degree of radioactivity was proportionate to the amount of sample material present. Temperature, sunlight or other chemicals did not affect her measurements. This simple observation led Curie to conclude that radioactivity was a fundamental property of the atomic structure of radium *itself*, not the result of some chemical reaction.

By placing radioactivity *within* the atom, Curie set off a chain of events that would prove the fundamental instability of atoms.[50] This was a total paradigm shift. The planetary model of the atom we know today was still very much in the making and wouldn't be established until 1911.[51] Isaac Newton had proposed the model of a billiard ball – 'solid, massy, hard, impenetrable particles'.[52] But scientists were still trying to reconcile this stable structure with the existence of wave-like phenomena, like light. In 1932, the physicist James Chadwick would finally explain radiation in terms of the subatomic particles, with the nucleus of a particle consisting of positively charged protons and neutrons, orbited by negatively charged electrons. Atoms of the same element can deviate in their number of neutrons, making them unstable. The nucleus stabilises when it emits subatomic particles. This is radiation.[53] The Curies had inadvertently challenged the very foundations of physicists' understanding of matter and sparked a technological revolution too.

The idea that the atoms in radium were themselves somehow responsible for the radioactivity was a conceptual leap that foreshadowed the discovery of atomic energy in the twentieth century. And yet, Curie didn't linger on the implications of the shift in thinking. Unlike Lovelace, she was more interested in the painstaking practical task of isolating the elements she had proposed to discover to prove they existed. She became, once again, engrossed in the focused laboratory work. Her view grew more precise, filling in the minute details that would eventually reveal a much bigger picture.[54]

Radium does not exist freely in nature. As we have seen, like uranium, Curie uncovered it in the mineral pitchblende. To isolate pure radium, you have to treat vast quantities of pitchblende in a laborious chemical process.[55] Curie took on the brunt of the momentous practical task, while Pierre did the theoretical work. She isolated, purified and characterised her elements, working from a cold and dusty shed provided by the Sorbonne. Later, she admitted she hadn't realised what she was getting herself into: 'Would we have insisted, despite the scarcity of our means of research, if we had known the true proportion of what we were searching for, no one can tell; all that can be said now is that the constant progress of our work held us absorbed in a passionate research.'[56]

The Curies had divided their roles for efficiency's sake, but in the sexist scientific establishment, that easily lent itself to the idea that Curie was Pierre's assistant. Pierre received a letter, surprisingly, from the monarchist mathematician and Nobel committee member Gösta Mittag-Leffler warning Pierre that the French Academy had put forward the names of Pierre Curie and the French physicist Henri Becquerel for the Nobel Prize in Physics.[57] The usually anti-institutional Pierre pushed for the Nobel committee to recognise Marie's work: 'If it is true that one is thinking about me [for the prize] I very much wish to be considered together with Madame Curie with respect to our research on radioactivity.'[58] In 1903, half the prize went to Becquerel. Curie shared the other half with Pierre 'in recognition of the extraordinary services they have rendered by their joint researches on the radiation phenomena discovered by Professor Henri Becquerel.'[59]

Curie first won the prize when it was in just its third year.[60] To this day, the prize represents the very pinnacle of scientific achievement. It has become central to our understanding, as a society, of what science is important, and *who* can participate in science. The Nobel Prize is how we sanctify our scientific giants.

The prize was established by the Swedish inventor and businessman Alfred Nobel. Nobel produced his most impactful invention by harnessing the explosive power of the compound nitroglycerin to make dynamite. He went on to develop a series of even more powerful explosives, as well as a detonator, and built numerous armament factories. His intellectual commitment to war proved lucrative. He filed over 300 patents during his life and, by the time he died in 1896, he owned the equivalent of $300 million.[61] And still, Nobel was paradoxically focused on peace for a man who had hedged his bets on warfare: 'When two armies of equal strength can annihilate each other in an instant, then all civilized nations will retreat and disband their troops,' he once wrote.[62] Realising this is not how the developments of the twentieth century were heading, he left a will dedicating his life savings to a prize that would honour 'the greatest benefit on mankind' in physics, chemistry, literature, medicine and peace-making.[63]

From the beginning, the prize was about consolidating (or in Nobel's case, rectifying) a personal legacy. The rationale behind the prize, with its emphasis on the benefits of scientific discovery to society, as well the very public-facing announcements of the award (made in October and, from the beginning, far more widely publicised than the ceremony itself), are all geared towards making public heroes.[64]

The prize committee has historically chosen institutionally acceptable heroes; largely white men from countries aligned with Sweden's political and economic interests. To this day, the discoveries most likely to gain Nobel recognition are those made by the sons of business owners, doctors or engineers in the world's richest countries. There have been modest improvements in representation since Curie's time. In 1901, the average winner's family background belonged to the top ten percent of earners; today, it's the top twenty percent.[65] Only four women have been awarded the Nobel in physics since Curie – though many have unfairly missed out. There's Chien-Shiung Wu, who overturned fundamental laws in physics, who we'll meet later, and also Emmy Noether, who didn't receive the prize for her 1915 theorem of conservation laws that underpin every physical law, from the turning of the wheel to the orbits of planets. Or

Lise Meitner, undeniably a vital part of the team who discovered nuclear fission; the 1944 prize went to her male colleague Otto Hahn. Or Jocelyn Bell Burnell, who discovered radio pulsars in 1967; the Nobel Prize for that discovery went to two men, in 1974.

This discrepancy reflects the fundamental flaw in the Nobel narrative. Science isn't done by heroes but, as the stories in this book show us repeatedly, by talented people who still require the right access, support and collaborations to advance (and so we could argue that the prize shouldn't recognise individuals at all but networks and teams).[66] A prize that doesn't take into account the barriers that prevent so many from participating in science, let alone receiving credit, will reinforce a false history of science based on genius rather than circumstance. And so, it will never shine light on the conditions that are needed to write an inclusive narrative.

The prize was just background to the Curies though. They had become preoccupied by the discovery of a new, otherworldly property of radium. It glowed in the dark. The mysterious light captivated them. They arranged their 'precious products' around the lab, so that they could admire them from all sides. 'These gleamings', Curie reflected, 'which seemed suspended in the darkness, stirred us with ever new emotion and enchantment.'[67]

Not only did radium glow: it spread, infecting nearby objects with radioactivity too. This was more troubling. Curie was concerned about the contagiousness disrupting her measurements. She took precautions to protect her equipment, but nothing was completely isolated. 'Dust, the air of the room, and one's clothes all become radioactive.'[68] An inconvenience, though it enhanced radium's magic.

Curie was increasingly enraptured and had intensified her mission to isolate the element. Almost immediately following their announcement of the discovery of radium in 1899, the Curies approached the scientific instrument-maker Société Centrale des Produits Chimiques (Central Chemical Products Company) to help them develop a semi-industrial method of processing pitchblende.[69]

The couple lived and breathed radioactivity; both were starting to feel the effects. Pierre in particular was suffering from the symptoms of what they didn't yet know was radiation sickness: chronic fatigue, anaemia, painful 'burns' on his hands and mobility problems in his limbs; aches that kept him up at night. In the end, though, it wasn't radiation that took his life. Another tragedy intervened. On a rainy spring afternoon in 1906, Pierre rushed across the street on his way from his publisher and was hit by a horse-drawn wagon carrying military uniforms. He was killed instantly. When his father heard the news, he lamented, 'What was he dreaming of this time?'[70]

His wife, of course, knew. In her diary, Curie describes losing him: 'it is the end of everything, everything, everything'.[71] But despite the depths of her grief, Curie pulled herself back to the lab. 'Crushed by the blow', she later recalled, 'I did not feel able to face the future. I could not forget, however, what my husband used sometimes to say, that, even deprived of him, I ought to continue my work.'[72] Curie would honour her husband by enacting their shared dream.

An opportunity came swiftly after his death, when the Sorbonne offered her Pierre's vacant professorship; she became the first ever woman to teach there. Curie took the post and doubled down on the task of isolating pure radium, succeeding in 1910. The following year, she won the Nobel Przie – in chemistry this time – for her discovery of radium and polonium and further work on radioactivity. This time, the prize was entirely her own.

Still, the prize was just currency to her. When German troops homed in on Paris in 1914, Curie made her way swiftly to the bank to donate the gold to the war effort. France had given her an education, a marriage and a career; when the bank officials refused to melt the medals down, she used the money to buy war bonds instead.[73]

Curie understood the threat of occupation; her first impulse was to protect the source of her deepening knowledge. She gathered her stock of radium, sealed it inside a lead-lined container, and took it to Bordeaux by train. There, she deposited it in a box at a local bank. With her most valued possession secured, she turned her attention to the war, though the destructive potential of radioactivity couldn't have been further from Curie's mind. She was going to harness her science to continue the work that had compelled her late husband: its potential to save lives.[74]

The war called for a different approach to science; less in-depth, more pragmatic. Curie returned to Röntgen's X-rays for the spark of an idea. X-rays were being used in city hospitals to identify everything from fractures to foreign objects inside bodies, but they were needed on the battlefield. Curie designed a 'radiological car', a 'little Curie' that carried both the X-ray machine and the darkroom to the army surgeons in the field. Her invention even incorporated a dynamo to generate the required electricity. Curie herself was an all-encompassing scientific machine; she invented but also did the crucial work of sourcing funding for the cars. The French military were too slow to respond, so she turned to the Union of Women of France, who financed the first car, which supported the Allies in blocking the Germans from entering Paris, at the Battle of Marne in 1914. Curie leveraged her scientific status again to increase the production of cars, persuading Parisian women to donate their cars. Her twenty sourced vehicles required operators; she recruited twenty women volunteers and educated them in the theoretical physics of electricity and X-rays, as well as practical anatomy and photography. The first cohort included her own daughter, Irène, just seventeen.[75]

Together with her daughter, Curie traversed the battlefront. She had to learn to drive, change tyres and fix the mechanics when the car broke down: a far cry from experimental physics, but a skillset that freed her to zoom around the male-dominated zone, mitigating the damage of their war. At the same time, she oversaw the construction of 200 stationary radiological rooms at the field hospitals behind battle lines.[76]

Irène became her mother's first assistant; she helped her train the further 130 women who followed. There are some remarkable photos of the two women from the period: one, taken just after the war, in 1919, with Curie at the front of a classroom, confidently demonstrating her science to the men in uniform, and Irène at the centre of the class of the American Expeditionary Forces, looking a little lost, perhaps, and yet remarkably, powerfully present, in her dress, at the centre of the picture.[77]

The frontline experience continued an already unique education for Irène. She was taught by six professors of an academic cooperative that Curie had initiated and participated in. Each taught the children their respective area of expertise. Irène had always been interested in science, which obviously delighted her mother (with a mother like that, a bit of a *fait accompli*, you might say, but Irène's younger sister, Ève, became a concert pianist and journalist, as well as her mother's biographer). After the war, Irène continued her studies in her mother's lab at the Institut du Radium.

Curie opened her Institut du Radium in 1918. The laboratory was adjacent to a hospital for the express purpose of collaborating with medical doctors.[78] The institute consisted of two labs housing two research groups, one devoted to physics and chemistry and headed by Curie. The other focused on biomedical research, and was led by the French doctor Claudius Regaud, who was pioneering radiation treatment in medical diagnosis and cancer therapy.[79] They developed 'curietherapy', which involves placing radium directly inside a tumour to kill it. The technique, today known as brachytherapy, is still used with less radioactive, artificial replicates.[80] Irène herself, along with her husband and collaborator Frederic Joliot, synthesised the first artificial radioactive atoms. In 1935, she followed in her parents' footsteps in more ways than one and shared the Nobel Prize with Joliot for the discovery.

The First World War revealed the hope and the danger of radiation. Millions of soldiers had been treated with radium and X-rays; it had helped, even saved, lives. But their doctors and technicians had also been exposed. They reported alarming symptoms, ranging from skin irritation to gangrene.

There were also troubling reports coming from lab workers. Madame Artaud, a member of the Radiochemistry Society, who Curie knew personally, died after spilling radioactive substances on herself. The engineers Maurice Deminitroux and Marcel Demalander had both died of leukaemia and anaemia a few months after they'd prepared radioactive substances in a factory outside Paris in 1925.[81]

Many of the disasters, though, came from outside the world of medicine. From the beginning, industry had been eager to capitalise on the radioactive glow. The public hype around X-rays faded as companies started mining and selling radium. Radium had huge cosmetic potential. Radium could be used to burn away imperfections on the skin. It was promoted as a fountain of youth – the Botox of its day. A whole range of products incorporating the wonder product filled the shelves of department stores on both sides of the Atlantic. There was even O-Radium, a radioactive wool for baby clothes.[82] Wristwatches with luminous, radium-daubed dials became all the rage. The first factories opened in America, where the delicate hand-painting of glowing watch dials was considered a women's chore. The workers, some as young as fifteen, painted about 250 dials a day, smoothing their brushes to a fine point between their lips. The first health problems emerged in the 1920s, when some of the women began suffering from fatigue, anaemia and trouble with their teeth. Dentists found jawbones so diseased that chunks of bone came out when they extracted the bad teeth. Young women died of what became known as 'radium jaw'. Their bodies turned radioactive, the radium absorbed into their bones, and they developed bone cancer.[83]

In 1921, the first radiation protection committee was established in Britain. Curie followed the standard safety practices for handling radium in her laboratory but went no further. She responded to a 1925 report that warned of the dangers of handling radioactive material, saying that she was unaware of 'any grave accidents due to radium' either 'among the personnel of other factories' or 'among the personnel' of her own 'Institut'.[84] She was unconvinced by the evidence. Her daughter Irène, who worked in the lab (and who herself died from leukaemia at fifty-nine, due to radiation exposure), wrote to her mother about an accident in a tone that captures the attitude at the lab. She casually dismissed the stomach troubles and hair loss one of their long-time employees, Sonia Cotelle, experienced after a vial of polonium exploded in her face: 'her present ill health perhaps has no connection', she wrote.[85] Even by 1931, when seven of twenty workers at her Institut du Radium had abnormal blood test results, and just four years before her own death from blood disorders connected to radiation exposure, Curie wasn't convinced it was lethal.[86] She wore her own radiation burns with pride and encouraged her workers to do the same. There were more important things.

By 1921, Curie seemed, to the outside world, at the height of her powers. She had just returned from a successful fundraising tour to America, organised by the journalist Missy Meloney. She had been shocked to hear that Curie could no longer afford the very radium she had discovered because she had never patented it. This while her discovery was being sold at $100,000 per gram and fuelling cancer treatments and military developments across the world. Meloney sensed a human-interest story, and she worked hard to shape the figure of Curie, increasingly detached and isolated since Pierre's death, into someone women would want to donate to.[87] Meloney wrote about Curie as a benevolent scientist who had always been driven, not by the creed of detached science, but to heal people. Meloney tuned Curie into the 'Jeanne d'Arc of the laboratory', the very face of 'suffering and patience'.[88] Curie was unconvinced by the narrative but duly went on the tour to collect her gram of radium, presented to her by the president himself.[89]

On returning home with her vial of precious radium, Curie was in the worst health of her life. Her eyes were full of cataracts. She couldn't read properly and had to paste coloured signs on her instruments to be able to identify them. She was, at some level, aware that radiation was the cause. Perhaps for that reason, she kept the multiple surgeries on her eyes a secret. She only fleetingly admitted the possible connection in a letter to her sister: 'I am very worried about it,' she confesses in writing about her health. 'Perhaps', she speculates, 'radium has something to do with these troubles, but it cannot be affirmed with certainty.'[90] On 4 July 1934, Curie died from aplastic anaemia, aged just sixty-six. She was buried alongside Pierre.

Curie was no Jeanne d'Arc; she was a radioactive woman. Curie helped open the door to a universe of unstable atoms. She laid the foundations of modern physics, cancer treatments and nuclear energy, as well as the atomic bombs that would obliterate Hiroshima and Nagasaki just over a decade after her death. Curie was not responsible for what became of her findings, but her life and her contributions contain the irreconcilable tensions that make up the history of science: creation and destruction, death and new beginnings. Curie's surviving notebooks remain radioactive to this day, kept, like her own precious radium, in a lead-lined box. They remind us that scientists are unstable atoms: expelling discoveries that ripple out through the world. It is for us to turn to their charged archives, always aware of the responsibility discovery brings.

7

Sabina Spielrein (1885–1942)

Psychoanalyst

Many of the women we've met so far advanced their careers, by necessity, through their associations with men. Few of them were defined by it. Despite the obstacles, we remember them for their science, not the men they knew. Ada Lovelace wasn't merely Lord Byron's daughter, nor is Marie Curie remembered solely as Pierre's wife – quite the opposite.

Sabina Spielrein was not so lucky. This Russian psychoanalyst made foundational contributions to psychology, an exciting, emerging science in her day, but she is remembered, above all, for her relationship with the much better-known Carl Jung, as well as the founder of psychoanalysis himself, Sigmund Freud. Much like Hypatia, Spielrein was not just erased from history but treated abominably by it: first ridiculed and degraded and, even in recent retellings, remembered primarily for her murder by the Nazis, rather than her pioneering career. She struggled with the twin 'deficiencies' of being both a woman – inferior to men – and a patient – inferior to doctors. Despite all this, she persevered by putting her own experiences at the heart of scientific inquiry. A marginalised story holds explosive potential. Move it to the centre, and it shows us what we were never allowed to know.

*

Spielrein was born Sheyve Naftulovna Spielrein on 7 November 1885 to a wealthy Jewish family in Rostov-on-Don in Southern Russia. The Russian Empire had never been a hospitable place for Jews, consigning most to the impoverished Pale of Settlement at the empire's western edge. But from the 1880s, Alexander III enforced a new wave of anti-Jewish policies, sparking violent pogroms. Over two million Jews fled Russia over the next four decades, the vast majority heading to the US.

At home, the troubles only magnified. While her grandfather, a rabbi, had allowed her mother, Eva (born Khave) Lublinskaya, to pursue an education and train as a dentist, he soon forced her to marry the Jewish Nikolai (born Naftul) Spielrein after she fell in love with a Christian. They formed an unhappy household. Nikolai, a merchant, was aggressive and volatile. When the sixteen-year-old Sabina told him that she'd prefer to spend time with friends over family, as most teenagers do, he reportedly threatened to take his own life.[1] Meanwhile, her mother beat her five children incessantly.

Spielrein's personality reflected all these shades of her upbringing. At five, she was sent to school in Warsaw, where her father's family lived. Here, she learned German and French, and these languages offered unique escape from her abusive home life. Spielrein began to develop a special connection with an angel who spoke to her in German and Russian.[2] This was a friendly spirit: one who understood the meaning of her words, often before she had spoken. Spielrein had been raised religiously; her mother had taught her to believe in God and angels as much as she beat the fear of God into her. Spielrein's powerful imagination resonated with the image of the angel and found in it a vision of the kind of safety, connection and hope her parents couldn't give her. Learning and spirituality would converge as a coping strategy throughout Spielrein's life. At ten, Spielrein returned home to attend a girls' grammar school, and graduated with a distinction in 1904. She loved the natural sciences and wanted to study medicine, but at fifteen disaster struck. Spielrein lost her six-year-old sister, Emily, to typhoid. The pain broke her.

At 10.30 p.m. on 17 August 1904, Spielrein was escorted to Burghölzli Clinic in Zurich, Switzerland by the medical police. The time of admission suggests it was an emergency, though it isn't clear what happened.[3] Spielrein herself complained of a headache but asserted she was merely upset, not 'mad', to the hospital's deputy senior physician. Dr Carl Jung took note of her symptoms, which included 'laughing and crying' in a 'strangely mixed compulsive manner', and numerous tics, like rotating her head, sticking out her tongue and twitching her legs.[4] This was a familiar cluster; neurologists in France had categorised the symptoms of hysteria in the late 1800s. Jung diagnosed her accordingly and arranged for her to stay in the women's ward with a private nurse.[5] Her parents paid the 1,250 francs a quarter for her accommodation: the equivalent of the salary of a senior physician.[6]

While Emily's death had been the trigger for Spielrein's breakdown, it was clearly the final drop in a traumatic upbringing. Spielrein was part of an epidemic of 'hysterical' women from bourgeois families who had all been exposed to chronic, often sexual, violence at home. Since the dawn of Western civilisation, their distress had been attributed to everything from misplaced wombs to dysfunctional ovaries to their feminine brains. Hysteria is derived from the ancient Greek for 'womb' after all. Anything but the domestic abuse that kept them beneath their patriarch's feet.

The French neurologist Jean-Martin Charcot strove to find a cure. He employed a method he called 'anatomo-clinical'. He would observe his female patients, noting patterns in their behaviour, like paralysis, impaired vision or periodic fits. When the patients died, he dissected their brains in the hope he would find a physical cause. However, the brains of hysterical patients were completely normal.[7]

Charcot initially prioritised female case studies, but he identified cases of 'male hysteria' too. Jewish men seemed particularly prone; he guided his student Henry Meige closely to develop a thesis based on a series of clinical cases that linked another dubiously neurological condition of 'neuropathic travelling' to the antisemitic myth of the 'Wandering Jew'. Meige and Charcot found clinical meaning in the Christian story of a Jew condemned to roam the Earth for eternity for refusing Christ water during the Crucifixion. The tale helped them classify a new hereditary condition of a morally and intellectually fragile people unable to defer to the recommended treatment, doomed to roam hospitals for all time in search of a cure.[8]

Jung had caught wind of a different approach to treating hysteria. The method, designed by Dr Sigmund Freud, was not neurological. Freud had studied under Charcot but became more interested in the 'talking cure' proposed by his colleague Dr Josef Breuer. Breuer had, in turn, taken the term from one of his patients, Bertha Pappenheim (known in her case files as 'Anna O'), to describe the cathartic effect of talking about her traumatic experiences – what a revelation, I know.[9] Based on Pappenheim's observations, Freud developed a method that included free association, dream interpretation and talking therapy.[10] While some of these components eventually became more removed from women's actual experiences (turning them into 'fantasies' rather than realities), initially, at least, this was a treatment that addressed the real root cause of 'hysterical' women's symptoms. A patient's own observations had set doctors on a new path.

Spielrein would join this lineage of influential patients. Though, in the end, Jung's treatment wasn't central to her recovery. While Spielrein was at the Burghölzli, Jung only had three sessions of something nearing analysis with her.[11] His notes remain a valuable source on Spielrein's childhood. We learn how, at thirteen, after one of her mother's punishments, she 'hid and doused herself in ice-cold water in winter in order to catch her death of cold'. Or how, at fifteen, she 'tried to starve herself', because she 'had made her mother angry'.[12] Jung also wrote of visitations Spielrein received at night, recognisable as flashbacks of sexual assault, likely by her father. Jung reports: 'In the night a great fright: there might be a cat or someone else in her room, someone was suddenly speaking in her ear. She felt something moving on her back like a snail, and something grasped her side like a hand.'[13] Jung viewed these flashbacks as further evidence of Spielrein's hysteria. For modern readers, however, his notes paint a picture of a young woman traumatised by parental violence.

Jung didn't make much of this family violence. It was his boss, the hospital director, Eugen Bleuler, himself an enlightened psychiatrist who reimagined the mental asylum as a therapeutic community, who understood that Spielrein would feel a lot better with a safe distance from her family. Bleuler took over correspondence with her father. While the set-up usurped Spielrein's agency and the reports were condescending in tone, it maintained a safe distance from her abusers. And it had another advantage. In his letters to Mr Spielrein, Bleuler framed education as in the best interest of Spielrein's recovery. He reports, for example, how they had succeeded 'in stimulating Miss Spielrein's interest in scientific pursuits so that she can be distracted for hours at a time from her pathological obsessions'.[14]

The letter makes it sound like the doctors enticed Spielrein into medicine, but as we saw, her interest started in childhood.

By the second half of her stay, Spielrein had moved beyond the remit of a patient and involved herself in scientific studies, assisting Bleuler in research, as well as Jung himself in a project on word associations.[15] Spielrein also decided herself that she wanted to continue her medical training at the University of Zurich. Her father reluctantly agreed but had her oldest brother enrol in the same year to keep an eye on her.[16]

Spielrein joined a tide of Russian women travelling to Zürich in pursuit of an education. For all the political upheaval in Russia, the 1905 revolution had established the country's first constitutional monarchy. The infamous Bloody Sunday on 22 January that year saw demonstrators massacred by the soldiers of the Imperial Guard. The event sparked a public consciousness and a wave of protests that would eventually lead to the definitive overthrow of the monarchy. Things were changing for women too. Serfdom had been abolished in 1861, and, though women were still banned from Russian universities, Nadezhda Suslova, the daughter of a serf, had set a precedent by enrolling at the University of Zurich and earning her doctorate there in 1867. Other women started to follow her lead and formed 'Russian colonies' – communities who supported each other and remained involved in the politics in Russia from abroad. Jewish men and women also sought refuge from the discrimination they faced in Russian universities, as well as the opportunities they were denied by the strict quotas limiting Jewish enrolment in higher education to as little as three percent in cities like Moscow and St Petersburg; other schools banned Jews altogether. [17]

After being discharged, Spielrein moved into her own apartment in June 1905. Now, she had to pave her own way far from the patriarchal but protective confines of the hospital. And it terrified her. On 24 April 1905, the day before her lectures started, she wrote: 'Somehow I'm afraid to get close to people.

I fear for my freedom. The only thing I have now is my freedom, and I'm protecting this last treasure with all my strength.'[18]

In the end, safety and a new-found sense of purpose, not psychoanalysis, allowed Spielrein to recoup agency over her own life (as well as the once again unacknowledged support of nurses, who provided the bulk of the care). Still, in the first of a succession of false claims to credit, first for her recovery, then for her ideas, Jung argued she had him to thank. He staked his claims in a series of letters to his friend (though they hadn't yet met in person), Freud.

He wrote the first in October 1905, a few months after Spielrein was discharged from the hospital. Spielrein, at this point, had developed a crush on Jung, which she wrote about to her mother (she had a habit of crushing on her superiors: schoolteachers, uncles and another doctor at the sanatorium).[19] Her mother quickly travelled to Zürich to ask Jung to refer her daughter to another doctor to cure her of her infatuation. With her in the room, Jung wrote a note to Freud marked 'For use if the occasion arises' and handed it to her. In it, he diagnosed Spielrein and her mother with hysteria. He also included intimate details of his former patient's symptoms, including how she masturbated and her sexual fantasies. He described them to a man who, at this point, he had never met.[20]

It is impossible to know exactly why Jung would do such a thing. But it is in this context that he perhaps tried to flatter Freud, by describing how 'using your method I have analysed the clinical condition fairly thorough and with considerable success from the outset'. In doing so, he much overstated his role in her treatment.[21]

Over the course of medical school, Spielrein's diaries show that her infatuation subsided. She was doing well there and had her own social circles. Their relationship could have become professional. Spielrein continued to attend his lectures and ward rounds and even assisted him in the lab. They understood each other intellectually; they were, as Spielrein put it, 'good at reading each other's minds'.[22] Together, they developed the idea that they would one day conceive a son named Siegfried who would be 'the greatest genius'.[23]

It was only at this point, outside the context of her formal treatment, that Jung thought it was appropriate to try out the in-depth psychoanalysis he'd claimed had been fundamental to her recovery. He analysed her wish for his child (as if it had been her own fabrication). He did it casually, after lectures or in social situations. It was concerning enough for one of Spielrein's friends to warn her that he was experimenting on her. Spielrein wrote to him herself, asking him to stop, wary that his behaviour would deepen her infatuation for him. And yet, he continued, he encouraged it and found himself developing feelings too. Over five months, they developed an emotional and physical relationship. The affair was doomed from the start; Jung himself admitted to his fickleness when it came to romance. But Jung's wife beat him to the chase. She wrote to Spielrein's mother to expose the affair. Jung cut off contact without so much as a note and only allowed her to see him in booked meetings at the hospital. Spielrein felt understandably betrayed; in her third appointment with him, things grew heated, and she assaulted him with a letter knife. Jung didn't see himself as guilty of any transgression; at first, he even argued the affair was his right, since Spielrein had never paid a fee for her (unsolicited) analysis with him and so had chosen not to keep it professional. Jung's ridiculous defences only amassed as he turned to justifying himself to Freud. He presented Spielrein as his first psychoanalytic patient: a 'test case'. It had taken him a while to realise she was trying to seduce him. Even as Spielrein wrote to Freud with her own, competing account, Freud didn't see much harm done. He reassured his friend: 'In view of the kind of matter we work with, it will never be possible to avoid little laboratory explosions.'[24]

Spielrein was reduced to an experiment. But she wasn't a naive victim. The scandal blew over, and she and Jung resumed their physical relationship for a while. But Jung was too hot-and-cold for her; she grew bored of the 'Don Juan' act.[25] In 1911, she earned her doctorate in psychiatry from the University of Zurich with a dissertation on schizophrenia. She turned her attention to the theories that would help her make her own sense of what had happened between them.[26]

Spielrein's own case had been used as a case study in early clinical discussions on schizophrenia.[27] Psychoanalysts were particularly intrigued by the neurological condition and the mechanisms involved. Now she was the first woman to have contributed a dissertation to the emerging field.[28] In late 1911, she presented a paper at the Wednesday night meeting of the Vienna Psychoanalytic Society (VPS).[29] Freud attended the meeting, alongside some of the most pre-eminent figures in the field, and her arguments left a mark.

No stiff-lipped, early-twentieth-century man would have picked her research topic. The speech, titled 'On Transformation', was based on Spielrein's dissertation, and would later become her best-known paper: 'Destruction as the cause of coming into being'.[30] In it, Spielrein explored a seeming paradox: the tortuous feelings associated with romantic love seem to contradict the biological drive to procreate. She went about trying to explain how the emotional experience of sex might be connected to the biological drive. Spielrein asked in particular why 'anxiety and disgust' as well as sexual desires associated with death often accompany sexual attraction.[31] She makes a clear statement, radically against the sensibilities of the time, that these feelings have nothing to do with the morality of sex. Spielrein saw sex as driven by a desire, however unconscious, for something beyond yourself. She argued that feelings of anxiety accompanying sex were tied, especially for women, to a sense of disappointment, as they asked themselves: 'Is this all?' This did not stem from their partner's sexual performance (although this would undoubtedly have been part of the picture too; women's pleasure was hardly a priority). The problem was, instead, that the experience had fallen short of the kind of transformation people were reaching for through romantic love.[32]

Spielrein took an interest in death but, unlike many of her contemporaries who saw it as the incommensurable opposite to life, connected it to a spiritual form of survival. Spielrein saw the drive of attraction and sexuality as embedded in an ongoing cycle of life that also required death. Merging with another is the ultimate expression of your own aliveness, of self. At the same time, Spielrein saw the risk of losing your own integrity and independence. This drive to lose yourself, to share your suffering with another and be held, when expressed too literally, led to suicide or sadomasochism.[33] But the paradox itself expressed a human desire for the metaphorical death we need to change or grow. We satisfied that need, figuratively, through a physical union with the object of our affection. She pointed to the work of poets and philosophers to show how people had long drawn the connection between sexuality and death. She even drew on biological arguments to make her point: an egg and sperm are both destroyed in the process of creating a new human being. More than this, procreating also means facing your own mortality, and, as Spielrein observed, for many mothers at the time, childbirth meant death.[34] Spielrein saw how there is a fundamental conflict between self-preservation and species-preservation that is at the heart of much of the meaning-making humans do.[35]

Spielrein connected the drive to destruction to self-preservation *and* transformation: change requires a death of an old 'self' to change and become 'new'. This was how people made sense of the world symbolically. Spielrein was well placed to know. She had transformed from a patient into a doctor, a caged woman into someone relatively free. Her deep and incisive observations reflect how she herself coped with change in her life.

But Spielrein also learned from her patients. Their experiences helped her develop her theory of personal transformation. She named the dimension people reached for in love, the 'We'. There is, she theorised, a level of the psyche that 'knows no "I," but only its summation'.[36] In the schizophrenic patients she had observed, the 'We' had become overwhelming, meaning that the 'I' could no longer be distinguished from the collective. The same, she said, was true for highly sensitive people.[37] At the same time that we don't want to lose ourselves to it, the 'We' is our gateway to a pool of resources that will allow us to transcend ourselves, and grow.[38] Though she gained her insight mainly from the dreams of people experiencing psychotic episodes, Spielrein argued later that we can all access this other realm through the images in dreams and stories provided by ancestors. This other, unconscious realm, in this sense, resolved the paradox between the need for self-preservation and self-destruction.[39]

Spielrein went against the grain of her time to develop theories based on how she and other patients had made sense of their experience. She used methods, like analysing symbols in culture and religion, that were considered unscientific by many psychoanalysts. This approach wasn't just innovative but humane. Spielrein took experiences, like female sexuality, sadomasochism and schizophrenia, deemed pathological by psychoanalysis and society alike, and returned meaning to them. Following in Freud's footsteps, she turned what had been deemed pathologies and aberrations into a set of meaningful human experiences.

Spielrein extended her insights more widely. We all make meaning through symbols, she argued: through stories, ideas and ritual. Her efforts to develop a generalisable theory inspired Jung's 'archetypes', which set out how dreams, art and religious literature have included the same symbols throughout time. This became evidence for his theoretical collective unconscious.[40] Jung's model continues to offer psychotherapists and readers a rich and challenging window onto human experience today, though Jung never acknowledged the extent of Spielrein's influence.

Freud and Jung generally downplayed Spielrein's impact on their thinking. Spielrein's influence was so obvious in Freud's work on the 'death instinct' that he briefly stooped to acknowledge her influence in the notes of his 1920 book, *Beyond the Pleasure Principle.*[41] But Freud also made a point of disagreeing with Spielrein, even when this seemed to contradict his entire philosophy. He argued that there is a biologically inexplicable drive to destruction in humans but refused to accept any of the symbolism that Spielrein had postulated.

Jung, too, took his turn to academically dissociate himself from Spielrein. In 1912, he published *Symbols of Transformation*, now one of his best-known texts, and where his work diverged from Freud and the psychoanalytic tradition. He makes a very similar argument to Spielrein in regard to the experiences of schizophrenic patients. Spielrein saw that Jung had plagiarised her and confronted him by letter. He replied in March 1912: 'Your study is extraordinarily intelligent and contains splendid ideas whose priority I am happy to acknowledge as yours. The death tendency or death wish was clear to you before it was clear to me.'[42] Off to a good start; he acknowledged not just Spielrein's influence but also her primacy in coming up with the concept of the death instinct. But then, he continued, with a masterly sleight of hand: 'I express myself so differently from you in my work that no one could imagine that you had borrowed in any way from me.'[43] So here, he shifts the focus to his own influence on Spielrein. Then, he continues: 'Perhaps I borrowed from you too; certainly I have unwittingly absorbed a part of your soul as you doubtless have of mine. What matters is what each of us has made of it. And you have made something good… I hope you will be able to represent my new ideas.'[44] Suddenly his intellectual theft was the inevitable outcome of the merging of souls, followed by vague platitudes on how it was all means to a noble end, namely, the continuation of *his* intellectual legacy.

As a woman and a patient, it was easy for Freud and Jung to discredit Spielrein. Even today, critics argue that Spielrein drew mainly on her own clinical experience, rather than client observation.[45] Her patient status combined with her gender has allowed historians to reduce her theories to romantic outpourings.[46] Spielrein's psychoanalytic theories on destruction in love, they say, are the outcome of her failure to lock down Jung in a romantic partnership or just a sign of her own unresolved penis envy. While Freud's theory of the death instinct is regarded as an academic contribution and a generalisable theory about the human condition, Spielrein's is a personal tragedy. All this disregarding the fact that Spielrein actually cut off her association with him because she considered her work to be more important, and that she went on to publish for thirty years after their love affair; none of which mentions Jung or his work.[47]

While Spielrein's outsider status prompted derision, it was always also the basis for her insights. She brought a new perspective. Freud and Jung saw people as individuals and emphasised the importance of becoming an autonomous and independent self. They looked into individual minds for explanations and solutions to mental suffering. Spielrein argued that people develop and change in relationship, through one another. In doing so, she foreshadowed ideas that are now at the heart of modern psychotherapy. Her death drive, for example, evokes what the existentialist philosopher Martin Buber (1878–1965) calls the 'I-Thou relationship'.[48] Buber did not see people as separate from the surrounding world. He argued that people live and change in and through relationships. This is not dissimilar to Spielrein's theory of the sexual act.[49] This relational way of understanding identity today gives mental health practitioners a different way of relating to patients, as people who depend on healthy relationships for their well-being.

Despite all the efforts to discredit her, shortly after presenting her dissertation at the 1911 meeting the 26-year-old Spielrein was invited to become a member of the VPS. She was one of the first women to have been admitted to the society and put women's experience unapologetically on the agenda, presenting to the all-male society on female masturbation. She was an active participant in these meetings alongside many of the field's notables. She challenged their entrenched views, like the idea that women are less sexual than men. All this in an era when women's sexuality was notoriously suppressed. The dominant medical belief at the time was in fact that women didn't enjoy sex at all and merely submitted to it for their husbands' pleasure. If they did enjoy it, this was most definitely a symptom of nymphomania, in turn a sign of hysteria.[50] Spielrein argued against these beliefs, presenting as an expert on the very disease she was purported to have.[51] Brilliant.

Spielrein continued to unsettle the boundaries between domestic and intellectual when she had her first child in 1913, having married another Russian Jewish physician, Pavel Nahumovitch Sheftel, the year before. She found innovative ways to fold motherhood into her career. Besides her interest in analysing mythic motifs in folktales, which was pioneering research in the field, she had also developed a specialisation in child psychoanalysis and developmental psychology. Freud's daughter Anna (1895–1982) is often considered to be the founder of child analysis, and the psychoanalyst Melanie Klein (1882–1960) is given the credit for 'object relations theory', which explores the development of the 'self' in relation to others in childhood. But Spielrein pre-empted many of these ideas when she began to observe and write about her own child Renata's development in 1913.

Spielrein published almost forty papers, offering both empirical observation and original theories about child psychology, as well as psychoanalytic theory. She went on to contribute foundational work on the origin and development of children's speech and thought. She was the first analyst to investigate the links between child development and language. She combined Freud with language theory to argue that infants learn through relationship to those around them, foreshadowing the theories of the later, very influential paediatrician and psychoanalyst Donald Winnicott (1896–1971). Again, Spielrein was interested in how people, now children, made sense of their experience in relationship. In all of this, she was guided by an intuitive understanding rooted in her own experience. She described case studies, including her own, to demonstrate and expand Freud's theories as they applied to childhood. Her own trauma had shown her powerfully how early life shapes the adult. It had also always inspired her with faith in the human capacity to learn, grow and overcome.[52]

Spielrein found early success, but by 1923, after years of teaching, publishing and practising psychology in Europe, primarily in Geneva, she had hit a wall. She was struggling to find work and had lost financial support from her family after they'd been devastated by the 1917 revolution. She decided to return to Russia.

While her family had lost their wealth, they were optimistic about the future of their country, and the Bolsheviks still supported psychoanalysis. Her brothers both held prominent positions at state businesses.[53] One was able to secure his sister a professorship.

Politics had brought Spielrein back into her parents' orbit. But things were different now. She wasn't their hysterical daughter anymore but an established doctor. Spielrein joined Moscow's State International Psychoanalytic Institute, the first in the country, at thirty-seven. She was the only woman there. Nonetheless, she taught and was even appointed director of the clinical programme. This was possible under Soviet rule, which saw a role for women in the state's crucial scientific advancement. One of the first acts of the 1917 revolution was to grant women equal rights to education at any level and in every field.[54] Women were permitted to work as professional researchers and university teachers, although they remained in lower-ranking positions, as assistants or lecturers.[55] Spielrein had undergone an unlikely transformation, not only given her start in life, but her political context too; still, anyone would now recognise her as a highly regarded leader in the Russian psychoanalytic movement.

And yet, brewing unrest would once again engulf her. Spielrein's institute had tried to secure itself in an environment that was increasingly hostile to intellectuals and especially psychoanalysts. They had initiated a project to find a way to reconcile the country's communist collectivism with the radical 'bourgeois' individualism of Freud's psychoanalytic principles.[56] By the mid-1920s, though, the Bolsheviks had turned on psychoanalysis; it was dangerous to encourage individuals to explore their inner worlds. The institute and the discipline slowly disappeared under Stalin's regime; teachers were fired, budgets slashed and psychoanalysts increasingly accused of treason.[57] Under these conditions, Spielrein was treading a tenuous line. She continued to work in a school in her hometown, Rostov-on-Don, as a paedologist. This was considered a properly 'Russian' science of 'transforming human nature during childhood', in line with Soviet ideology, but was eventually also slashed when the Party began to suspect subversive messaging.[58]

Despite the danger it posed, Spielrein wouldn't abandon psychoanalysis. She published until 1931, finally stopping to protect her anonymity, but continued to see patients at home. Work brought meaning in the face of an increasingly nihilistic world. Her entire family was disappearing around her. In 1937, her three brothers were all executed in the gulags, then her husband and parents died in quick succession. And yet more tragedy loomed. In 1941, Nazis invaded Rostov-on-Don – the Red Army managed to fend them off for a year, but in July 1942 the city fell to the Germans. Life was bleak. It is understandable, and also commendable, that she persevered with her work in the face of these dehumanising conditions. As the Holocaust survivor and existentialist psychiatrist Viktor Frankl would later put it as part of his now renowned approach of logotherapy: 'Those who have a "why" to live, can bear with almost any "how."'[59]

Many authors paint Spielrein as a wretched figure in her fifties, desperately clinging to her practice in deadly circumstances. Some even twist these decisions into a kind of self-sacrifice connected, protractedly, to an ongoing romantic obsession with Jung. There is no evidence that she gave any thought to Jung in her later years. She had other concerns, given she'd spent two decades enduring an oppressive, communist regime, the spread of fascism through the continent and the death of her closest family members.[60]

In the end, it wasn't Spielrein's identity as a woman or a patient, but a Jew under Nazi occupation that dominates her story. The precise details of her death are disputed, but the most widely accepted version is that she was killed along with her two daughters in August 1942.[61] Twenty-seven thousand people, primarily Jewish, were murdered by the Einsatzgruppen at Zmievskaya Balka from 1942 to 1943, a ravine three miles from Rostov. On 11 August 1942, 13,000 Jews were directed to gather, strip and were then shot, their bodies immediately covered with clay.[62]

Most accounts depict this as the painful final chapter it was, but some, disturbingly, warp it into proof of Spielrein's self-destructiveness, painting these horrific historic conditions as some kind of enactment of her 'death instinct'. It is a remarkable act of victim-blaming, suggesting how far some historians will go to squeeze women into the straitjacket of hysteria.

Psychoanalysis is today the basis for a whole range of modalities in psychotherapy and psychology; these disciplines have grown the seeds for a humane response to human suffering contained in the work of the psychoanalytic pioneers. Today, these disciplines enact many of the principles Spielrein understood were necessary for recovery or development (in children and adults alike): that people require supportive, safe relationships in which to grow and make sense of their lives. She had experienced the harm and the difference another human could make. She drew on that knowledge from beyond the realms of sanctified science to shape her theories. She brought a subversive empathy. Only she could have thought to ask the questions she did: as a woman, a patient and an interdisciplinary scholar.

8

Barbara McClintock (1902–1992)

Cytogeneticist

When we think of corn, we tend to picture shiny yellow sweetcorn, peeled and trimmed in the vegetable aisle. Not all corn looks like this; ancient, heirloom varieties grow in a kaleidoscope of yellow, red, purple, black and white, even blue. No two are alike. At first glance this might seem like a neat quirk of nature, beautiful but scientifically uninteresting. Most scientists working on corn in the 1950s saw it this way; their aim was to find ways to increase yield or resistance to pests, not fuss over aesthetics.

Not so for the cytogeneticist Barbara McClintock. To her, the colours of corn revealed an all-important question. Nothing about these random patterns fitted with contemporary understandings of heritability. Working, first in wartime, in seclusion from Cold Spring Harbor Laboratory, Long Island, New York, she asked: 'how?' The question unravelled into the second half of the twentieth century, showing her a whole new dimension of heredity. With her deceptively simple question, she made a discovery that overturned the most fundamental assumption in the genetics of her day.

*

McClintock's childhood primed her for life as a pioneer. Her mother, Sara Handy McClintock, was an American who traced her lineage back to the Pilgrims aboard the *Mayflower*, and her father, Thomas Henry McClintock, was the son of British immigrants. Sara's father, a Congregational minister, disapproved of Thomas; not only was he a foreigner, he was also still in medical school and in no position to support a family. But the pair were determined and married against Sara's father's wishes. Without any support from their families, they slowly paid off Thomas' medical school debts and settled in Connecticut, where they had four children. On 16 June 1902, their third child, Eleanor McClintock, was born.[1]

From the beginning, McClintock was a loner. As a baby, her mother would leave her on a cushion with a toy for hours and she'd be perfectly content. Her parents even changed her name from Eleanor, which seemed too delicately feminine, to Barbara, which they thought better captured her independent spirit.[2]

In extended interviews conducted by the physicist and historian of science, Evelyn Fox Keller, in 1983, McClintock paints a vivid picture of her childhood, but her most treasured memories aren't exactly conventional. She recounts, for example, how after the family moved to Brooklyn when she was six, she discovered a special form of running. She would jog with a very straight back so that she felt like she was floating and didn't get tired. She discovered later that this resembled the style of running practised by Buddhist monks in Tibet, known as the 'Running Lamas'.[3]

I love this image of Barbara floating through the streets of Brooklyn like a Buddhist monk. That this is among her favourite recollections of childhood tells us something: moments of solitude stood out to her. She also enjoyed reading and just sitting, 'thinking about things'.[4] And, while her inclinations sometimes worried her mother, for the most part her parents accepted her. They let her skip school when she didn't like her teachers. They even instructed teachers not to give her homework, because they imagined school as just a small part of growing up. McClintock was encouraged to develop a clear sense of her own interests and to trust her intuition.[5]

And yet, as she grew older, McClintock's independence became a problem. She was 'tomboyish', enjoyed playing sports and was persistently interested in science. She challenged her science teachers by solving difficult problems in unconventional ways.[6] Her mother became increasingly concerned that McClintock would become a strange college professor who didn't 'belong to society'.[7] McClintock realised that she would have to come to terms with being different. She had no choice, once she set her mind on something, she said: 'I had to go in that direction.'[8]

McClintock's parents had cultivated the very attitude that led her to rebel against them.[9] Just like her mother had done when she'd married her father, McClintock ignored their objections and was determined to go to college. Her father was supportive of her education, but McClintock had graduated in the thick of the First World War. She had to wait for him to return. While waiting, she didn't waver: 'I just knew what I wanted to do.'[10]

In 1919, McClintock enrolled at Cornell University in the college of agriculture. Her father had been worried about her being treated badly as a woman on the programme, but the environment proved surprisingly welcoming. The roaring twenties saw women breaking with convention; the 'flapper' rose as a symbol of changing social roles. Women gained the right to vote in 1928. And in education, too, middle- and upper-middle-class women were steadily moving into science. Cornell was at the forefront of this movement. It was the first private coeducational institution in the country. A women's newspaper was founded in the decade, and with the increasing presence and success of female students, they constructed a new women's wing of the building (it wasn't until 1943 that women were allowed to share all cafeterias and lounges with male students).[11] When the increasing presence of women triggered a backlash, with male students protesting against coeducation, the president of the university publicly defended women's right to study there.[12]

At Cornell, McClintock was free to come into her own. As it is for many of us today, university was her time for self-discovery. On a mundane level, this meant that, following a 'philosophical talk' with her local barber, she decided to cut her hair short. She swapped her dresses for a pair of cropped trousers, or 'golf knickers', better suited to working in the cornfields.[13] None of this caused as much of a stir on the progressive campus as it would have elsewhere. Besides, bobbed haircuts were *the* look *du jour* for fashionable women.[14] Nonetheless, McClintock was expressing her independence. She also befriended a circle of mainly Jewish people, who were isolated as a group on campus. She even learned Yiddish. When her friends were rejected from sororities, she declined her own invitation to join.[15]

McClintock was exploring the levels and forms of intimacy she was comfortable with. For a while, university life tempted her into the world of human relationships, but she gradually reverted to the autonomous way of being that was her default. She was popular in her freshman year, even elected president of the women's class. And yet, it didn't last. She tried romantic relationships, but they weren't for her.

Increasingly, she was turning to what would become the true love of her life: biology. She particularly loved zoology and genetics. Her approach was guided purely by her interests. She took on a heavy course load but, as had been her way since school, she would drop courses that fell short of her expectations. She didn't care that it set her back and prevented her from graduating early. She trusted her instincts to guide her.[16]

McClintock chose autonomy, repeatedly, and in doing so she was starting to discover a different kind of intimacy. As McClintock progressed on to become a graduate student, she was closely instructed by Lester Sharp, a cytology professor in the botany department.[17] Cytologists study the structure, life cycle and pathology of cells. McClintock was particularly interested in chromosomes, the entities inside cells that had been hypothesised to carry the theoretical 'genes': the name for the as of yet unidentified units of heredity. As the evidence for the connection between chromosomes and genes mounted in the early 1900s, the field of cytogenetics was emerging. This would become McClintock's specialism.

The development of cytogenetics in the first decades of the twentieth century reanimated a field that had been dormant since the Augustinian monk Gregor Mendel first established the basic rules of heredity in the mid-nineteenth century. From 1856 to 1863, Mendel grew around *28,000* pea plants in his monastery's garden, recording the frequency of different traits of various crosses – what happened, for instance, when you crossed a plant with purple flowers with one with white flowers? Or a plant with smooth seeds with one with wrinkled seeds? What he discovered upended the popular theory of 'blending'; crosses between different coloured flowers didn't result in lighter purple flowers but an entirely purple crop. His experiments showed that offspring presented with traits of *one* of the parents, not both. However, when you bred these 'hybrid' white–purple offspring, roughly a quarter of their offspring bore white flowers. From this, he theorised a system based on dominant and recessive '*elementen*'. Purple flowers, tall stems, smooth seeds and green pods were all dominant; white flowers, short stems, wrinkled seeds and yellow pods were recessive. Mendel's theories allowed him to deduce which of the parent's traits would be expressed in their offspring and at what frequency and allowed him to postulate how these instructions are stored and carried from one generation to the next.[18*]

Mendel's conclusions were all based on observing the appearance, or 'phenotype', of pea plants. The underlying physical mechanism for this system remained unknown.

By the mid-nineteenth century, scientists had located the hereditary material of cells in the nucleus. But still, they hadn't seen it. In 1880, the German anatomist Walther Flemming used stains to darken the elements of the nucleus, so that he could see them under a microscope. He found a fibrous network inside the nucleus, which he called chromatin, or 'stainable material'. He noticed that this chromatin clustered into bodies, later known as chromosomes, during cell division. Based on his observations of cells in the various stages of division, Flemming identified the sequence of chromosome movements during the cell divisions that allow for growth, repair and replacing old cells, now called mitosis. He noticed that chromosomes split along their length during mitosis and hypothesised that these halves would form daughter cells carrying identical genetic material. In jellyfish-like diagrams, he connected the theory of heredity to a nuclear mechanism. At the end of the decade, embryologist Theodor

---

* Most traits, especially in humans, are *not* Mendelian. The properties he was examining were controlled – unbeknownst to him – by one gene with just two alleles (alternative forms or mutations of the same gene), thus why they were so predictable. While some traits in humans, like colour blindness or sickle cell disease, *are* inherited this way, most, like height (affected by over 12,000 genes, in addition to environmental factors), hair (at least 124 genes) and eye colour (at least 16 genes) are more complex.

Boveri added that chromosomes are carried by the germ cells, the predecessors of the sex cells, between generations. He also noticed that the chromosome number was halved in the sex cells, though he still lacked the evidence to explain the other fundamental type of cell division: meiosis. At the turn of the century, the American graduate student Walter Sutton filled in the picture. He identified individual chromosomes in the testes of the lubber grasshopper. In a 1902 paper, he described the configurations of chromosomes at various stages of meiosis. Sutton suggested that chromosomes have a stable structure that is maintained through generations. He showed how each new cell receives a paternal and a maternal chromosome, and that a reducing division, meiosis, ensures that the daughter cells carry half of each parents' chromosomal material. Sutton also identified eleven pairs of chromosomes that he distinguished according to their sizes, as well as one that he suggested, correctly it turned out, was the chromosome that determined the sex of the animal.[19]

The geneticist Nettie Stevens consolidated the theory for sex determination. Though her work is another casualty of the Matilda effect, her research in 1905 proved for the first time that there are two classes of reproductive cells. Working with male mealworm beetles, she showed that some sex cells carried larger chromosomes (the 'X') and some small ('Y') and that females only carried Xs while males carried an X and a Y. She inferred that a female offspring forms when an egg is fertilised by a sperm carrying an X and a male when an egg is fertilised by a sperm carrying a Y. Stevens concluded that biological sex is determined by chromosomes, weaving together the threads, in the words of one historian, 'of more than two thousand years of speculation and experiment on how animal, plant, or human becomes male or female'.[20] Stevens filled out the mechanism for sex inheritance, offering important experimental proof that chromosomes were the vessels of inheritance.

In the stories of modern genetics, however, we more often hear about the work of Thomas Hunt Morgan. In 1910, Morgan and his colleagues at Columbia University found a mutant strain of *Drosophila melanogaster* with a different eye colour to the rest.[21] Lilian Vaughn Morgan, Thomas' wife, suggested that these exceptions happened when replicated chromosome halves failed to separate during the second stage of meiosis. This was strong evidence that chromosomes carried genes. The exception proved the rule.[22] The findings brought together the findings on chromosomes and sex determination elegantly. Morgan acknowledged his intellectual debt in a 1912 paper; he even mentioned Stevens, though she remained a 'Miss,' rather than 'Dr', while Edmund Wilson, who made a similar discovery slightly later that year, is consistently 'Professor'.[23]

Most of the evidence in cytogenetics over the centuries had come from studying organisms with specific, visible traits, like pea plants and 'crossing', or mating, pairs, allowing the scientist to correlate the traits of successive generations with the presence of a particular chromosome.[24] McClintock was so good at the work, which required manual dexterity as well as mathematical aptitude, that she quickly became Sharp's first assistant. He wasn't a practical person, more of a thinker, and he gave her free rein to do whatever she wanted. That suited her well.[25]

Corn was one of the two model organisms typically studied by geneticists at the time. The other was *Drosophila melanogaster*: the fruit fly. *Drosophila* remains a workhorse for genetic experiments due to their genetic simplicity, short life cycle and high number of offspring, which allow geneticists to study their genes over multiple generations. Roughly seventy-five percent of disease-causing genes in humans can also be found in fruit flies, putting them at the forefront of medical research. But the maize plant had its own advantages. The colours of kernels on a cob of maize are a perfectly legible, almost diagrammatic, expression of genetic traits. The chromosomes for maize hadn't been identified and studied. Scientists only knew that it had ten pairs of chromosomes; for comparison, *Drosophila* has just four pairs. McClintock demonstrated that the genetics of maize, like *Drosophila*, could be studied not only through breeding but by examining the chromosomes under the microscope.[26] Using a new staining technique, she identified the chromosomes, ordered them according to size, labelled them and followed them through corn's life cycle. The rough shape of a chromosome resembles an X, composed of two thread-like strands, crossed at the centre. But McClintock revealed their individual differences. Every chromosome had its own length, shape and structure. Some had longer arms or had a pattern of bead-like structures (chromomeres) during the first stage of division where the DNA is more tightly packed, and their centromeres that link two halves of the X-shape fell in different places. These chromosomal features also varied considerably between strains – in one strain, the end of the second-smallest chromosome had a visible, deep-staining knob. Features like this became genetic tags, allowing genetics to distinguish between the chromosomes and their associated genes.[27] This level of detail was unprecedented for any organism studied by a cytologist and established McClintock as a forerunner in her field.[28]

Completing her PhD at twenty-five, McClintock was set on her next task. In *Drosophila*, geneticists had established that there were genes inherited together, called linkage groups, that were carried by specific chromosomes. These genes are located close together on the same chromosome.[29]

Genes are usually separated through recombination during meiosis. Meiosis, we have seen, is gene division that forms the sex cells, or gametes, with half the number of chromosomes. At the very beginning of meiosis, homologous chromosomes carrying the same genes (though possibly different versions, or 'alleles') line up, side by side, in the nucleus. When chromosomes make contact, they sometimes exchange fragments in a process of crossing over, or homologous recombination. When genes are far apart on the same chromosome, crossing over can put new alleles together on the same chromosome that will then travel into the gamete in the subsequent stages of meiosis. When the genes are closer together, there is a smaller space between the genes, so it is less likely that they'll cross over. These genes are more likely to travel into the gamete together, as linkage groups.

McClintock wanted to do for plants what was already underway with *Drosophila*: to connect particular traits of maize to the genes on the chromosome. It was hard to garner support for this new specialism, but McClintock didn't particularly care. She was intensely focused on the work. She prized the science over human relationships. She had no reservations, for example, about beating her colleagues to the interpretation of their results and including her name on the publication.[30] Her relentlessness served her well. By 1931, with the help of a PhD student, McClintock had published conclusive evidence of the chromosomal basis of genetics in maize, to wide acclaim in her field.[31]

McClintock had established herself as an expert. And yet it wasn't clear at this point where her career was going. It was still uncommon for women in the US to progress onto research careers; usually, they became assistants.[32] McClintock maintained a pragmatic attitude to the gender barriers she faced. When the male PhD students in her department received job offers and she didn't, she turned her attention to finding a place to do her research elsewhere. She eventually landed a job at the University of Missouri in 1936. While her research progressed happily there, it was less accepting than Cornell. She was repeatedly excluded from meetings and kept in the dark about positions at other universities. That year, news of a Barbara McClintock's engagement hit a newspaper. Mistaking the woman in the paper for Professor McClintock, the chair of the botany department called McClintock into his office, and told her, 'If you get married, you'll be fired.'[33] When asked to describe her time at the university, McClintock responded: 'awful, awful, awful'.[34]

When another job opportunity came up at Cold Spring Harbor, a key centre of genetics research on the picturesque North Shore of Long Island, McClintock took it – even after she learned that she'd likely be offered tenure at Missouri. She rejected security for a position that gave her what she most wanted: freedom. It was a difficult decision, but one that made sense to her.[35]

That is how McClintock found herself, on the eve of the Second World War, in a position that suited her well. She was secluded in a remote location with her own lab, a place to grow her corn and no other responsibilities. The culture at the lab was both casual and fanatic: everyone wore jeans and worked eighty hours a week.[36] In the summer, the number of investigators tripled with scientists visiting to exchange ideas in a relaxed holiday setting. During that time, McClintock mingled and tended to her cornfields. The winters were different; wartime rationing and the dwindling of researchers really meant there was nothing to do but work.[37] But that wasn't the worst thing. McClintock worked there seven days a week and entertained friends from the lab. She worked on a long surface made of several desks pushed together. In a side room, she stored boxes of dried corncobs, each carefully tagged with information on the lineage of each strain of seed.[38]

McClintock's scientific reputation continued to grow. She became the third women to be elected to the National Academy of Sciences in 1944 and was elected the first woman president of the Genetics Society of America in 1945. The war, for all its atrocity, broke barriers for women scientists. In 1947, McClintock rejoiced, 'Opportunities for women have never been greater than they are at the present time. There is no question in my mind that these opportunities will become increasingly better and at a very rapid rate. The restrictions in opportunity...are being steadily removed.'[39]

McClintock wasn't alone in her enthusiasm about the wartime advancements in her field. Some amazing discoveries emerged from Cold Spring Harbor and surroundings during the war. Some of the old work continued, with McClintock, for one, fleshing out the complexities of inheritance. But there was also a different branch of research that was telling a very different and enticingly simple story about the same process.[40] In 1953, James Watson and Francis Crick, with no small help from Rosalind Franklin (who we'll meet later), identified the structure of DNA. The double helix, which could be neatly unzipped during cell division, offered a simple explanation for how traits were passed between generations. DNA soon dominated research, overshadowing any other findings. It provided an elegant solution, distilling inheritance into the passing of a simple code. The theory was tantalising for its promise of mastery, and it paved the way for an era in which altering the genomes of plants, animals and even humans is increasingly the norm. And yet, in doing so, inheritance has also proved to be more complicated than DNA pioneers initially thought. The simple theory of a static code, it turns out, missed much; and, for a while, McClintock's work was part of the collateral.

McClintock had developed her own distinctive method during the war years. One formative experience came when her former colleague and friend George Beadle invited McClintock up to Stanford to help him with a problem he couldn't solve in his organism of choice, *Neurospora* (a red mould found on bread). The chromosomes were so small no one could spot them. In fact, the entire meiotic cycle was a mystery; was there a typical process for fungus at all? He thought that if anyone could do it, it was McClintock. McClintock didn't just manage to identify and count the chromosomes; she figured out *how* fungi transmitted their genetic material by following them through a whole cycle of replication. How did she do it?[41]

The answer might sound bizarre and perhaps a little Newtonian. McClintock went and sat under a eucalyptus tree. She was, she said, 'thinking'. Though not in the usual, scientific, purely analytical way. It was a kind of 'intense, subconscious thinking' that allowed her to grasp the problem in a new way.[42] While at first, she'd been unable to follow the chromosomes through the cycle, once she had performed her meditation under the tree, she could see. In her head, she reconstructed the process from the separate slides she had viewed under the microscope, so that she could 'watch' the process in real time. She saw how the nuclei from the two parents moved towards each other. How the chromosomes, tiny at first, became longer and fatter after they paired up, like railroad tracks. Big enough now to distinguish and organise according to size. Slowly, she went on, assimilating the data into a process that she knew but that could also be changed by the data specific to the fungus.[43] The chromosomes of *Neurospora* followed similar stages in meiosis to higher plants and animals. When she returned to the microscope, her mental picture allowed her to see the chromosomes in front of her more clearly.[44] Within five days, she'd solved the problem of how the fungus passed on its genetic material.

This was the second time in her life that McClintock had accessed something akin to a meditative state. Just as she had tried to transcend her body with the 'running lama' method as a child, here she had found a way of transcending her conceptual limitations to access some broader form of understanding that then allowed her to see the fungus from within this expanded frame. Her thinking then allowed her, in turn, to 'merge' with the organism she was studying to know it more intimately. As she put it, 'As you look at these things, they become part of you. And you forget yourself. The main thing about it is you forget yourself.'[45]

McClintock is sometimes accused of being a 'mystic' because of statements like the above, in which she is candid about the aspects of the scientific process that are difficult to explain. Since Francis Bacon, the father of empiricism, established the foundations of credible modern science in the seventeenth century, scientists have been reluctant to admit to any kind of processing or thinking that doesn't fall within this paradigm of observation and logical deduction. And yet, McClintock isn't alone in her descriptions of the process of scientific insight. Many scientists have made similar statements, not least Einstein, who said: 'The state of feeling which makes one capable of such achievements is akin to that of the religious worshipper or of one who is in love.'[46] This is different from the spirituality of the other scientists we have seen: Nightingale's 'calling' or Spielrein's angels that implored them to pursue their path. This is a form of transcendence that describes an approach, a method; it is akin to what we might in today's psychological language call lateral thinking, or perhaps that state of absorption we find on the cusp between current ability and challenge, sometimes described as 'flow'. McClintock describes accessing a perspective not through logical, step-by-step thinking but through a creative process of cognitive modelling over time, allowing her to access insights on the brink of her awareness.

The experience under the eucalyptus tree gave McClintock the confidence she needed to take the next steps in her research.[47] She returned to Cold Spring Harbor and began to pursue what would be her career-defining work.

McClintock had noticed an interesting anomaly in the appearance of one generation of maize. Given what she knew about how traits were passed on, the maize should have been colourless. But there were clearly distinguishable coloured patches. There was something going on that was causing this change that could not be explained in terms of the transmission of genes in fixed locations. She studied the patterns of inheritance of the colour of the kernels over several generations and managed to identify two locations on the chromosome, which she named *Dissociation* (*Ds*) and *Activator* (*Ac*). She found that the *Ds* seemed to cause chromosomes to break at a particular location and had additional effects when the *Ac* was also present, like disrupting the normal patterns of inheritance of certain traits. She then discovered that the *Ds* and the *Ac* could switch locations. She did meticulous, microscopic analysis, gathering more evidence than she could publish because the results were so outlandish and she knew she'd need a rigorous evidence base. But she'd become confident of her finding that when the *Ds* moved a gene that coded for colour, which the *Ds* previously suppressed, was suddenly activated, so that it coded for pigment synthesis of the cells in the maize kernels.[48]

Between 1948 and 1950, McClintock consolidated a theory that described how mobile elements, like *Ds* and *Ac*, regulated the genes by deactivating or regulating their action.[49] This mechanism of gene regulation, she believed, had implications far beyond explaining the patterns of inheritance in maize. It could explain a question that was more pertinent for embryologists than geneticists: namely, how cells differentiate into their various functions.[50] How does each cell of the body, containing the same set of genes, know whether to become a skin cell or hair cell or brain cell? Just knowing that a gene codes for a trait, after all, does not tell us *how* this process happens. To make sense of that chain of events, scientists needed a more complex mechanism like the one McClintock proposed for how genes get switched on or off. But this was not the thinking for geneticists at the time.[51]

McClintock argued that individual genes did not work as a simple code. Genes didn't work independently but as part of a complex system, acting differently at particular times in an organism's development.[52] That meant that the function of genes changed depending on their position on the chromosome. This understanding of gene action is standard for geneticists today. But at the time, this question had been sidelined and relegated to the 'less exciting' field of embryology.

The cleft between McClintock's work and the leading science of her day was widening. James Watson, of the DNA trio and just a student at this point, attended Cold Spring Harbor summer schools while she was there. He played baseball in the fields neighbouring McClintock's cornfield. Flippantly, he recalled how the ball 'all too often' landed in her fields. The anecdote is symbolic of the growing rift between the scientists' respective fields, as well as their respective attitudes to their work: Watson, as a boisterous member of a fraternity of men with his bold and catchy theories, and McClintock, an 'old' and isolated woman doing the unglamorous manual work of unfurling an increasingly complex picture.[53]

McClintock's breakthrough took six years. It was highly laborious work that involved breeding corn, staining chromosomes and meticulously recording data. Though it was equally theoretical. For one, a gene, at this point, was still just a theoretical entity; she was trying to develop a theory of an entire mechanism based on something that was presumed to exist. And of course, she was trying to make sense of data that fell outside the accepted paradigm of heredity. She was proposing that mobile elements, what we now know as DNA sequences called 'transposons', moved and copied themselves to new locations in the genome, changing how genes are expressed. To arrive at her conclusions, she had used the same method as she had years earlier, under the eucalyptus tree, integrating the data into a working theory; her mind working, in her own words, like a 'computer' to assimilate this data into a conceptual model that she would then flesh out using observable evidence.[54] It was highly abstract stuff. You had to grasp her logic to be able to visualise the system she was trying to describe. Communicating her theory proved impossible. In the summer of 1951, after years of gruelling work, she presented her work at the annual symposium at Cold Spring Harbor.

She was met with silence.

Her presentation had been dense, full of statistics and proofs. There was some mumbling, even laughter. People didn't understand. 'It fell like a lead balloon,' her graduate student Harriet Creighton recalled. A leading molecular biologist called her 'just an old bag who'd been hanging around Cold Spring Harbor for years'.[55] And yet, McClintock knew what the reception was really about. She'd challenged the theory on which her esteemed colleagues' careers rested. She had, as she put it, 'collided with the stable chromosome'.[56]

Most people complained about the density of McClintock's arguments. And yet, it was hard to avoid. It is always difficult to visualise a new reality. The response she received reflected scientists' bafflement but also an unwillingness to entertain the contradictory data. One biologist admitted to his wilful ignorance: 'I don't want to hear a thing about what you're doing. It may be interesting, but I understand it's kind of mad.'[57]

McClintock tried again in a series of seminars and a detailed publication published in 1953.[58] Only three scientists outside the field of maize genetics read the article. McClintock gave up on publishing for a decade and wrote up her data in large notebooks instead. She submitted summaries of her work for publication in the annual reports of the Carnegie Institution of Washington that only a few libraries bought. She also stopped teaching or talking to people about her work. Until then, McClintock had often chosen solitude, in which she found the freedom to follow her interests. Now, in her own recollections, it had turned to loneliness.[59]

Though McClintock was by no means the old bag rotting away in Cold Spring Harbor that Watson made her out to be. She remained relevant. She received numerous awards throughout the 1960s – from Cornell University, the National Academy of Sciences and the National Science Foundation. Just not for the work that mattered to her.[60]

Even when it came to transposition, scientists continued to discuss her work in the hope that someone would make sense of it. When asked about her work in 1951, the eminent geneticist Alfred H. Sturtevant, who constructed the first genetic map of the chromosome, replied, 'I didn't understand one word she said, but if she says so, it must be so.'[61] Many graduate programmes continued to teach her work; textbooks included descriptions of even her most challenging findings (today, transposons are even an established part of sixth-form biology!). And yet, students couldn't make head nor tail of it. The Nobel Prize-winning biologist David Baltimore recalls how he struggled: 'growing up as a student in the sixties…all of us tried to read McClintock's papers… But a lot of us gave up.'[62]

McClintock's story shows how difficult it is to prove anything outside the accepted theories in science. It takes an accumulation of findings to build a different worldview. The supposedly objective observation of the greatest scientists actually relies on the process of assimilating knowledge into an ever-developing collective model of reality: the very process McClintock described herself doing on an individual level under the tree. Her failure to make herself understood, combined with her frankness about her process, challenge us to look at science not as an authoritative body of knowledge, but as an evolving understanding. That understanding will be as good as our tools for communicating and evaluating our most widely held theoretical models, as well as our capacity to make sense of unfamiliar perspectives and new ideas. The geneticist James Shapiro put it this way:

> If she says something has happened, she has seen it in dozens and hundreds of cases. One reason that people don't read her papers is because the documentation is so dense. So first they said she's crazy; then they said it's peculiar to maize; then they said it's everywhere but has no significance; and then finally they woke up to its significance.[63]

Shapiro was one of a wave of geneticists who, in the 1960s, began to discover how gene expression was regulated in bacteria. Their descriptions resembled McClintock's for maize. But still, there was a resistance to the idea of transposition among most geneticists. It also took a long time for bacteria geneticists to draw a link to McClintock's work. While there was increasingly talk of 'jumping genes', it was difficult to see how this translated to the more complex, multicellular organisms, which underwent a whole process of development the simpler organisms didn't.[64] Eventually though, geneticists described similar mechanisms in yeast, fruit flies and even antibodies in mammals, and in 1976, a paper made the first explicit analogy to McClintock's work.[65] Slowly, her work gained traction. In 1983, McClintock, at eighty-one, received the Nobel Prize in Physiology or Medicine – on her own.[66]

Transposition has since helped explain many mutations, inherited birth defects and even how cells acquire resistance to antibiotics. Scientists put transposition to work in genome editing; transposable elements can be cloned and used to carry genes to new host cells.[67] McClintock's work has added to the theory of evolution based on random selection, a short-term, adaptive form of inheritance.[68] Increasingly, we are seeing that genes are far from the stable, isolated units that geneticists once thought they were but are actually in constant dialogue with their environment.

Scientists, scholars and readers of Keller's biography have debated extensively about why it took three decades for McClintock's work to be recognised. Some say it was her isolationist way of working, her complex way of communicating and, also, her 'unscientific' descriptions of her process.[69] Some point to the ways in which these qualities were subversive for her gender and became the target of sexist dismissals. Some say that the language McClintock used to describe the science itself – in terms of intuition and feeling – are culturally coded 'feminine', as opposed to the rationality of scientific thought, and are easily dismissed in a system built to promote male expertise. Some say it was simply the state of knowledge at the time that made it impossible for her theories to be accepted and that other crucial pieces of evidence, like the actual structure of DNA, or the techniques required to intervene in parts of the genome to test the theory, were missing.[70] I think all of these played a part.

What is important is that McClintock's story and her thoughtful descriptions of her process help us to build a more realistic picture of how we arrive at new insights about our world. She reminds us that science is ever evolving and entirely dependent on the ongoing efforts of individuals to include more of that reality. More than that, though, McClintock arrived at her way of working because of who she was. She was unapologetic about the ways in which she made her scientific practice her own. The closer we get to scientists, the more we see how personal and subjective science can be; people are *doing* the work, not hovering objectively over it. People find ways to experiment and theorise based on how they are in the world; people can only see from within the limitations of their own perspective. This view on the scientific process gives us another argument (though we shouldn't need one) for including people and perspectives historically excluded from science. Progress requires that we explore ideas that appear unfamiliar or even antithetical to established fact; they might just turn out to be fundamental.

9

Dorothy Hodgkin (1910–1994)

Chemist

'Oxford housewife wins Nobel' read the headline of the *Daily Mail* in 1964, the year Dorothy Hodgkin received the prize for distilling the chemical structures of penicillin and vitamin B12. 'British woman wins Nobel Prize – £18,750 prize to mother of three', the *Daily Telegraph* told its readers. The *Observer* saw fit to comment that the 'affable-looking housewife Mrs Hodgkin' had won the prize 'for a thoroughly unhousewifely skill: the structure of crystals of great chemical interest'.[1]

Dorothy Hodgkin is not a name we've heard as often as Marie Curie's or Florence Nightingale's. She is the first, and only, British woman to date to have received the Nobel Prize in science, but very few know her name or what she did. Besides the mandatory sexism of her time, it didn't help that the work she did was not much flashier than washing the dishes. Crystallography, a branch of science concerned with the structure and properties of crystals (not just the shiny kinds but any solid material with a very ordered microscopic structure), is highly technical. Unless they translate directly into medical breakthroughs, the findings can be difficult for a non-expert to grasp. Both Hodgkin and crystallography, though, are especially interesting to anyone looking for visions of a more inclusive and collaborative approach to discovery. Hodgkin, herself a brilliant bridge-builder and communicator, and her field show us that egalitarian science is also effective science. With a personality and a skillset so very different from McClintock's, she shows us a vision of another unique approach to science and another dimension of what it takes to get it done.

*

For Hodgkin, science always coexisted with a sense of social responsibility. She was born Dorothy Mary Crowfoot in Cairo on 12 May 1912, sharing a birthday with Florence Nightingale. Her father, John Winter Crowfoot, worked for the Egyptian education service, while her mother, Grace Mary 'Molly' Crowfoot, was a botanist, archaeologist, trained midwife textile historian or, in the words of one historian, a 'polymath and superwoman'.[2] In 1916, when John was posted to Sudan, Molly joined him there. She took up weaving and spinning to get to know the local women and later published papers comparing the Sudanese methods to 11th-dynasty practices in ancient Egypt. She was shocked to learn from the women about the local tradition of female genital mutilation (FGM). She vocalised her horror at a 1921 dinner party with the British governor general – to his embarrassment, he later said. Still, that same year, the colonial government sanctioned the setting up of the Omdurman Midwifery Training School; part of its remit was to tackle the practice of FGM.[3]

Molly had been educated at home and encouraged her daughter to follow her interests. Besides this, Molly also imparted a commitment to peace. She lost all four of her brothers in the First World War and impressed the futility of war on her children. Hodgkin metabolised this belief into a strong sense of social justice and pacifism.[4]

In 1914, when war really seemed like it would break out, Molly moved her children back to England, leaving them with a nanny near their grandparents' home in Sussex before she herself returned to Egypt. As the oldest, Hodgkin had suddenly to become quite independent. She didn't see her parents again until 1918, when they returned with a fourth baby daughter. The family then set up home near Beccles in Norfolk, but her parents were always in and out. Hodgkin's mother taught the children and their cousins for one year, before returning to her father in Sudan.[5]

Life in England, though, continued to be full of exploration and inspiration. Hodgkin attended a tiny primary school run by the Parents' National Educational Union, an organisation set up to train governesses. The science syllabus extended beyond that taught by mainstream schools of the time. In chemistry, Hodgkin got to make solutions of alum and copper sulphate, which, in time, turned into beautiful, blue crystals, like sapphires. 'I was captured for life by chemistry and by crystals,' she reminisced.[6]

Hodgkin took on some extracurricular learning too. She set up a laboratory in her attic at home and bought chemicals from the local pharmacist to start growing her first crystals. When Hodgkin was twelve, a family friend who was also a government chemist, Dr A. F. Joseph (aka 'Uncle Jo'), helped her set up experiments to identify local minerals in Sudan, on a trip to visit her parents.

When the time came to start secondary school, it was crystal clear to Hodgkin what she was going to study. Only the standard education for girls didn't include any science subjects. Hodgkin had to push to be allowed to pursue even just one science. But she was eventually enrolled and could receive a more structured education in chemistry. She even learned about a new cutting-edge technique that allowed scientists to see the atoms in a molecule and how they were arranged in space: X-ray crystallography.[7] One textbook, Sir William Bragg's *Concerning the Nature of Things*, explained how X-rays could be used to reveal the three-dimensional patterns formed by atoms in crystals. 'Let us realise', he wrote, 'that in the last 25 years or so we have been given, so to speak, new eyes… The discovery of x-rays has increased the keenness of our vision a thousand times, and we can now "see" the individual atoms and molecules.'[8] The moment was decisive for her. 'I really decided then that this was what I would do,' she recalled, gleaming at the memory, in a BBC interview in her 70s.[9]

It was clear to Hodgkin that she wanted to go to university to study chemistry. Only, again, the limited education meant she didn't meet the entrance requirements. Since her father was responsible for state education in Sudan, he felt it important that his daughters also attended state schools in Britain. Hodgkin had attended a state school that wasn't used to preparing students for Oxford or Cambridge. Hodgkin had her one science, but she needed two, as well as maths and Latin, to get into university. Luckily, she'd finished school a year early.[10] Hodgkin's parents were idealistic about their daughters attending a state school, but they could afford private tuition. Hodgkin spent a year with a tutor preparing for Oxford.

Her father's work offered other, unique forms of education. He was interested in antiques and later in life worked as director of the British School of Archaeology in Jerusalem, carrying out excavations on Mount Ophel, as well as at Jerash, Bosra and Samaria.[11] At eighteen, Hodgkin accompanied him on an expedition to modern-day Jordan. On the dig, they uncovered perfectly preserved Byzantine mosaic pavements that Hodgkin drew meticulously in her notebook. She was as interested in their geometry as their beauty. 'They are nearly all geometrical interlacing patterns,' she wrote of the experience to her grandmother.[12]

Hodgkin's archaeological experience became the scaffolding to her work at Oxford. Studying there was still a rare privilege for a woman in the 1930s. There were only five women's colleges and a strict quota restricted women to less than a quarter the number of men – it would not be repealed until 1957. Hodgkin attended Somerville College, suitably named for Mary Somerville. Hodgkin made the most of it. She worked with a dedicated intensity. In her spare time, she sat in her room sketching the tessellated pavements from her father's digs. 'I wonder she didn't go blind doing all the tiny squares,' wrote her friend Betty Murray in a letter to her parents in her second year at Somerville.[13] Hodgkin was making a dedicated study of pattern. Her illustrations of the mosaics for one dig, today kept at Yale University, formed part of the official record of the excavation. She became the third woman ever to be awarded a first in chemistry at Oxford.[14]

After finishing her degree, Hodgkin was itching for more. Again, her family's connections helped her. Uncle Joseph stepped in and contacted a friend of his at Cambridge, a professor who had hired J. D. Bernal as the first lecturer in structural crystallography at Cambridge.[15] Bernal agreed to have Dorothy join his laboratory.

Scientifically and politically, Bernal's lab was Hodgkin's kind of place. Bernal was a leading crystallographer and among the first to use X-ray crystallography to study the complex molecules that constitute the bodies of living things.[16] He received all sorts of crystalline samples from across the world to investigate – vitamin D, sex hormones and the digestive enzyme pepsin.[17] In his lab, collaboration sat at the heart of the science. Crystallography is not the kind of work that can be done alone; to this day, it relies on a varied skillset to both collect and interpret the data and requires the help of mathematicians, chemists, biologists and physicists. There is a culture of sharing findings and filling in each other's gaps that is rare in the competitive atmosphere that pervades many other fields of science.[18] One student, Max Perutz, later described a scientific utopia: 'The laboratory space was transformed into a fairy palace by Bernal's learning and enthusiasm, the fascination of his conversation and the great interest of the problems on which they were working.'[19]

The spirit of egalitarianism in Bernal's lab was distinctly political. Bernal was a member of the Communist Party. This might sound radical to those whose minds go to the anti-communist fever that reigned in the West in following decades, but communism was common among Cambridge's academics in the 1930s. The Great Depression had brought the terrible poverty of Britain's working classes into sharp relief, and, combined with the rise of fascism, socialism offered hope to many of Cambridge's more politically minded. This also came with a distinct view of the purpose of science: something done not simply for the pursuit of knowledge but for the good of society.[20]

Bernal was somewhat of a social pioneer and unusually supportive of women scientists. He took on Hodgkin and many other female students who went on to shape the field. Crystallography remains one of the most gender-equal fields of science; from very early on, equality was written into its DNA (not least thanks to Hodgkin's own advocacy for women). Bernal worked closely with her – in 1934, he passed over his experiments on the enzyme pepsin to her. He had been sent the crystals from Sweden. Hodgkin, importantly, showed that X-ray radiation did not inhibit the activity of the enzyme. This meant that it could be possible to visualise – to *see* – the structure of the molecule.[21]

Cambridge was a formative and inspiring time for Hodgkin. The university was a scientific firecracker at the time. In the Cavendish Physics Laboratory, James Chadwick discovered the neutron in 1932, John Cockcroft and Ernest Walton split the atomic nucleus for the first time and proved experimentally that $E=mc^2$.[22] In 1936, Alan Turing published his study, *On Computable Numbers*, proposing a programmable computer. Hodgkin herself had laid the groundwork for her science and developed techniques that would carry her through her career. A method that became particularly important was one in which she made artificial versions of the natural molecule she was studying, only containing a heavier element, like mercury. This allowed her to compare X-ray photographs to fill in gaps of the structure that she'd missed.[23] Besides all this, she'd found a set of values – humanitarian and socialist – that directed her for the rest of her life. 'My research with Bernal formed the foundation for the work I was to do during the rest of my career,' she later said of the time.[24] She wanted to stay, but her mentors at her old college in Oxford were trying hard to get her back. In the end, they made her an offer she couldn't refuse: a rare offer of job security in the midst of an economic depression.

Oxford gave her many things but not a decent laboratory. From 1934, Hodgkin worked from the basement of the Oxford Museum of Natural History, surrounded by dinosaur bones and cases of dead beetles,[25] yet she didn't let the conditions distract her. Here, she took on cholesterol, well known to us today as the fat linked to heart disease. To deduce its structure, she only had convoluted methods at her disposal. X-ray crystallography requires great practical and mental aptitude. The process starts by growing crystals, then shining X-rays on them to capture a pattern that needs to be interpreted using complex mathematics to, finally, deduce the shape of the molecules. The process is complicated, moving from a hypothesised model of the crystal pattern to a prediction of the resulting diffraction pattern produced by the X-rays. And this gets more onerous the bigger the structure you're looking at. Usually, the hypothesised structure didn't match the diffraction pattern, and so you'd have to repeat huge sets of calculations until you landed on the right formula.[26] When Hodgkin started her work, she had to do all this without the benefit of computers. By uncovering these structures, crystallographers help to reveal the connection between a chemical's structure and its properties; they help show us how life 'works'. This, in turn, creates the possibility of synthesising the molecule as a drug.

The day Hodgkin finally cracked the structure of cholesterol, she danced around her lab, weaving between the dinosaur bones. This was the largest, most complex molecule yet analysed by crystallographers and the first to have its structure deduced by the technique alone. In 1937, Hodgkin received her PhD for her highly skilled and technical work.[27]

Hodgkin had never been afraid of a complex project. But the outbreak of the Second World War brought her an unfamiliar kind of challenge in the form of penicillin. Until now, Hodgkin had done her painstaking work in the spirit of deepening a fundamental understanding of the molecules without any real-world applications in mind. This is perfectly legitimate; many of our scientific breakthroughs of the previous century came from scientists pursuing their curiosity in this way (including, well, penicillin).

And yet, there is also something to be said for more targeted efforts. The outbreak of the Second World War shifted the emphasis from fundamental science to real-world applications on an unprecedented scale. Modern warfare continued to present the same challenges it had in Nightingale's day. The cause of death hadn't changed; soldiers still died predominantly from infectious disease, though scientists' understanding of the causes of disease had changed. Germ theory was now well established. Unlike Nightingale, scientists now understood that germs, not vapours, spread pathogens.

Doctors now treated battle wounds with antiseptics to prevent infection.[28] And yet, the Scottish microbiologist Alexander Fleming had noticed during the First World War that, while antiseptics may prevent further infection, they also killed the white blood cells that help prevent it. When the war ended, he returned to St Mary's Hospital in Paddington, London, to look for other substances that might treat infections without the counterproductive side effects. He got as far as identifying the antibacterial effect of fungus after one of his samples of the bacteria *Staphylococcus* he was testing in the lab was accidentally contaminated. He noticed a ring around the fungus without any *Staphylococcus*, which led him to deduce that it had contained some kind of antibacterial chemical. In March 1929, he named the antibiotic penicillin.[29]

Fleming never harnessed penicillin to cure disease. His research sat dormant for over a decade. It took another world war to put penicillin at the centre of scientists' agendas. This time, the research didn't just deliver the first broad-spectrum antibiotic; it revolutionised the way science was organised. At the beginning of the Second World War, a group of Oxford scientists had come across Fleming's paper and established an unusually interdisciplinary group of doctors, epidemiologists, microbiologists and biochemists to explore the antibacterial properties of penicillin more thoroughly. They made progress in isolating the antibiotic and even testing it on human patients, but the issue they faced was production. In 1940, the team had watched powerlessly as a patient began to recover after taking the drug for five days, only to die due to lack of supplies. They needed large quantities of fungus as well as the equipment and labour to produce penicillin in sufficient quantities. With UK drug companies increasingly involved in supplying the war, they turned to the US for the partnerships they needed.[30]

The US had recently started the Office of Scientific Research and Development (OSRD) as a new way of funding medical and scientific research related to national defence. Howard Florey, the pathologist representing the Oxford group, met with the chairman of the medical division (and an old friend), pharmacologist Alfred Richards. Their friendship went back a long way – despite lacking the clinical data on penicillin, Richards was prepared to support the group's efforts. This is an early example of what we'll see in the following chapters becoming the old boys' club of the American pharmaceutical industry – sometimes sidestepping inconveniences like missing clinical data with disastrous consequences. In this case, though, the leap of faith proved lifesaving. Richards assembled the representatives of the drug companies Merck, Squibb, Lederle and Pfizer and appealed to their patriotism, gaining their support for an unprecedented form of contract between industry and government. In the past, government had commissioned companies to make, for example, aeroplanes for the military. Richards was proposing more of a research partnership, with a committee of pharmacy and government representatives coordinating on supplies and production. The collaboration between government and industry laid the foundation for what became a military advantage as the war progressed. In June 1943, penicillin research was brought under the umbrella of the War Production Board, who started the Penicillin Producers Industrial Advisory Committee to really ramp up production. Companies grew exponentially. By 1943, there was enough penicillin to supply an active military; by 1945, anyone could buy a pack for ten cents from their local pharmacy. When people discuss the most important invention in history they tend to reach for large, tangible ones: the wheel, the combustion engine, electricity. But none of those have had quite the impact on our quality of life as antibiotics; they've increased the *global* average human lifespan by twenty years.[31] The new government contracts also became big moneymakers for pharmaceutical companies. Over the course of the war, the pharmaceutical industry went from the sixteenth largest industry in the US, below areas like furniture and cars, to the most profitable.[32]

This booming industry brought scientists closer together than ever before. This international, interdisciplinary way of coordinating scientific research towards a real-world goal brought a new kind of ethos to scientific work. Even the most fundamental science of crystallography was conducted with a human application in mind. This outlook overlapped with Hodgkin's. As a left-leaning scientist, she believed that science should serve the collective goals of a society. Unlike McClintock, who used the war as an opportunity to retreat further into her isolated interests, Hodgkin responded to the wartime need; she was willing to be flexible.

The Oxford team needed to understand the structure of penicillin to be able to produce it. They had noticed that the chemical broke down into crystals. This was Hodgkin's domain. Even though so much of the chemical formula of penicillin was a mystery to scientists, Hodgkin was fairly confident she'd be able to deduce the molecular structure. She told them to send her a sample.[33] It was difficult work. The forms of the penicillin crystals varied, making the crystallography even more challenging than usual. It took Hodgkin four years, but she solved the puzzle in 1946.[34] Her work on the structure helped chemists create synthetic penicillin that turned out to be better than natural penicillin at attacking some bacteria.[35] Hodgkin had claimed her place at the centre of a new scientific–industrial complex in response to an urgent societal need. Hodgkin herself said in her Nobel lecture that 'Our urgent reason for doing the crystal structure [of penicillin] was to try and plan its synthesis.'

That outward focus also led her to emphasise relationships as they took on new significance in an age of international collaboration. Take her chummy relationship with James Watson, who pestered McClintock at Cold Spring Harbor. Hodgkin was one of the first people to see the model of DNA that he and Francis Crick had built in 1953 (drawing on the X-ray crystallography of another scientist with whom Watson had a tense relationship…but more on her later). Hodgkin was an accepted and respected figure at the centre of these growing, powerful male networks, providing the technical expertise that they couldn't do without. Both her science and her approach were about building bridges and filling in gaps, *responding* to the world around her. She was so good at both that she became the fourth woman to be elected to the Royal Society, that pinnacle of the British scientific establishment, in 1947.[36]

With a stellar reputation as a crystallographer, Hodgkin found herself repeatedly at the cusp of a knowledge gap. In 1948, the drug company Glaxo sent Hodgkin a vial of wine-coloured crystals that they had produced and asked her to work out their structure.[37] This was vitamin B12. People with pernicious anaemia whose bodies could not extract vitamin B12 from their food needed supplements, and Glaxo wanted to learn its structure to produce it at scale.[38]

This was a much larger molecule than penicillin and even less known. Despite her immense capabilities, this one, Hodgkin realised, needed a computer. Hodgkin worked with the scientist

Kenneth Trueblood from the University of California, in Los Angeles, who had programmed an early computer.[39] Hodgkin gave her data to Trueblood for analysis, deducing the structure, in her own words, by 'post and cable'.[40] This technologically advanced, international collaboration was just another repetition in the pattern that made Hodgkin's career.[41] Synthesising vitamin B12 not only meant it could be used to cure anaemia, but solving its structure suggested that large proteins could be next. The breakthrough proved that the young field of crystallography was a serious science with profound applications for human health. The crystallographer and Nobel laureate Lawrence Bragg likened Hodgkin's work on vitamin B12 to 'breaking the sound barrier'.[42] In 1964, she became the first British woman to win the Nobel Prize in any science.[43]

All this was just a prelude to Hodgkin's biggest achievement. Insulin is the hormone that helps cells convert sugar into energy and is today known to most predominantly for its function in diabetes. This, Hodgkin would always say, was her most satisfying problem to solve, because it was the hardest. Insulin's 777 atoms proved a monumental challenge (compared to vitamin B12's 181). Hodgkin spent more than thirty years on the molecule. The work required analysis of no fewer than 70,000 X-ray spots. The challenge wasn't just in the calculations but also in the experimental challenge of growing large enough single crystals suitable for the X-ray technique.[44]

Hodgkin used the same techniques as she had to sketch the mosaics in her teens. Just as she had pieced together the individual tesserae to reconstruct the floor, she positioned the atoms of insulin in three dimensions. Her first photo of insulin, produced in 1935, was an outline that she filled in with further data. She understood that the final solution would be constrained by the repeating pattern of molecules in the crystal, in the same way that the mosaic pattern was limited by the symmetries of its geometrical form. The molecule, she showed, has trigonal symmetry; it looks identical when rotated around a central axis by 120 degrees. She used this aptitude for patterns to model the possible solutions, testing them against the data of her photographs.[45] It had taken her entire career to develop the technical acumen required, but she finally cracked it in 1969.[46]

And while she was doing all this scientific trailblazing, she was having to innovate in the home too. Hodgkin had met her husband, Thomas Hodgkin, in 1937 and had three children with him. A communist vicar, according to one biography, 'quoted Marx over the couple on the altar steps'.[47] But married life brought new hardships. The third pregnancy triggered an acute attack of rheumatoid arthritis, which never left her. Not just this, but Thomas worked away from home. She had to juggle childcare with her demanding career. An unusually supportive employer, Somerville College, awarded her what would today be called maternity leave – unheard of at the time.[48] Dorothy spent the early months of 1939 at home but used the time to continue her work on insulin. She presented a key paper at a major meeting at the Royal Society in 1938 while she was eight months pregnant.[49] Her colleague and fellow Nobel Prize winner Max Perutz said of the experience: 'Dorothy lectured in that state as if it were the most natural thing in the world, without any pretence of trying to be unconventional, which it certainly was at the time.'[50] Dorothy seemed to have a pragmatic, and unapologetic attitude to the necessary blurring between work and life. She showed women that motherhood, that all-defining cipher of femininity, in no way undermined your right and ability to contribute to science at a time when perceptions were very different.

Besides multitasking, she found other workarounds. During the war, she and her husband housed refugees who helped with childcare. In 1958, the Hodgkin family and Dorothy's sister Joan and her family set up house together in a large Victorian villa in Oxford. The house was lively and chaotic, always full of relatives' children and friends from all over the world.[51] All this proves the difficulty of balancing working life and caring responsibilities, but it also offers models for how women can be supported to pursue careers.

Hodgkin was so successful and well supported by her university that it is easy to forget how unusual her situation was for women at the time. We get a reminder, though, when we look at those headlines reporting her scientific achievements: 'Oxford housewife wins Nobel'; 'affable-looking housewife'; 'prize to mother of three'. Apparently, it takes more than a Nobel Prize to be recognised as a scientist rather than a mother.[52]

Once the inane commentary was done with, though, the Nobel Prize gave Hodgkin a new, more public platform to voice her political ideas. Hodgkin had never formally joined the Communist Party (though she had founded the Oxford branch of the Association of Scientific Workers in 1936 – a decidedly leftist organisation).[53] She had been relucent to define herself as a communist. After her Nobel, though, she became increasingly politically vocal.[54] She publicly opposed the Vietnam War. She pushed to include Chinese and Soviet scientists in the International Union of Crystallography, which she helped to found at the start of the Cold War. She maintained relationships with Soviet scientists throughout the Cold War and was even denied a visa to the US in the 1950s due to her communist involvement, continuing to exchange crystals with them and with China – she even attended a dinner in the presence of Ho Chi Minh and Mao Zedong. In 1987, she was awarded the Soviet Union's International Lenin Peace Prize, which she gladly accepted.[55]

In addition to her work on disarmament, she was also appointed the president of the Pugwash Conferences on Science and World Affairs in 1975, an organisation founded in 1957 to seek peaceful solutions to global security threats. She took the same approach here as she did in her science; she built relationships, drawing on the growing network of scientists and students establishing themselves internationally. She remained the president for thirteen years, proving herself astute at facilitating an exchange of perspectives to build a shared vision in yet another arena.

All the while, Hodgkin upheld her politics in the lab too. She fostered the same environment of inclusiveness and equality she had experienced with Bernal. She trained and encouraged many women in their scientific careers; many went on to lead their own research teams. One of her less scientifically successful mentees was Margaret Thatcher. Their relationship is emblematic of Hodgkin's approach to science and politics. Thatcher, from her university years, was already a dedicated member of the Conservative Association. This wasn't an issue for Hodgkin, who was very supportive of her as a student, and even helped her get funding to continue her studies. Thatcher worked under her for a year on a crystal of the molecule gramicidin S, which had been grown in the Soviet Union and sent to Hodgkin for analysis; I wonder if Thatcher dwelled on the fact that her work at that time was made possible by such a political collaboration.[56]

Thatcher always had the highest respect for her mentor, was even intimidated by her, and supposedly kept a portrait of Hodgkin in 10 Downing Street throughout her premiership – again, despite their differing politics.[57] This ongoing bond meant that Hodgkin was included in a discussion with Thatcher and her advisors on the UK's relationship with Eastern Europe in

1983, where Hodgkin pushed for nuclear disarmament. After the meeting, Thatcher agreed to visit Hungary and negotiate with the then general secretary of the Communist Party, Mikhail Gorbachev, to deescalate East–West tensions.[58] Hodgkin had her own political leanings but she did not let these become a barrier to dialogue.[59] The biochemist Tom Blundell described his former mentor's approach perfectly: 'Dorothy saw politics as individual personalities, in terms of the people, not in terms of dogma or political convictions the way most people do.'[60] Hodgkin relied on relationships as a vehicle for her science, and this approach travelled with her beyond the lab. It was her style in everything she did. It was *how* she practised science, and it was *how* she was.

Both Hodgkin and the field she worked in were centrally about building relationships. Hodgkin was there, as the field established itself (in part, thanks to her contributions), growing and consolidating an attitude based on dialogue and collaboration. Showing women that you could be mothers and scientists. Despite the newspaper reports that diminished her science based on her gender, in another way Hodgkin also *demonstrated* that the skills traditionally associated with motherhood, like caring, listening and relating, were essential to the work of science. Hodgkin suggests another way for us to dismantle sexism in science. We can celebrate women as professional scientists, not mothers, while also valuing the skills traditionally associated with motherhood as integral to the work of any scientist.

The numerous portraits of Hodgkin focus in on her frail, arthritic hands, including one that hangs in the Royal Society in London, by Henry Moore.[61] We could read into these hands the kind of gentle femininity of a Nightingale, or the supposedly saintly Curie's commitment to the manual work of experimentation. But neither of these images describe Hodgkin's style. Her hands, to me, capture the passion of a woman whose life was about using her hands to see. Using them to develop new tools, reveal structures in unprecedented depth, build bridges between disciplines and people, and to see through the collective gaze of collaboration. At a time when science was becoming international, scaled-up and streamlined for innovation, Hodgkin was a scientist who modelled the best a collaborative spirit could do. She embodied an ethos based on social ties and social responsibility. 'I think with my hands,' she used to say.[62] Her hands may represent her 'feminine' vulnerability: they were also her greatest strength.

10

Chien-Shiung Wu (1912–1997)

Physicist

In December 1956, Chien-Shiung Wu was stationed at the US National Bureau of Standards in Maryland, in the midst of a pivotal experiment. Her vision of the universe hung in the balance; it all depended on the direction the cobalt atoms in her experiment would go. She had cooled them to absolute zero, then used a powerful magnet to align them along the magnetic field. Now, she had a fifteen-minute window to measure the direction the nuclei would emit their electrons. If the particles came shooting out symmetrically, the universe was as physicists had believed it to be. According to the principle of parity, the laws of physics applied equally to the left and right. If not – and the alternative was almost unthinkable – this fundamental law was wrong. The universe was lopsided.

The unthinkable turned out to be true. The universe does, against everything that was previously thought, on some levels, distinguish between left and right.[1] The discovery was seismic. It overturned everything physicists thought they knew about matter, and how we come into existence.

The discovery won Wu's two male colleagues the 1957 Nobel Prize in Physics. Wu received nothing.

It is no wonder that Wu was open to the notion of asymmetry. Her exclusion from the Nobel Prize was the symptom of various imbalances that shaped her life and career from the start. Sexism, racism and even factional hierarchies between theorists and experimentalists in physics intersected, creating obstacles. Even today, she is sometimes remembered by the moniker 'the Chinese Marie Curie', as though her being of Chinese nationality is what makes her contributions remarkable. At the same time, comparing her to Curie implies that the comparison to a Western scientist somehow 'elevates' her contributions, while also appropriating her achievements into the Western scientific canon. The sleight of hand, where Wu was simultaneously celebrated and denigrated, colours many of the stories told about her. I'm going to try to do better.

*

Wu was born on 31 May 1912 in a small fishing village in Jiangsu province in eastern China. Her name, which translates to 'courageous hero', was a big one to live up to.[2] Unusually for a girl at the time, Wu was allowed an education. Born on the cusp of the Xinhai Revolution, which overthrew the last Chinese dynasty and established the new Republic of China, Wu grew up as a new generation of leaders challenged traditional norms, some very oppressive. Hitherto, girls had been banned from schooling, raised to serve their husbands and their subjugation exemplified by the debilitating practice of foot-binding. Luckily for Wu, her father, Zhong-Yi Wu, was at the forefront of this cultural shift, diverging from his engineering background to open Jiangsu's first school for girls. Her mother, Fu-Hua Fan, was also an educator, and she helped Zhong-Yi persuade other families to enrol their girls in school.[3]

Wu was one of the first girls of her generation to obtain a formal education. And yet, despite her parents' efforts, Wu quickly reached the limits of what they could teach her.[4] Her parents enrolled her in a girls' boarding school, fifty miles away, when she was eleven. The selective school trained its students, tuition-free, to become teachers – a role that would be achievable and acceptable. But Wu wanted more. Already she was particularly interested in physics, borrowing classmates' textbooks to read more at night. Her hard work earned her a place in the physics department at the National Central University in Nanjing, where she began studying in 1930.

Wu continued to bump up against the limits of what her education could teach her. She graduated in 1934, top of her class, but there was nowhere in China for her to continue her studies at graduate level. There was, on the other hand, so much happening in physics in Europe and the US that fascinated her, especially on the workings of atoms. And so, at twenty-four, an uncle paid for her to cross the Pacific to continue her studies at the University of Michigan. It was a good time to leave. Japan, at this point, was threatening an invasion. Wu had protested for Chinese leaders to prevent fully fledged war. She was aware of the threat but couldn't have anticipated that the escalating violence would prevent her from returning home for over thirty years. The US was meant as an interlude to a career in China.[5]

And the US was an exciting place to be for a burgeoning physicist. Ernest Lawrence and M. Stanley Livingston had just blown apart the atom – quite literarily. Remember how Curie hinted that the atom might not be the stable building block scientists took it to be? She turned out to be right. Over the first half of the twentieth century, scientists turned their attention to the atomic nucleus to understand its properties, for sure, but also with a particular interest in the energy released by subatomic elements. By the 1930s, Lawrence had designed an accelerator that used a magnetic field to collide particles in a spiral path, producing more energy than ever before. The cyclotron was a powerful new tool that opened the subatomic world to physicists. It was also the birth of a new kind of 'Big Science', involving international teams and research conducted in government-funded laboratories, much in the way we saw for penicillin.[6]

After landing in San Francisco in 1936, Wu decided to pay Berkeley a visit on her way to Michigan. Once there, she immediately fell in love – twice. First with the futuristic particle accelerator, which was so fascinating that, with a little encouragement from Lawrence, Wu switched to Berkeley to complete her PhD. She also met another Chinese physics student, Luke Yuan: her future husband.[7]

According to most biographies, part of the reason Wu abandoned Michigan was that she'd heard reports that they barred women from entering the student union through the front door.[8] Physics was, and continues to be, one of the most male-dominated of the sciences. Women were generally excluded entirely or trained solely as assistants. This was no different in Berkeley. Wu stood out there, due to both her gender and ethnicity. Male colleagues talked openly about her appearance, calling her 'the belle of Berkeley'. Admirers followed her around campus.[9]

Nonetheless, it was the right decision to go there. Berkeley was crackling with invention. The man who would become synonymous with the atomic bomb, J. Robert Oppenheimer, had recently taken a post there. Having earned his doctorate at the University of Göttingen, he taught the new European quantum theory – explaining the particle physics that made the accelerators possible. His eminence attracted all the most ambitious scientists to Berkeley. Wu wanted to be part of that. She wasn't intimidated. 'I didn't see a man there who could compare with my father,' she later said.[10] These American men could never compete. Her confidence remained her unshakable foundation as her hope of returning home slowly dissipated in the fallout of Berkeley's atomic science.

At Berkeley, Wu learned that one of Lawrence's assistants, Emilio Segrè (who shared the Nobel with him in 1959), was studying the atomic nucleus. Wu promptly joined his group. Segrè had a reputation for grouchiness, but he warmed to Wu and even tried to protect her from a tendency he'd noticed in her. 'Miss Wu, you have to spend some time reading physics. You have to stand back from things and see the whole picture,' he would tell her. It was all in vain; her single-mindedness became her hallmark.[11]

With Segrè as her supervisor, Wu developed a two-part thesis. First, she studied the electromagnetic energy emitted when particles slow down as they move through matter. For the second half of her thesis, Wu focused on the radioactive gases emitted when the uranium nucleus splits.[12] She was building on the work of Lise Meitner. Meitner had provided a theoretical explanation of the process of nuclear fission – where a neutron strikes the nucleus of a heavy atom, causing it to split and release energy. Meitner had worked from Sweden, while her German colleagues, Otto Hahn and his junior colleague Fritz Strassmann, had conducted the experiments from their lab in Berlin (as a Jew, she'd had to flee Germany in 1939). Meitner and her nephew, Otto Frisch, published their theoretical interpretation of Hahn and Strassman's experiments in the February 1939 edition of *Nature*, and coined the term 'nuclear fission'.[13] Still, the 1944 Nobel Prize in Chemistry 'for the discovery of fission in heavy nuclei' went only to Hahn. Hahn was wary of his association with a Jewish researcher while living under Nazism and didn't credit her in his publications; maybe that explains why Meitner didn't get the prize. It does not explain why she was nominated another forty-eight times, including by Hahn himself in 1948, in both Chemistry and Physics, without a single victory.[14] We can add her to our history of giants who succumbed to the Matilda effect.

Wu became *the* fission expert at Berkeley. After she completed her PhD in 1940, Danish physicist Niels Bohr invited her to give a talk to explain the process to the other scientists on campus. Oppenheimer himself called her 'the authority'.[15] Wu nicknamed him 'Oppie'. And yet, despite her accomplishments, American racism remained inescapable. When the *Oakland Tribune* reported on her PhD work on nuclear fission in 1941, they recalled 'a petite Chinese girl' who 'looks as though she might be an actress or an artist or a daughter of wealth in search of Occidental culture'.[16]

Even Berkeley didn't take her seriously. After she completed her newsworthy PhD, they refused to hire her. For Wu, these were the convoluted reverberations of an event that had loomed over her entire PhD career. Less than a year after arriving in Berkeley, in 1937, Wu had opened the morning paper from her student dorms to the headline 'Japan Invades China'.[17] The war meant that Wu was stranded in the US. She couldn't even contact her family and wouldn't hear from them for eight years. All the while she received a stream of horrific news about the 42,000 murders and rapes in her old university town. The Japanese soldiers invading Nanjing took photos of their handiwork. Chinese shop clerks ordered to develop their films smuggled copies to the newspapers, exposing the atrocities to the world.[18] Wu could only watch the horror unfold.

China was at war with Japan, but in the US all Asians were the same.[19] Japanese–Americans were contained in internment camps, but any Asian person was suspect.[20] There were ever-stricter anti-Chinese immigration laws. While her home country fought a common enemy, while she contributed to the American knowledge base and suffered the excruciating silence of her family, a twisted, hateful logic had made the talented Wu an untenable candidate for a position she was more than qualified for. Her mentor Segrè was outraged. 'They would have had a star,' he told her.[21]

The uncertainty and injustice would be too much for anyone to bear; there was no choice but to split off, to escape into another world. Wu threw herself even more totally into her lab work. Tellingly, Marie Curie became her self-appointed idol at this time. Wu said that she resonated with Curie as a determined foreigner who had relied on her own hard graft.[22] Wu certainly shared Curie's dedication. But her situation was more precarious. A failed exam could leave her destitute. She worked late, often till four in the morning. University officials worried about her safety when she went home at night. They appointed a fellow graduate student, Robert Wilson, to coax her out and escort her home each morning. He observed her emotional cycles, how she became 'willowy' in the lead-up to every exam.[23] How, when she passed (with flying colours, of course), 'there'd be a triumphant celebration at a Chinese restaurant'.[24] An unstable atomic nucleus emits subatomic particles to return to equilibrium. Wu had found her own coping strategy in the form of periodic scarcity and occasional release.

The war provided Wu's next big opportunity. This was, in some respects, a 'physicists' war' (though as we have seen, it was a biologists' too). Radar had helped Britain face the German air force; the atomic bomb would end the war.[25] After a few years of teaching (part of it as the first female lecturer at Princeton), Wu was invited for an interview with the Division of War Research at Columbia University. In a famous anecdote, two physicists questioned her about physics but withheld all information about the lab's secret projects. At the end of the day, they asked, 'Now

Miss Wu, do you know anything about what we're doing here?' She smiled and replied, 'I'm sorry, but if you wanted me not to know what you're doing, you should have cleaned the blackboards.' It was, of course, her physics, not her dry wit, that landed her the job on the Manhattan Project. They asked her to start the next morning.[26]

Wu, now a Columbia professor, worked from a converted Nash automobile warehouse. She developed radiation detectors and performed research on 'gaseous diffusion', a process used to enrich uranium to increase the supply of fuel for the atomic bomb later dropped on the people of Nagasaki and Hiroshima.[27]

The bombs saved Wu's family and killed many others. They ended the conflict that had severed her from her family and left a death toll that weighed on many of the scientists involved. Like her friend Oppenheimer, Wu had complicated feelings about her role.[28] Though not enough to refuse, as Meitner did when Oppenheimer invited her to join the Manhattan Project. Like Marie Curie, she preferred not to talk about it. She opted for vague pronouncements about the goodness of humanity when asked about the potential the bomb posed for global annihilation. In 1959, she said: 'Do you think that people are so stupid and self-destructive? No. I have confidence in humankind. I believe we will one day live together peacefully.'[29] Of course, she was right to frame the bomb's capacity for destruction as a human responsibility rather than a quality innate to the bomb. But she did miss an opportunity to use her platform to speak in more concrete terms about the role scientists could play now that they had been situated at the heart of global politics (like Hodgkin and others did in the Pugwash movement). Wu chose to remain detached. In 1965, she did advise the Taiwanese president, Chiang Kai-shek, against developing an atom bomb when he consulted her, but not explicitly on moral grounds.[30]

On a personal level, the Manhattan Project hadn't just saved Wu's family from ongoing conflict, it had accelerated her career. She was one of the few physicists on the project asked to remain at the university after the war. She was appointed senior investigator on one of Columbia's largest federal research grants. At the same time, China wanted her back. The National Central University offered Wu and her husband full professorships. They suggested that they spend another year gathering equipment, then return home to them. Wu might have been torn but her father told her not to do it. A civil war had erupted in China. The communists were emerging as the victors and were taking control. Her father's advice turned out to be sound, though it also meant they'd never see each other again.[31]

The world was splitting once more. The People's Republic of China was founded in 1949, and by 1950 the US had joined South Korea in the war against the communist China, Soviet Union and North Korea. Chinese scientists in the US were under intense scrutiny. Notoriously, the MIT rocket scientist Xuesen Qian was accused of communist sympathies and detained for five years.[32] Wu became a US citizen in 1954 so was spared the suspicion to some degree, but the US blocked all travel to and communication with China. When she was finally allowed to travel back, in 1973, her parents and brothers had been killed in the Cultural Revolution; even their tombs had been desecrated.[33]

Wu now held an esteemed position at the top physics departments in a country at war with her homeland. She had supported the US war effort in part, according to one biography, with a vague hope of saving her family.[34] But the science she had helped to produce had only given the world powers ammunition to point their weapons at those dearest to her. While she grappled with these contradictions, Wu had landed on an inconsistency in the subatomic universe too. Only this one, she could and would make sense of.

In the midst of world tensions, Big Science was thriving. New accelerators were shattering particles into ever more parts, and, with it, the fundamental laws that physicists revered. One subatomic particle, the K meson (or kaon), was behaving particularly strangely. Wu's colleague at Columbia, Tsung-Dao Lee, had been working on it with the theoretical physicist Yang Chen-Ning at Princeton. They had noticed that, when the K meson decayed radioactively, sometimes two particles, sometimes three, were produced. Other physicists assumed they were unwittingly observing two different particles, but Lee and Yang had shown that the K meson consistently weighed the same and had the same properties. There was something else going on.[35]

What was at stake here was the principle of parity. Left–right symmetry, or parity conservation, stated that the laws of physics applied equally to the left and right; particle processes like collisions or decays should happen in exactly the same way when mirrored. The principle had been proven to hold when it came to most objects: from atoms to molecules, planets to baseballs.[36] But not, it turned out, inside the nucleus. The K meson's decay seemed to be lopsided. They wrote up their findings and published a paper outlining the experiments that could put the question to bed.[37]

No scientists responded to the call. The maths was there – nobody believed that such a fundamental law wouldn't hold. The quantum mechanics pioneer Wolfgang Pauli wrote to his colleague about the proposition: 'I do not believe that the Lord is a weak left-hander.'[38]

But to Wu, the principle's fundamentality was an argument *to* test it. A law this important, she reasoned, should have the experimental evidence to support it. It took her some time to decide. Once she did, she was hit by an incredible urgency. She didn't want anyone else to get there before her. She was booked to go on holiday on a luxury ocean liner, the *Queen Elizabeth*, that spring, to celebrate the anniversary of her arrival in the US with her husband. He sailed alone.[39]

Wu accelerated. She slept four hours a night, and she hounded her team to keep up. Her colleagues secretly called her 'the Dragon Lady' (comments tainted with the self-same racism that followed her everywhere, even as they expressed a kind of deference).[40] But they did it. As the Christmas lights were going up in 1956, Wu conducted an experiment at the US National Bureau of Standards in Maryland to test Lee and Yang's theories. The experiment she had planned was complex even by today's standards; with the equipment available in the 1950s, it was near to impossible.[41] Timing was everything. Wu decided to use atoms of the isotope cobalt-60. She had to eradicate the random heat energy that kept the atomic nuclei in constant motion so that she could measure the direction the electrons were ejected during radioactive decay. She brought the cobalt-60 atoms down to absolute zero temperatures so that they came to a near-standstill. She then used a magnet, many times stronger than the Earth's magnetic force, to align the nuclei along the magnetic field.[42] That gave her a window of fifteen minutes to measure the direction the electrons were ejected. If the cobalt atoms emitted beta particles asymmetrically, it would prove Lee and Yang's idea. [43]

The results of her experiment were astonishing. The left side of the nuclei emitted more electrons than the other; no matter which way she spun it, she got the same asymmetrical effect. The universe, at the nuclear level, skewed definitively to the left. It is hard to express the full ramifications of this finding. It could, for example, help explain how matter came into existence. It could have been through a break of symmetry that matter came to outweigh antimatter. Monumental theories like this flowed from Wu's work. The experiment was named after her.[44]

Against all expectations (even her own) Wu had disproven the sacred law of parity. She had got there before anyone else. Wu barely had time to celebrate. She and her colleagues drank champagne out of paper cups in the lab, then set to publishing the results. Wu whipped up her team to write up a paper, capturing nine months of work, in one afternoon. Her colleague Raymond Hayward had his own interpretation of her sense of urgency at this time: Wu, he said, was 'very, very, very competitive – because of some deep-down insecurity perhaps. It was as though she was afraid she wasn't going to get credit.'[45]

Wu's anxiety was grounded in a reality. Academia had not always treated her kindly. In this case too, her fear was to some extent realised. She was bitterly disappointed the day she heard that the Nobel Prize for the discovery had gone to Lee and Yang, not her. This came as a surprise to many of Wu's colleagues, including Oppie.[46]

Most of her supporters put the decision down to another implicit hierarchy in the field of physics that places theorists above experimentalists. The Nobel committee made a similar decision in 1914, when Max von Laue theorised the principle of X-ray diffraction (in a conversation, not even a published paper) and two university colleagues performed the gruelling experiment proving that the technique worked. Still, the committee gave von Laue the prize. Theory is valued over the practical work of proving or disproving ideas. This reflects, again, the genius narrative woven into the Nobel's DNA. The way that the prize bolsters individual legacy and values moments of inspiration and a haughty objectivity, over the manual labour required to bring science to life. Also notice how the characteristics we associate with 'proper' science in this scheme are once again culturally connected to masculinity (rationality, mind) and those devalued to the feminine (affect, body). In the case of parity conservation, the committee also explicitly refused to credit anyone on the experimental side.[47]

The deliberation behind the Nobel committee's decision will remain sealed until their documents are released after fifty years, and then only if all the concerned parties have died.[48] Only then will we be able to hold the committee accountable for any discrimination. We do have statistics, though. Since its inception in 1901, only four women have won a Nobel Prize in Physics. Lee and Yang were the first physicists of Chinese nationality to receive a Nobel. It is difficult to discern how social attitudes worked against Wu getting credit for her experiment, but, given the widespread reality of sexism and racism at the time, the question is more one of how than if.

Wu defied the laws of the universe to find meaning and even recognition in conditions that were distinctly nihilistic. She didn't receive the Nobel Prize but won multiple other prizes, and achieved many 'firsts': she was the first woman to receive the Research Corporation Award, the first woman to win a Comstock Award from the National Academy of Sciences, the first woman president of the American Physical Society, the first woman to earn an honorary doctorate of science from Princeton University – the list goes on. Columbia fully embraced her and granted her tenure in 1972.[49] In 2021, the US Postal Service issued a commemorative stamp in her honour.[50] She has become an icon for women in STEM and a national treasure in China, featured on various memorabilia.[51] Until her retirement in 1981, Wu remained one of the world's most respected experimentalists, influential in various fields, from magnetism to nuclear fluoroscopy.

And yet, we have to do more than celebrate. Wu's career is equally a window onto the fear, hatred and disavowal immigrants navigate to survive in a non-native country. To be rejected by a society you contribute to, that you come to call home. To be celebrated by a culture that continues to diminish you. As her granddaughter points out, Wu's stamp brings the total of Asian American women featured on stamps to just two, alongside chef Joyce Chen, who popularised moo shu pork.[52] We can celebrate Wu's overcoming, but we have to recognise the real obstacles she faced. Perhaps we *can* make the comparison to Curie (especially considering Wu herself did), but we have to acknowledge the differences too. It took an internal splitting and accelerating momentum for Wu to maintain her balance amidst the morbid geography of her world. Her tenacity was remarkable but was also sadly necessary. Truthfully, perhaps she never quite found the order she was hoping for. Wu herself joked it was the first time anyone had won a prize in science 'not for establishing a law, but for overthrowing it'.[53] Wu's career would not be defined by a stabilising law; her presence in the scientific establishment was fundamentally subversive.

11

Frances Oldham Kelsey (1914–2015)

Pharmacologist and Doctor

The 1950s. Women found themselves confined to the home once again after the fleeting freedom of the wartime years. The feminine ideal proved stronger than ever, and along with it came the pressure to endure childbirth and domestic confinement, all with a smile on your face. It was an era that plagued women with what the feminist writer Betty Friedan, in *The Feminist Mystique*, dubbed 'the problem that has no name' – feelings of listlessness, purposelessness, depression – which doctors eagerly recognised as the latest incarnation of hysteria. The ground was ripe for a wonder drug to ease the burden; while drug prescriptions for barbiturates soared for housewives, where was a drug offering to ease the burdens of pregnancy?

Enter thalidomide. In the late 1950s and early 1960s, the drug, initially developed in Germany, was lauded around the world – a sleeping pill that avoided all the usual side effects of barbiturates. Moreover, it seemed particularly effective at easing morning sickness in pregnant women.

However, this 'miracle' drug was indeed too good to be true. Over the four years that thalidomide was circulating the globe, thousands of babies were born with shortened or missing limbs, no ears or damaged hearts. Some were never born. Some died. The scandal swept through Europe, Australia and Canada.[1]

But in the US, the number of babies born with defects was in the teens, not the thousands.[2] And that was due to the work of the pharmacologist and doctor Frances Kelsey. Kelsey stood her ground at the Food and Drug Administration (FDA) when no one else would. Kelsey didn't discover something new, but she prevented significant harm. She shows us a different and crucial side to the rapid scientific advancement we have seen: how scientific findings are often an argument to *de*celerate.

*

Kelsey was born on 24 July 1914 on Vancouver Island in British Columbia. Known as 'Frankie', she was one of three children. Her father was an officer in the British army who had retired to Canada. They travelled around a bit, and she attended schools in Victoria, British Columbia and England. In her early years, though, she picked up reading and writing by listening to her mother teaching her older brother. She learned to navigate male-dominated spaces from a young age. Her first school also wasn't meant for her: a small, private, all-boys school in British Columbia. Kelsey was the only girl there for several terms before more girls were admitted. 'I started off in the atmosphere of boys,' Kelsey writes in her memoir. 'I was used to being in a class with men.'[3]

Kelsey did very well at school. She was an honours student and weaved her way through the educational system, all the way up to a doctorate. When she received the offer from Eugene Geiling, professor and chairman of the pharmacology department at the University of Chicago, she was addressed as '*Mr* Oldham'. She asked her tutor if she should write to Dr Geiling to correct him. Her tutor told her not to be ridiculous, to sign and put 'Miss' in brackets. Which she did. She never discovered whether, had her name been Elizabeth or Mary Jane, she would have received this life-changing position as a PhD student. Apparently, she did try to ask Dr Geiling throughout their long friendship and collaboration whether her gender would have made a difference. He would never answer.[4]

Kelsey started her degree in pharmacology in 1936. During her PhD, she witnessed her first disaster. In 1937, Elixir Sulfanilamide arrived in Geiling's lab. The drug was an effective treatment for strep infections but was also killing people: over 100, including children.[5] Geiling had been asked to identify the toxic substance in the elixir. His team found that the drug contained antifreeze. It was distressing work, feeding lab rats the pills and watching them die. The experience imprinted on Kelsey the real harm seemingly benign drugs could do.[6]

The event left an impression on the public too. The scandal precipitated some of the most fundamental legislation in the pharmaceutical age. Congress responded by passing the Federal Food, Drug and Cosmetic Act in 1938, which required manufacturers of drugs to provide evidence that a drug was safe *before* going to market.[7] Today, this seems common sense. But at the time, it wasn't required. Even after the new legislation was passed, there was a loophole that allowed drugs to be distributed as part of a clinical trial for 'testing purposes'.[8] Drugs could still be distributed to patients as part of a sanctioned, human experiment with little or no oversight. This flaw in the regulation would prove fatal.

Kelsey continued to work with Geiling after she completed her PhD in 1938. When the Second World War broke out, the lab joined in another urgent medical project: combating malaria. Malaria was one of the infectious diseases killing more American soldiers than combat in the Southwest Pacific.[9] Kelsey worked with the well-known antimalarial drug quinine to investigate how rabbits broke it down in their livers. But the team had done something radical in their experiment (that definitely wasn't required at the time, especially not in a wartime project meant to save the lives of male soldiers) – they had included female, not only male, rabbits. Some of these rabbits were pregnant. Kelsey made several key observations. Quinine, she found, passed from mother to foetus through the placental barrier. Once in the foetus' system, the rabbit in utero processed the drug differently to the adult mother. The mother's liver broke down quinine effectively, while the foetus didn't at all. When a drug isn't broken down, it will accumulate in the body, potentially to toxic levels. This meant that a drug could be harmless to a parent but hurt the developing foetus.[10] Kelsey's research drew attention to a dimension of drug safety that wasn't top of the list of priorities in the new war-powered pharmaceutical machine fixated on male soldiers. The findings would become crucial to her work on thalidomide.

Kelsey married her lab-mate, Ellis, in 1943 and they formed a household centred around science. They never watched TV, instead reading scientific journals together in the evenings.[11] At dinner, they exchanged work stories about mismanaged drug trials.[12] Moreover, Kelsey didn't just remember the lessons of her quinine experiment; she *lived* them. When they had their two children, Kelsey refused to take any drugs during her pregnancies.[13]

And it made sense; through their pharmacological work, Kelsey and Ellis were getting a very different picture of what was becoming the era of the 'wonder drug'. Across the world, penicillin had sparked optimism about drugs as one-stop fixes for a healthier, happier life. From antibiotics to treat everything from meningitis to diphtheria and syphilis, to the new anti-psychotics and antidepressants that promised to make your problems go away. The new wave of chemicals promised veritable miracles to a public who had reasons to want to boost their moods in the aftermath of war.

It wasn't that these drugs weren't incredible; many of them did alleviate symptoms and sometimes targeted the disease itself. But none of them did so in a vacuum. Penicillin had been successful not just because the chemicals themselves worked but because governments had coordinated efficient production lines to ensure the demand was met. That production line had served its purpose for quickly supplying soldiers at war, and then for producing profits, but it was less suited to meeting the needs of a complex society over time. Streamlined production left insufficient process for safety. Kelsey had already witnessed the impact of unexpected side effects. This was the shadow side of the new, rapidly growing pharmaceutical industry. Producing drugs in larger quantities than ever before also amplified any harm they did, not just adverse effects, but also the risk of dependency (as in the case of barbiturates) or antibiotic resistance that left scientists out of their depth. Drugs were rushed enthusiastically to market without any real regulation to ensure they reached their consumers safely. In the background, nuclear radiation was a galvanising force for scientists realising that in a new, large-scale, powerful science, governance was everything. It was clear to many that the wonder drug story was overly simplistic. Kelsey was a scientist, investigating the effect of chemicals, but what she was really seeing was a failure in the human structures to deliver them safely.

Kelsey and Ellis had contributed equally to the quinine project at the University of Chicago and supported each other in their careers. And yet, when it came down to it, Ellis' career came first. The university had a policy of not hiring two members of the same family. Now that they were married, either Kelsey or Ellis would have to leave. The couple decided that one of them would attend medical school, not because they intended to practise but because it might bolster their qualifications. They chose Kelsey to go. To explain their reasoning, Kelsey said, 'I needed the extra credentials. Let us face it, [as a woman] I needed all the help I could get to obtain a job.'[14]

Kelsey finished her medical training in just four years and graduated in 1950. She had a job waiting for her as an editorial associate at the *Journal of the American Medical Association*, through Geiling's contact. Her job was to review research papers. Many of them weren't very good. And yet, curiously, Kelsey and her colleagues noticed repeatedly how the papers they rejected appeared in other journals. When Kelsey started at the FDA, she again spotted many of those names on the clinical studies in new drug applications (NDAs).[15] Something other than academic rigour was opening doors for these scientists.

In 1960, Kelsey started the job that would make her career. She was working as a reviewer of NDAs at the FDA. Her job built on her skillset as an editor, but this time with higher stakes: she was assessing if there was sufficient evidence in each application to show the drug was safe for humans.[16] It was an unglamorous job. As a doctor, you could earn a lot more in private practice, and it was generally considered a transitional job for newly qualified doctors.[17] Besides this, there was a high turnover of doctors leaving to join drug firms, where the salaries multiplied.[18] At forty-six, Kelsey was much more experienced than any of her colleagues – a doctor, researcher *and* a medical reviewer.[19] The job was arguably below her paygrade. That didn't bother her.

All things considered, the FDA wasn't the worst place to be a woman at the time. The agency was one of the oldest US regulatory agencies and had overseen food and drug labelling since 1906 when it was established under the Food and Drugs Act. But its culture was more progressive than most universities'. When it opened its first Food Research Laboratory in 1908, it was directed by the female chemist Mary Engle Pennington (known for discovering that refrigerating foods helps to preserve them).[20] Women held prominent positions in FDA laboratories from the outset.[21] The organisation had made strides in professionalising, with a team of competent pharmacologists running its labs, and yet the 1906 Act had already proven inadequate in facing the scientific challenges of the evolving century: poisonous foods, deadly cosmetics and, of course, cases like Elixir Sulfanilamide that shocked everyone with the dangers of untested drugs.[22]

It was a dynamic time to be working for the agency. Things weren't perfect but were rapidly improving. New amendments were constantly passed. After the Second World War, testing procedures were introduced for antibiotics like penicillin. In 1950, two pharmacists pushed for addictive drugs to be available on prescription only. It was a time when the taken-for-granted regulations governing medical drug use today were being established, although, sadly, often in response to human tragedies.[23]

Kelsey was stationed in a Second World War temporary prefabrication building, or 'pre-fab', on 7th Street, Washington, DC. The application for Kevadon, the US brand of thalidomide, was the second to cross her desk.[24] At this point, thalidomide had been selling like hotcakes across Europe under other names. The established US pharmaceutical firm William S. Merrell was hoping to cash in on the trend. They were pitching Kevadon as an exceptionally safe sedative compared to the addictive barbiturates that people overdosed on. Kevadon didn't have these undesirable 'side effects', the company claimed; you could use it as a sleep aid or prescribe it to pregnant women for morning sickness. You could take as much as you wanted and be absolutely fine.[25]

The application was sloppy. Because the drug had already been used widely around the world for several years, Merrell, and likely the FDA, expected this to be a straightforward case.[26] Though there were other reasons that Merrell may have expected their product to be approved easily. The last woman to work at the organisation, Kelsey's predecessor, was a whistle-blower. Barbara Moulton had spoken out about the chummy relationships between the FDA reviewers and big pharma. She had testified in Congress that these politics undermined the FDA's procedures and meant that dangerous drugs were going to market before they were proven safe.[27] Moulton had tried to raise the issue at work, but they had only dismissed her. [28]

Kelsey wasn't part of the boys' club either. She walked in fully aware of the challenges of working in a male environment – she'd even read Moulton's testimony before joining the FDA.[29] Kelsey would fight the battle by insisting on scientific rigour.

Kelsey was in charge of reviewing the human clinical trials that formed part of the Kevadon application. Her colleagues would review the parts on the chemistry of the drug and animal studies. They then compared their results, and if there was any sign of harmful effects, or if they weren't convinced that the drug's components were safe or the manufacturing consistent, they'd reject the application.[30]

This all sounds systematic and sensible, but the amount of paperwork is humongous. According to Kelsey herself, it amounted to about four volumes – a volume is about the size of a phone book.[31] On top of that, the reviewers had just sixty days to respond to an application. Otherwise, the drug would *automatically go on sale*. This sounds irresponsible to modern ears. And it was. Employees lived in fear that an application would fall by the wayside and unintentionally default to going to market.[32]

The system of review, although somewhat improved, definitely still skewed in favour of profits, rather than human safety.

It didn't take long for the team to find gaps in the evidence. For one, the chemistry component had been translated from the German imperfectly, which had been spotted by a reviewer who had trained in Germany.[33] This led them to wonder if the results were being misrepresented, though it proved nothing. The animal studies also seemed suspicious, even unrealistic. Merrell had reported that the drug was entirely non-toxic. No matter how high the dose they gave the test animals, they didn't die. This is almost unheard of. Besides, if Kevadon's sedative power didn't increase with a higher dosage, it was likely because animals weren't absorbing the drug, in which case the findings were moot.[34]

But Kelsey's own part of the review was the most concerning of all. The clinical evidence was either lacking or irrelevant.[35] There was no information on the patients who had received the drug: their age, sex, prognosis, the dosages they were given or any description of any side effects. There was no actual data. All Kelsey had to review were statements from the thirty or so doctors running the trials – testimonials, rather than scientific studies.[36] It was, all in all, a sub-par application offering nothing but a doctor's word. That Merrell submitted it expecting instant approval said a lot about how their applications had been treated in the past. Not this time.

Kelsey was not going to let this pass; that much was clear. But she also didn't have a solid basis for a rejection. She was missing the very data she would need to make that decision. Kelsey was under growing pressure. The Merrell liaison was pushing for the FDA to approve their application. It was all taking a lot longer than expected and Merrell resented the delay. They already had brochures printed, and enough raw material to make 15 million pills for the Christmas season.[37] Who wouldn't want to be pleasantly sedated through the holidays?

Not just that, but time was running out on the sixty-day window; the drug would soon go to market. Kelsey consulted her predecessor for advice on how to navigate the deadline. Moulton shared a helpful loophole she had found in her time at the FDA. If you marked the application as incomplete (which, of course, it was) and requested further information, the deadline would be extended by another sixty days.[38] That's what Kelsey did. She sent Merrell a letter spelling out the holes in the application: the incomplete animal studies and clinical trials, the lack of data on long-term safety. It was a long list.

Merrell wasn't used to demands. Gone were the days of special favours. When they came and presented their updated evidence the following month, Kelsey still wasn't convinced. The company got so frustrated that they called Kelsey's boss, though he didn't concede.[39]

Under the mounting pressure, Kelsey finally came across some real evidence of the dangers of thalidomide. The Scottish doctor Leslie Florence had described a side effect of the drug in a letter to the *British Medical Journal* (*BMJ*) titled: 'Is thalidomide to blame?'[40] The doctor reported peripheral neuritis – a painful inflammation of the peripheral nerves that caused a prickling, cramping and coldness in the arms and feet – in patients who had taken thalidomide for more than a year.[41] When the patients went off the drug, their symptoms subsided.

Kelsey was disturbed by the findings – nerve damage was not the kind of 'side effect' you should be experiencing from a sleeping pill.[42] More than that, the article had been published weeks before, and Kelsey had heard nothing about it from Merrell. All this while their representative had been harassing her throughout February to hurry the application along. Kelsey called him again to see if he would mention it now. He didn't. When she asked the rep directly, he admitted that he *had* seen it, though he still downplayed it and assured her that there was nothing to worry about.[43]

The truth was that he'd sent a panicked letter regarding the study the previous week to the British distributor of thalidomide, Distillers, asking if they knew anything about the report. Distillers, it turned out, had known about the peripheral neuritis for *two years*. The *BMJ* findings had forced them to put warning labels on their pills that same year. The producer in Germany, Chemie Grünenthal, had its first reports of adverse side effects even before the drug went on sale as early as 1959. The neurological side effects of thalidomide had been reported when Kelsey was assessing the US application, and yet companies were pushing their products as non-toxic, lying about side effects.[44]

Merrell tried every tactic to convince the FDA. Their rhetoric was tasteless at best. From April 1961, they started comparing their drug's safety to that of barbiturates. Barbiturates had a known suicide risk; you didn't kill yourself with thalidomide, they reasoned. They released their first publicity campaign just after Marilyn Monroe died from a barbiturate overdose. 'If Marilyn Monroe had taken thalidomide, she would still be alive,' was a common refrain.[45] A lack of suicide attempts seems a low bar for safety to me. With questions about other side effects still unanswered, it was also entirely beside the point. It definitely wasn't going to persuade Kelsey.

In the end, though, the symptoms in adults turned out to be the least of the companies' worries. Kelsey wasn't aware in early 1961 that doctors in Germany and Australia were starting to publish on the link between mothers who had taken thalidomide and a growing epidemic of babies born with severe malformations: blocked intestines, heart defects and missing ears. The

Australian doctor William McBride had even called Distillers to warn them, but, again, they did nothing.[46]

Kelsey was totally unaware of the emerging evidence. Her research career, though, had primed her to ask this question. She knew drugs passed the placental barrier and that foetuses responded to substances differently. Kelsey asked Merrell for studies on the drug during pregnancy. They responded with some actual evidence but only for late pregnancy. Kelsey wanted to know the drug was safe throughout. We can see now how crucial this research is, knowing that thalidomide disrupts foetal development; it makes sense that it would do more harm the earlier it enters the foetus' system.[47] Merrell refused. They agreed to add a warning to its labels indicating that thalidomide's effects on pregnancy were unknown,[48] but beyond this, they were still only concerned about getting their drug to market ASAP.[49]

In these negotiations, Kelsey stood her ground. Despite the ticking clock and the daily abusive phone calls, she remained stoic.[50] She had a confidence around men, cultivated since childhood. But she was also grounded in a scientific code. Her daughter, Christine, recalls the dictum at home was to 'question everything...don't believe it until you can test it, or until you know it's true'.[51] Kelsey's commitment to the scientific method, as for so many of the scientists in this book, anchored her in the face of personal onslaught. The drug hadn't been proven safe: it wasn't going anywhere.

It is important to remember, though, that the clinical trials meant that, while the drug wasn't on sale in the US, it was being prescribed to women by doctors as part of this research. Hundreds of pregnant women in the US were already taking thalidomide while the drug review was in process. Legal or not, it is pretty shocking how Merrell conducted its studies. One piece of research, for example, was published under a doctor's name when the whole thing had actually been written by his buddy at Merrell based on quotes he gathered over lunch or a game of golf. This doctor prescribed thalidomide to expectant mothers. Their children died or went on to live with life-long disabilities.[52]

Merrell's unethical conduct continued. Towards the end of 1961, the epidemic of malformations in babies in Germany had been publicly connected to thalidomide. The paediatrician and geneticist Widukind Lenz had been carrying out his own investigations, noting the abnormalities in infants of mothers who had taken thalidomide, and had presented his findings to government. To the chagrin of the German producer of thalidomide, Chemie Grünenthal, the government banned the drug. The story hit the press, sparking public outrage.[53]

But it took another year before the drug was removed from market in the US. Merrell did tell Kelsey, but, in what we by now recognise as Merrell's evasive hallmark, they *still* tried to downplay the evidence. They told her that they hoped the connection between the drug and the malformations was just a coincidence, and that they were sending a representative to Germany to glean more information.[54] They also promised her they'd notify the doctors they sent the drug to – in the end they contacted just 37 of the 1,200 doctors involved in their trials.[55]

Kelsey may have taken them at their word, had she not heard from Dr Helen Taussig. Taussig was the founder of the field of paediatric cardiology. She had travelled to Europe after hearing about the babies affected there and had uncovered evidence of their limb defects, and also the congenital heart disease that was her speciality. She talked to drug manufacturers, to parents of the affected children, to scientists and to epidemiologists to build a full picture.[56] Once back in the US, she invited Kelsey, as a representative of the FDA, to her home in Baltimore, where she shared the evidence with her. She had photographs, interviews and case histories. She had a record of an affected baby born as early as 1959. She had proof of a steady annual increase of incidents in the following years. The mothers in these cases had all taken the drug. Taussig estimated that 6,000 babies had been affected in Germany alone.[57] This was the moment, Kelsey recalls, that convinced her completely that thalidomide harmed the children of women who had taken it. The association was indisputable. 'Therefore', she said, 'we would have to take all the measures we could to make sure that none of the drug was remaining in this country where it might be used.'[58]

Taussig's findings were reported in the *New York Times* on 12 April 1962. However, these reports didn't cause public outrage the way they had in Germany. Senator Estes Kefauver, though, took note, only for different reasons. He had been investigating pharmaceutical companies for price fixing. Thalidomide could help him push for the drug reform bill he'd been trying to pass for years. He needed a face to the campaign, a character with a story people could relate to.[59]

'"Heroine" of FDA keeps bad drug off market' read the headline of the *Washington Post* on 15 July 1962, accompanied by a portrait photo of Kelsey. Kelsey was thrown into the public eye and was suddenly bombarded with TV and radio interview requests. She was painted as a stoic, principled and determined heroine with refreshingly less of the sexist stereotyping you might expect; such was the power of the disaster she had averted.[60] 'She saw her duty in sternly simple terms, and she carried it out, living the while with insinuations that she was a bureaucratic nitpicker, unreasonable and even, she said, stupid.'[61]

Of course, there were also the many cracks in the government's regulatory processes to paper over. President Kennedy spoke out publicly about the 'urgency of providing additional protection to American consumers from harmful or worthless drug products'.[62] In 1962, the Kefauver–Harris Amendments were passed, followed by the 1963 investigational drug regulations. The new legislation established a bare minimum. Drug companies had to obtain informed consent from clinical trial participants, their advertising claims should be based on scientific evidence and they had to prove to the FDA that their drug was safe *before* starting clinical trials. It is questionable even today to what extent companies abide by these laws, but they nonetheless enshrined the idea that drug companies did indeed have an ethical responsibility towards their expansive consumer base.[63] Lest our miracle drugs become our poison.

Kelsey herself was presented with the President's Award for Distinguished Federal Civilian Service, the highest award attainable for a US civil servant.[64] Kefauver had submitted her name just two or three weeks before the event, and still, she won.[65] There were political agendas at play. That doesn't mean, of course, that Kelsey didn't deserve the recognition, but this *is* an invitation to be as discerning as Kelsey herself and to ask what this heroine's tale misses.

The victory against thalidomide in the US, it turns out, wasn't total. The narrative that Kelsey saved the US from disaster obfuscated the truth that many women had been given thalidomide by their doctors as part of clinical trials. Many of these cases went unreported for decades. Survivors had to lobby for recognition, let alone compensation. They had to prove that their mothers had taken the drug.[66] This was especially difficult because their doctors hadn't kept adequate records. It wasn't always the doctors' fault; many had been told by Merrell that the trial was more of a performative exercise, that the drug was about to come to market anyway, and that they didn't need to worry about keeping extensive records.[67] Only seventeen people have been recognised by the US government as thalidomide survivors, but there are likely many more. Even those recognised haven't been compensated. Perversely, the Kelsey narrative in some ways distracts from the ongoing impact of thalidomide on American people. This is the very absence of accountability that Kelsey would have been vehemently against. USA Thalidomide Survivors today are still campaigning to get the support they increasingly need as they age.[68]

Kelsey isn't an American heroine, in the same sense that she isn't a scientific giant. Her career wasn't about what she discovered; it didn't culminate in a scientific breakthrough. She shows us instead that scientists, public servants and representatives of institutions make decisions on the smallest and largest scales about *how* they will conduct their science. McClintock, Hodgkin, Wu – all the scientists in this book have had their own answer to that question, their own style. In all cases, their *how* reflected who and what they valued. Kelsey valued the lives of the people taking the drugs she reviewed. It was her how, not the what, that shaped her contribution; the how that prevented the dangerous drug from reaching patients. In an era where miracle drugs and silver bullets race each other to market, the sky's the limit but the risks remain the same. In a history of scientific discoveries, Kelsey invites us to tally our progress anew.

12

Marie Nyswander (1919–1986)

Psychiatrist

Since the Sumerians etched their prescriptions for opium poppy into tablets, opioids have presented societies with a potent tension. They effectively mitigate pain, suppress coughs and treat diarrhoea. They are also highly addictive.

That tension has long played out on the political stage. In the 18th century, it brought Europeans to the shores of China, who traded Indian opium for silks and tea. The Chinese government moved to block British opium in 1839, sparking the first of the Opium Wars. At the same time, in the West, opium was a widely used painkiller. In 1804, morphine was isolated and used to treat sick and wounded American soldiers in the Civil War. Thousands of veterans became addicted, but, with the medication easily available, so did their doctors and countless middle- and upper-class housewives in what became known as America's first opioid crisis. New prescriptions replaced the old, but opium-smoking remained popular among Chinese immigrant labourers who had arrived with the California gold rush – this changed the face of opium use. Tougher legislation in some municipalities followed, now that users were Chinese, gamblers and sex workers supposedly taking Americans' jobs.[1] The publicity campaign lent Roosevelt convenient credibility as he tried to stymie the opium trade in the Philippines, then under US control.[2]

Towards the end of the century, heroin was synthesised from morphine. Again, the chemical had been imagined as a medicine, but by the 1960s, heroin use in the US was rising faster than any drug that had come before – this time, mainly among young African American and Latino men.[3] The drug, once again associated with a minoritised people, was banned. Swathes of users were imprisoned. And yet, even as heroin became the prime target of the war on drugs, Dr Marie Nyswander and her team at Rockefeller University in New York proposed something truly radical. What if doctors used another opioid – methadone – to treat heroin addiction?

It would make her some powerful enemies and save countless lives.

*

Mary Elizabeth Nyswander was born in Reno, Nevada, on 13 March 1919, to a German father, a mathematics professor, and an American mother. Nyswander's parents divorced when she was just two, and her mother, Dorothy Bird Nyswander, became her role model. Dorothy was a professor in psychology and an important figure in international public health. She modelled a life of social purpose and fierce independence.[4] They moved together, first to Berkeley, where Nyswander watched her mother teach at high school by day and work on her PhD by night. They spent time around liberal thinkers, like the anthropologist Margaret Mead, whose study of Samoan sexual norms published in 1929 fuelled the sexual revolution with its evidence that gender roles are learned, not biologically determined.[5] Nyswander's mother went on to found the school of public health at Berkeley, and set up day care for mothers helping in the war effort. She later joined the World Health

Organization at sixty-two and travelled the globe to set up health education programmes, sleeping on cement floors in rural India into her eighties.[6]

Her daughter watched her mother and knew what was expected of her. When Nyswander started college, she told Dorothy she'd be changing her name to Marie; there were too many 'Marys' out there, she told her, and, speaking to her mother's outlook, 'she wanted a name with more character to it'.[7]

Nyswander decided to become a doctor like her mother. After graduating from the progressive Sarah Lawrence College in 1939, she applied to Cornell Medical School. She wanted to become a surgeon – lofty ambitions for a young woman in the 1940s. She applied for her residency with the U.S. Public Health Service in 1946, hoping to emulate Dorothy's life of travel and adventure. She tried the navy, but they didn't accept women as surgeons.[8] She was to be sent instead to the backwater of Narcotics Farm in Lexington, Kentucky. 'Narco' had been set up by the US government as an experiment that would treat former prisoners with drug addictions as patients – to rehabilitate, rather than punish them.[9]

This was a radical idea at the time. People with drug addictions in the US had, until then, been treated as criminals by doctors and politicians alike. In 1914, during Prohibition, the US government had passed an act that criminalised drug addiction.[10] Today, a medical professional might still fall short, but these doctors weren't even allowed to *try* to help people with addiction. By the 1920s, the strategy had proved unsustainable. Prisons were heaving with drug-related arrests. Narco presented a new approach: a prison *and* a hospital.[11] Here, addiction might still be considered a criminal offence, but it was also treated as a medical problem. The environment was designed, to some extent, to heal. The farm's wholesome activities distracted its patients from drugs. There was farm work, alongside art therapy and leisure activities. A third of the inmates came voluntarily. Narco even became a jazz hub, with the likes of Chet Baker and Sonny Rollins performing there; for some it was a career goal to travel there and practise alongside the jazz greats.[12]

Harry J. Anslinger, the racist commissioner of the Federal Bureau of Narcotics from 1930 to 1962, hated jazz. He enjoyed an analogy when stoking up American racial fears to justify his anti-drug policies; he likened the music to 'jungles in the dead of night'. It made sense that he wasn't a fan; jazz was a hectic nexus of all the forces he opposed. In the 1950s, as many as forty percent of jazz musicians used heroin – notoriously, Billie Holiday and Charlie Parker.[13] Jazz was an irreverent and avant-garde expression of counterculture. Many of the performances took place at prohibition speakeasies.[14] Anslinger employed agents to target jazz artists. He ordered them to 'shoot first' in their drug raids.[15] Anslinger shook a grim cocktail of criminalisation, racism and addiction theories that became the cornerstones for America's drug policies in the second half of the 20th century. While the place was a refuge to many, Narco was still a prison. Anslinger was against community healing; he wanted control. The place existed against the backdrop of Anslinger's increasingly punitive drug policy.[16]

Nyswander saw that Narco was no utopia. The place was still a far cry from the liberal ideals she'd grown up with. The medical staff were often abusive, and she was shocked by the racism of the Deep South.[17] All this, she saw, wasn't helpful to people's recovery; treating people as addicts reinforced their identity and behaviour as addicts.[18] Narco was a veritable country club compared to the former prison system, but to Nyswander, the bleating of lambs and basket-weaving did not hide the bars on the windows. People were still sent there for minor drug offences.

Nyswander saw how the staff looked down on their patients.[19] It still looked like a prison to her.

It turns out that Nyswander was onto something: the attitudes she sensed were symptomatic of some abhorrent activities taking place at the facility between 1953 and 1962. Narco was eventually closed in 1975 when it became apparent that researchers there had recruited patients as test subjects for LSD treatments. The experiments were part of a CIA-funded project, MK-Ultra, to develop drugs that could be used to push individuals to breaking point in interrogations. Testing drugs on society's marginalised to develop methods of torture; it would be hard to come up with something more twisted.[20]

Personally, Nyswander also struggled at Narco. She'd arrived as a chain-smoking, New York intellectual who wore trousers; a single woman among male doctors all married with children. The doctors' wives, too, generally kept their distance.[21] Her social isolation coloured her impression of the place, but it also allowed her to bond with Narco's patients. These men welcomed her and tried to console her when they noticed how down she was. They played jazz for her, which she'd loved since adolescence, and Baby Laurence, a famous tap dancer, gave her a private performance.[22]

The patients gave Nyswander the human connection she was missing and challenged everything she'd been taught about addiction. In the theories of addiction she'd been taught at medical school, addicts were painted as psychopaths, inherently criminal (thus justifying their incarceration). That was not what Nyswander had witnessed at Narco. Her experiences led Nyswander to the completely obvious but highly subversive idea that 'addicts' were full human beings perfectly capable of relating, working, loving, playing; their addiction did not, in fact, define them, or indeed, make them immoral.[23]

Her time at Lexington led Nyswander to change the course of her career. She decided to switch from surgery to psychiatry. As Nyswander put it, she'd discovered that she wanted 'more closeness with the patient rather than with his kidney or his gallbladder'.[24] Her training in psychiatry, in turn, led Nyswander to Freud's psychoanalysis, which, in the 1950s, was considered the best theory of human behaviour by most US psychiatrists.[25]

But her training did not bring her closer to humanity. After Lexington, Nyswander completed a two-year residency at Bellevue Psychiatric Hospital in Manhattan, where she was faced again with the horrors of addiction 'treatment'. She was especially uncomfortable with the way doctors acted like prison guards, denying patients drugs even as they writhed with violent withdrawal symptoms.[26] Nyswander quickly became disillusioned with the whole system. She was done with it. She wrote a paper, published in 1950 in the *New England Journal of Medicine*,[27] offering her insights on withdrawal treatment, followed by a book, *The Drug Addict as Patient*, in 1956. She'd passed on the baton and shared everything she knew. Now, she fully intended to let others continue the disheartening work.[28]

And thus commenced what strikes me as a real head-in-the-sand era in Nyswander's career. Moving away from addiction, she started to see private clients for the somewhat less pressing problem of 'sexual frigidity'. Nyswander still saw people with addiction occasionally, but her focus had shifted to what she seriously believed was 'one of the gravest problems of our times'.[29] She became a proponent of female 'sexual surrender' for women who couldn't orgasm.[30] The cause, she offered, was a failure to accept the biological destiny of motherhood, often the result of developmental problems like having an absent father or a feminist mother (just like her…).[31] Not quite the progressive views we would have expected of her.

Nyswander burrowed herself in a quaint, middle-class life. She had remarried for the third time, remained childless and consistently prioritised her career over her home life. It was apparently possible for her to ignore the hypocrisy of the treatment she was advocating. She had chosen to abandon inconvenient realities – personal, professional and social – to embrace the most conservative views of her class uncritically.

Nonetheless, those realities simmered on the brink of her awareness. Nyswander herself describes how she was roped back into addiction work. Her 1950 'how-to' paper on managing drug withdrawal had set her up as an expert. And so, as she puts it, dryly: 'Far from being rid of the problem, more calls.'[32] The hospital she'd worked at kept calling her for help with people experiencing horrific withdrawal symptoms. Even in her state of Park Avenue retreat, it became too much to ignore. She recalls: 'How can you walk away when there's a woman in the hospital ...whom the intern has cut off from all barbiturates, and she's starting to convulse? You cannot not treat that woman.'[33]

Heroin addiction was on the rise. Initially, it had been envisioned as a medicine for tuberculosis and other respiratory diseases, brought to the market by the German pharmaceutical company Bayer in the late nineteenth century.[34] When the medical community eventually established the addictive and harmful effects of heroin, it was banned but already fuelling an underground market. By the 1920s, ninety-eight percent of all people addicted to drugs in New York took heroin.[35] The Second World War temporarily blocked supplies, but by the 1950s heroin was back and affecting different communities. In the 1940s, when Nyswander was at Lexington, users had been mostly white working-class men. In the 1950s, this shifted drastically to young, working-class Black and Puerto Rican men.[36] Harlem became the centre of heroin distribution, where, according to some accounts, the drug was easier to buy than cigarettes.[37]

Reluctant, but unable to sit by, Nyswander joined up with the East Harlem Protestant Parish to do what she could. She set up an addiction programme offering walk-in psychotherapy. Unlike other programmes, you didn't have to be an inpatient, and, importantly, you didn't have to quit drugs to be admitted.[38] According to newspaper reports of the time, her clients valued the service. They appreciated being talked to as *people* about what may have led them to addiction, rather than being policed.[39] One man told a *New York Times* journalist: '[She] didn't put us all in one box, she sorted us out, because she got inside. I could walk in next door and just blow my top if I want to; and believe me, she could blow her top too. I dig her because she swings. She's really alive.'[40]

Nyswander connected with her patients, but, if the aim was to stymie their addictions, the effects were underwhelming. The talking cure did not seem to work. At least, no better than the abstinence enforced at hospitals. Relapse rates were similar – that is, terrible – at around ninety percent.[41] Nyswander was, understandably, discouraged. But she *had* discovered something important. The outcomes of talking therapy may have been bad, but they were also no worse than other regimes. At the very least, this was evidence that the brutal measures Nyswander had witnessed in hospitals were unnecessary; talking about your troubled relationship with your mother or your masturbation angst was as effective and a lot more humane.[42]

Nyswander was making some progress. Enough to irk the old guard in addiction treatment. In the late 1950s, she presented on her work at the National Research Council. She explained to a room of male doctors how she offered psychoanalysis to patients still addicted to heroin. Anslinger was the first to stand up and leave. Nyswander's scheme undermined his long-established, singular drug policy of detoxify and punish.[43] In 1951, he had pushed Congress to pass the Boggs Act, setting mandatory sentences for drug convictions. Someone arrested for marijuana possession faced two to ten years in prison. His arguments relied on the same old racial profiling, this time linking marijuana to 'crazed Mexicans' and Black men supposedly corrupting white women.[44] Many others in the room followed Anslinger; they were wise to if they valued their careers. One of his agents later said, 'Anybody that came out with any academic work that could be critical of him, his Bureau, or his philosophy, had to go to prison, or be beheaded.'[45] When the American Medical Association (AMA) published a report debunking many of Anslinger's claims, he threatened to fire any of his agents who read it.[46] When another professor argued for compassionate care over incarceration, like Nyswander, Anslinger had him wiretapped and falsely told his university that he was involved with a 'criminal organisation'.[47] Nyswander herself told the historian David Courtwright that she found an official report confirming the Bureau of Narcotics' agenda to systematically intimidate anyone working in outpatient addiction with the aim of shutting their practices down.[48] Nyswander perceived the threat but couldn't get a lawyer to take her on; she had some powerful enemies.

Nyswander took it all in her stride. Nevertheless, she felt frustrated. She was doing well professionally, appointed as assistant professor at New York Medical College alongside her private practice. But her addicted patients were still relapsing and dying.

Until Vincent Dole called her. Dole was a researcher at Rockefeller University. Dole was a big name, not in addiction, but in obesity and metabolism research. He was on his way to winning a Nobel, but his work had started to bore him. He was frustrated with the inconsiderate obese patients who just couldn't seem to sustain weight loss (you can tell he had a narcissistic streak). But his frustration was really with a scientific problem; his patients' bodies seemed hardwired to an immovable metabolic default that he just couldn't shift.[49]

Dole commuted to work. He switched trains at Westchester at 125th Street, then walked several blocks through East Harlem to catch the train to Rockefeller. That walk took him through what he described as 'an epidemic sea of misery.'[50] Until then, he'd been totally protected from the heroin epidemic in his privileged bubbles. The new-found reality pressed on him. He wanted to do something about it. It wasn't his area of expertise, but he reasoned that addiction might have something in common with the metabolic processes behind food cravings.[51]

Dole had no experience, but he had status and the right connections. It took one conversation with the chairman of the New York City Health Department's Committee on Narcotics, Lewis Thomas, to make the switch. Thomas was going on sabbatical and didn't feel like the work was going anywhere – he named Dole his replacement on the spot.[52]

Dole knew he wasn't equipped for the role, so he read around a little and came across Nyswander's book. *The Drug Addict as a Patient* went further than Nyswander's article did. For one, it framed her recommendations for managing withdrawal with a social history of criminalisation. She spelled out the view of addiction she had observed, plain and simple, at Lexington. This, she wrote, wasn't a psychological problem rooted in a psychopathic personality. It was a physical disease.[53]

To a doctor like Dole, with all the necessary resources at his disposal, the disease model of addiction appealed. Here was something he could investigate and, theoretically, *cure*. It was the most interesting work he'd read on the subject, so Dole invited

Nyswander to join him in solving addiction through the body, not the mind.[54]

Nyswander started at Rockefeller in 1964. She worked with Dole and the clinician Mary Jeanne Kreek to conduct studies on addicted patients. Very quickly, they saw a pattern. When they interviewed patients, they described the driver of their addiction not so much as the high of the drug itself but the withdrawal, the unbearable feeling that led them to chase their next hit. They explained how they didn't feel physically 'normal' when they were off heroin.[55] Kreek, with her background in endocrinology and guided by her own experience of recovering from polio as a teenager, began to make sense of these patients' accounts in terms of a physical disease. Dole drew his own parallels to the metabolism.[56] Nyswander, of course, already saw addiction as a physical illness, but now they were gathering the biological evidence. The research team began increasingly to understand the problem of addiction as a biochemical imbalance. This meant, to them, that the 'cure' had to be pharmacological. Addiction, they argued, was a biological deficiency requiring constant maintenance. They frequently drew the comparison to diabetes. Just like a person with diabetes needs insulin, a person with an opioid addiction might just need opioids. As a kind of prescription.[57]

That being said, the pharmacological intervention Nyswander and her team came up with in the end didn't treat a biological 'disease' but a mode of delivery. Because heroin was illegal, people were injecting it on the streets using contaminated needles and inconsistent doses that were killing them. Skyrocketing prices were pushing people to commit crimes to afford the drug. Thus came the idea of maintenance. What if they gave people the addictive substance – for free, in a controlled setting? The team experimented with various drugs to keep their patients comfortable and manage their withdrawal symptoms. They tried morphine, Dilaudid, cough medicine, even regular doses of heroin itself, but nothing worked.

Until they tried methadone hydrochloride.[58]

Methadone was another powerful opioid and had originally been manufactured by a Nazi-allied German drug company.[59] Its origins were dubious, but methadone was known to stymie withdrawal symptoms, and, at a higher dose, it led to noticeable improvements in patients' functioning within a few days. When they returned home, they weren't tempted by heroin anymore. One of their first two patients graduated from college with a degree in aeronautical engineering.[60] The team continued to expand their studies and repeatedly found patients completely transformed – able to return to work and school, and get on with their lives. The team had found a medical treatment that actually worked.

Alongside Dole, Nyswander enjoyed a new-found prominence. Her name came second on the paper the team published proposing their maintenance approach in *JAMA*, the prestigious *Journal of the American Medical Association*, back in 1965.[61] Kreek's was entirely absent. This raises uncomfortable questions about the conditions for women's entry into science at the time. Nyswander's professional status was perfectly commensurate to her hard work and experience – perhaps her name should have been second (or indeed, first) – but Kreek's should definitely have been there too. Nyswander, though, was by this point romantically involved with Dole (she left her husband the same month the paper came out). Dole was the gatekeeper to scientific credibility, and his criteria were sexist. The gender policing continued as Nyswander began to appear in the media. Her face was splashed across magazine pages, with journalists celebrating her beauty and charisma over and above her science.

A 1965 *New Yorker* profile piece described Nyswander as a 'slim, attractive woman'. [62] In 1968, *Vogue* announced: 'Now Dr. Nyswander, who has disarming candour, enthusiasm, and enormous attraction, is married to the understanding, big, handsome, and craggy Dr. Dole.'[63] Nyswander was one half of a publicly palatable couple that, ever so conventionally, combined her beauty and Dole's status. It is hard to reconcile these depictions with the trouser-wearing, chain-smoking renegade at Lexington. Nyswander could have been the face of a burgeoning women's movement, but, in the popular press, the romance story overshadowed her subversive personality and science. It ensured that her name was erased from the history of science until historians of science unearthed her important contributions in the 1990s.[64]

However, the sheer contrast between *Vogue*'s reports and the earlier descriptions of Nyswander does point to a quality that was genuinely hers: a capacity for transformation. From changing her name, to her degree, profession and husband, she was unafraid to pivot to suit her needs. Her transition this time was swift and total: she didn't just change jobs but traded in her psychoanalyst circle for Rockefeller friends. She divorced her husband and married Dole.[65] Nyswander didn't quite have the status or freedom Dole did, but she moved easily between worlds and circles throughout her life. That was, of course, a luxury only upper-class and white people enjoyed. She was able to move between Narco and Park Avenue; Harlem and Rockefeller.

Nyswander had allied herself with a man, as well as a powerful institution.[66] At the same time, she still had some high-octane enemies. Other psychiatrists and physicians still opposed the approach. Maintenance, in their view, was dangerous, leaving the underlying addiction untreated (and, it was implied, criminals uncorrected). Psychiatrist Judianne

Densen-Gerber, a contemporary, was very vocal in her view that methadone maintenance was 'immoral'.[67]

In the press and media, the debate grew. A 1965 *New Yorker* article exemplifies Nyswander's response to the criticisms. The journalist describes Nyswander sitting at her storefront clinic in East Harlem, talking with a group of patients about the benefits of maintenance treatment. Most had positive experiences, but one patient, a man identified in the article as 'Pete', becomes the mouthpiece for the opposing argument. Suddenly, he gets up and protests: 'This is a synthetic heaven you're giving people. As long as you're on [drugs]', he says, 'you can't enjoy the things you can when you're normal…it's wrong.'[68]

Nyswander is unconvinced by the familiar objection. 'What', she asks, 'does "normal" mean?' From the outside, you couldn't tell if someone studying or working was taking methadone. Narcotics agents visiting her hospital couldn't tell the difference between the patients and the staff, she says. She evoked her favourite insulin analogy: 'Is a molecule of methadone more immoral than a molecule of insulin?'[69]

The government's surveillance also continued. By this point, the Federal Bureau of Narcotics had been monitoring Nyswander for years.[70] Now, they amped up their assault. They started with threats. When that didn't work, they turned to other tactics. As Dole recalls in an interview: 'They infiltrated clinics and stole records in order to get what they considered inside information' and 'floated various rumours' to 'discredit' them with legislators and to encourage them to pass 'prohibiting legislation'. They even pushed the National Research Council to condemn their work.[71] They wanted research and treatment stopped.[72] Narcotics agents frequently dropped in on her at work unannounced. They disrupted her meetings. It was only Dole's position that allowed Nyswander to continue.

Dole provided the financial and legal protection of a powerful research institution. He gave Nyswander all the protection he enjoyed as a Rockefeller scientist.[73] As Nyswander puts it, Dole was a 'member of a very exclusive club'.[74] His status allowed them to expand their treatment programme across the city and defy the government agenda. Nyswander had gone from disillusioned outsider to a powerful counterforce to government-sanctioned addiction treatment. And while it protected her work, it was this very same status that, in the end, set the limits of what she could do for her patients.

*

By the 1970s, methadone maintenance had become widespread. Maintenance treatment was already ongoing in Britain at the time. Still, in the US, the idea was radical and potentially illegal. There, the maintenance approach now conflicted directly with emerging government policy. In 1971, President Richard Nixon announced his plan to manage drug dependency in the population. His 'war on drugs' took a page from Anslinger's book and turned addiction into the 'enemy'. It would be necessary, he said, 'to wage a new all-out offensive'.[75] He called for a two-pronged approach that included both law enforcement and treatment programmes. For a while, he allowed some methadone treatment to continue, but, when underground markets emerged, he clamped down on regulations and made it impossible for patients to access methadone without strict monitoring. Under a new Drug Enforcement Administration that replaced the Bureau of Narcotics in 1973, there was a decisive shift towards criminalisation all over again.[76]

Nixon may have named addiction the enemy, but he was actually targeting communities. A famous 1994 interview with

Nixon's domestic policy advisor, John Ehrlichman, revealed as much. 'We knew we couldn't make it illegal to be…black,' he recalls, chillingly, but by 'getting the public to associate…blacks with heroin' and then criminalising *the drug*, they could target these communities. 'We could arrest their leaders, raid their homes, break up their meetings, and vilify them night after night on the evening news. Did we know we were lying about the drugs? Of course we did.'[77]

Nyswander stood firm as a counterpoint to this dominant and sadistic narrative about Black people as addicts. As the government pushed its abstinence models and bans, she argued persistently and publicly that addiction was a chronic disease. That, when medicated, supposed 'criminals' were restored to functioning members of society. She separated the disease from the person. Her aim was to have methadone readily available via prescription from a doctor, along with social support in finding housing or work. People with addiction could be functioning members of society, the only difference being that they were taking carefully dosed medication. Nyswander contested what was considered 'normal' as the measure of health and posed a solution based on what helped. This message directly undermined the government's. Nyswander stood firm in her approach: ever cool and unaffected. Her defiance in the face of power was admirable – less so, when confronted with criticism from the very people she was trying to help.[78]

Nyswander believed strongly that methadone maintenance empowered people to rebuild their lives. She opposed Nixon's view that Black people were inherently criminal. She gave people with addiction the tools they needed to live in the face of the hardships they faced. And yet, when it came to working with Black communities, she seemed blind to her own complicity in reinforcing their oppression.

Many members of the communities in Harlem that Nyswander visited shared Ehrlichman's analysis of the heroin problem. White doctors were promoting a treatment for a disease that was, they saw, the result of systemic racism. As Samuel Kelton Roberts, professor of history and sociomedical sciences at Columbia University, put it, this may have led to a line of reasoning that went something like 'you give us bad schools, you give us no jobs, you know, you give us bad housing, we're still being brutalized by police. And as a result of all that, you know, our children are, you know, using this junk. And what do you give us? You give us more junk.'[79] From that perspective, it's hard to see the sense in maintenance.

Doctors in the 1960s enjoyed a higher status than ever before. With a booming pharmaceutical industry and a widespread enthusiasm for miracle pills, this was the era of the 'Great Physician' – the expert, who prescribed miracle cures to ill-informed patients.[80] Nyswander's work ran counter to the mainstream of addiction treatment, but she was still at the heart of a science that elevated doctors to a new, god-like authority. She failed to see how her status sometimes further disempowered her patients to shape their own response to addiction in their communities, even as she acknowledged that the goal of treatment should be to empower people to work, learn and support themselves in a society that denied their basic needs.

Considering the experiences many Black communities in New York had of medical abuse, it was hard to imagine how they wouldn't be wary of new medical schemes. By 1972, while methadone maintenance was enjoying its moment in the sun, the sordid truth of the Tuskegee studies was hitting the press. Under the same public health service Nyswander had worked for, Black men with syphilis were deceived about their condition and denied treatment so that researchers could study how the disease killed them.[81]

Nyswander did not seem to consider the oppression she reinforced by posing herself, as a white woman allied with a powerful medical institution, as the solution to Black Americans' problems. To her, their resistance was a knife to the heart – these were people she had spent time with and cared for, whose trust she thought she had won. Their position also struck her as illogical. Nyswander denied her Black patients the reasonableness of their objections. There was a disease, and she had the cure. In this case, Nyswander saw herself as a doctor who knew best. This, to her, justified imposing treatment on people against their will. They were resisting, in her words, 'the only salvation they had'.[82]

The Harlem community's resistance to her treatment should have prompted Nyswander to do what she did so well during her early days at Lexington: to listen to people. Individual agency was the foundation to the maintenance approach she developed. Without it, methadone treatment is a medicalised version of the same control.

Nyswander continued to defend methadone treatment up to her death from cancer, at 67. She also continued to smoke heavily. She lived as she'd always been: defiant, committed to her cause and unfussed about her blind spots. In the end, it was her own addiction that killed her.

Today, methadone treatment in the US is largely privatised. There are usually many hurdles to accessing treatment, like assessments, that continue to control, patronise and dehumanise people with addiction while denying them help with accessing housing, food or safety. And obviously we haven't made much progress in tackling the problem of systematic inequalities that create and sustain addiction. Nonetheless, it is no longer considered ludicrous to treat an opioid addiction with medication.

Methadone maintenance exists today and is available across the world. Nyswander defaulted to a position of superiority at times, but, in a climate where addiction justified incarceration, she gave many their lives back. There is a humane vision of support at the heart of Nyswander's approach. And yet, Nyswander, unlike many of the women in this book, was never accused of being selfless or a saint (even the reporters recognised this as a stretch). She was an ardent individualist from a privileged background, driven by idealism at first and then by a reality that weighed on her. She wasn't called to action by angels or a god; she was dragged into the work by the force of a gaping void of compassion at the centre of a reality she saw had to be filled, however imperfectly.

13

Rosalind Franklin (1920–1958)

Chemist

When I was working on my PhD at Cambridge, I would often stroll down to the pub on a Friday night at the end of the street on which my department was located. I was aware, as I walked, that I was following in the footsteps of the American biologist James Watson and English physicist Francis Crick, who, in 1953, darted to The Eagle to announce that they had 'found the secret of life'. They had determined the structure of DNA, the molecule that encodes the instructions for building and replicating all living things. That pub is now a favourite of tourists, with a plaque commemorating the dynamic duo. A third name is missing, though, as I would tell the staff each time I ordered a beer (which I'm sure they appreciated).

It is, in fact, a story that's been told many times. First pieced together by friends and colleagues who were outraged by her sidelining in the famous account of the discovery of DNA written by Nobel Prize-winning, credited co-discoverer James Watson. Her first biography, written in 1975 by Anne Sayre, a writer-friend of hers, set the record straight in the silence left by Franklin's untimely death. The book became a manifesto for a rising women's movement; 'A vivid view of what it is like to be a gifted woman in an especially male profession,' the subtitle read.[1] Franklin, cruelly dubbed 'the dark lady of DNA' by her boss, became the emblematic shadow of the women excluded by the biggest, boldest names in science. Her shrouded figure exists alongside the shape of the double helix, as an iconic reminder of the disavowals that have made the science of our time.

Perhaps Friday night at The Eagle wasn't going to be my platform, but the story of Rosalind Franklin's erasure from the discovery of DNA is one that I felt everyone *had* to know.

I'll try here instead.

*

Franklin was born into a wealthy North London Jewish family on 25 July 1920. Her family were proud of their Jewish heritage and traced back their ancestry to King David, the founder of Jerusalem. They were bankers but also owned a publishing firm, Routledge & Kegan Paul, which offered jobs for the sons who weren't interested in banking. They were a civilly minded family and always animated with lively debate. Franklin didn't struggle expressing her ideas and happily took on Ellis, the family patriarch and his conservative worldview, in some epic showdowns. A civil servant who was working with Ellis on refugee applications at the Home Office witnessed the pair in action on a visit: Ellis was a 'domestic martinet', he said, and she was 'her father's daughter'.[2]

In line with her father's conservatism, there were some more traditional expectations of the women in the family. They were encouraged to cultivate their talents and excel in their education but always with an eye to performing community service, perhaps in education or politics.[3] Franklin, though, was interested in science. She discovered chemistry and physics at her high-ranking private school, St Paul's, in London. She was, as her Aunt Mamie wrote in a letter about her six-year-old niece, 'alarmingly clever' – that is, far better than any of the boys.[4]

While her aptitude unsettled her family, Franklin herself wasn't too concerned within school. She was more interested in photography. Her brother recalls how happy Franklin was when the image materialised; how at age twelve, she said, 'It makes me feel all squidgy inside.'[5] She was in touch with what she liked and wanted. She told her brother she chose to study science at school because it was so 'badly taught at school she thought there must be something more in it than that', he recounts. This was very characteristic of her dry, very British sense of humour: 'She was very amusing and she had a great sense of the ridiculous.'[6]

Franklin didn't waver from her interests. She enrolled for a bachelors degree in natural sciences at the University of Cambridge in 1938. Cambridge had admitted women since 1869 and Jews since 1871, but unlike Oxford they didn't grant women degrees. Women joined their designated colleges, Girton and Newnham, not the university itself. They completed the same honours-level work as men, attended the same lectures but didn't get the same credentials; only what was called a 'Decree Titular'.[7] Until 1930, the female students were segregated in the front rows of lecture halls and were forbidden the swooping black gowns that marked distinguished insiders. And yet, Franklin was part of a privileged minority – there was a cap on the number of women admitted each year to ensure that women never comprised over ten percent of undergraduates.[8] She was supposed to, and did, consider herself one of the lucky few.

The outbreak of the Second World War presented a new threat to Franklin's education. Franklin did not need her family to remind her of her duty to serve – it had been instilled in her from childhood. But so had her defiance. She responded defensively to her father's letters imploring her to show greater solidarity with her brothers at war: 'On one point you are quite unjust,' she wrote. 'When I first applied to do research here a year ago… I was led to believe that the first problem I had was war work. I soon found out that I had been deceived.'[9]

Franklin had genuinely hoped that she'd be given a war-related project when she began her PhD in the physical chemistry department. Unfortunately, the head of the department, the acclaimed photochemist Ronald Norrish was, by all accounts, bad-tempered and autocratic.[10] He was just one in a succession of men Franklin locked heads with during her career. And, as became an equally common refrain, he didn't like her for it. He'd given her a problem far below her abilities: the polymerisation of formic acid and acetaldehyde, or how these chemical compounds assemble into more complex structures (not so simple to most). Franklin was frustrated – her school friends were codebreaking at Bletchley Park, while she was wasting her time on dead-end research. Still, she did what he asked. She only confronted him when the scientific evidence proved his experiment was a dud. When she explained it to him, it led to, she confessed dryly to her parents, 'a first-class row – in fact, several'. She didn't regret a thing: the science was there. She did, however, finish his experiments. The incident revealed two unwavering aspects of Franklin's personality: she stood firm in the face of men in power, and she believed that science itself would reveal the truth. Franklin's Cambridge physical chemistry don, Frederick Dainton, put it this way: 'She once told me the facts will speak for themselves, but never fully accepted my urging that she must try to help the receptor of her messages…the logical sequential arguments meant everything to Rosalind.'[11] Franklin never forgave Norrish, but he invited her back to his lab after the war. Franklin wasn't surprised. She wrote to her parents: 'He's merely expressed willingness to have me work for him for a year as an unpaid stooge.' To her, he would always remain 'ill-mannered and tyrannical'.[12] Science revealed the truth, and people revealed theirs. Franklin erected the necessary defences to observe with a clear eye.

The Norrish feud only deepened her resolve to leave. In 1942, Franklin jumped at the chance for some war-related work. The first director of the newly established British Coal Utilisation Research Association (BURCA) was hiring graduate physicists directly from universities to study coal. Coal had taken on a new function during the First World War. This was a new form of warfare: chemical. Scientists frantically moved through iterations of gas masks that would filter out the fatal effects of the gases unleashed on their nation's armies, starting with canvas hoods treated with chlorine-absorbing chemicals and landing, finally, on a gas mask that incorporated charcoal. The charcoal made from the shells of fruits and nuts performed especially well. In the US, the government organised a public recycling project, with 'patriotic citizens' all over the country gathering peach stones and nut shells in a bastardised vision of the bucolic. As one 1918 popular science magazine reads: 'One of our pictures shows a mass attack by a company of boys on a horse-chestnut tree laden with a profusion of the most beautiful red-brown nuts enclosed in their spiked shells.'[13]

Gas masks became part of the fabric of daily life for regular citizens in the Second World War. There was a clear remit at BURCA, but students had far greater freedom to design their own research projects. Franklin investigated why certain types of coal were less penetrable to water and gas than others. She worked on bituminous coal (containing a tar-like substance called bitumen used in asphalt) and anthracite coal (compact with a high carbonic content) from across the British Isles. She ground the coals into fine powders, then measured their density using water, methanol, hexane or benzene. She compared these results with her density measurement using helium gas – helium, with its small molecule size, provided a 'true' density as a point of comparison. Franklin then compared the resulting structures under an electron microscope. The variation in the permeability of coals was related to a variation in the width of the constriction of the pores in the coal. She also heated the coal – up to 1,000°C – and found that heating increased the constrictions of the pores.[14]

Franklin wrote up her theory of 'molecular sieves' in a thesis and several landmark papers after the war. Her thesis, 'The physical chemistry of solid organic colloids with special reference to coal and related materials', earned her a PhD from Cambridge in 1945.[15] She went on to co-author a paper with her BURCA supervisor, 'Thermal expression of coals and carbonised coals', which was published in the esteemed British journal *Transactions of the Faraday Society* in 1946. She'd authored four more coal-related papers by 1949 and another twelve over the next decade. Together, these papers changed the way physical chemists understood the microstructure of coal and related substances. Before Franklin was the DNA expert, she was *the* leader on the less glamorous, but, as it turns out, foundational 'holes in coal'.[16]

Cambridge helped Franklin find her expertise. She had progressed, a little bumpily, because she didn't ingratiate herself to male superiors. As much as her parents begged her to pick her battles, Franklin refused to sit back when the truth was on the line. And while this trait seems combative, I think it is more helpful to understand it as Franklin's reaching for a different kind of intimacy: alignment around the facts. She didn't understand the fear others might have of new ideas and the struggle they might undergo to let go of comforting worldviews. *Her* assurance came from seeing things as clearly as possible. For a woman making her way among a people particularly committed to social niceties, this didn't always do her any favours. Had she been a man, of course, no one would have raised an eyebrow.

Franklin had found a culture that received her more willingly. During her time at Cambridge, she befriended the metallurgist and French Jewish refugee Adrienne Weill, who had introduced her to a group of French friends. Even before she was proficient in the language, Franklin felt immediately in her element around them: 'their standard of everyday conversation is vastly superior to that of any English gathering I have been in and they are all so much more quick-witted and alive'.[17]

The French circle included the crystallographer Marcel Mathieu. Mathieu was a communist with egalitarian attitudes to women scientists and Franklin moved to his laboratory in Paris in 1947. In his lab, Franklin started working with X-ray crystallography to further illuminate the structure and properties of coal. She was doing highly complex work, working with disordered matter that didn't have the regular patterns of atoms or molecules of simpler crystals. This was a highly desirable skillset for a chemist working at a time when the field was tackling increasingly complex atoms – not least, DNA.

Franklin was very happy in Paris. Here, she could be herself. She joined in the animated discussions in cafes every day at lunch time. She also swam, hiked in the Alps, rode pillion on motorcycles and danced. Franklin had always been focused on her work, but her time in Paris was, above all, about socialising. Her sister observes that the 'the pile of family letters' from the period 'talk far more about her holidays, her friends, or living conditions in post-war Paris, than about her work'.[18] In one letter, she alerts her family, who first initiated her into the culture of free debate, that this aspect of her personality was thriving:

'I've been told several times my French is best when I'm angry. I had a glorious row this a.m. in a shoe shop that sold me a pair of straps of wildly different length. I enjoyed it immensely.'[19] Paris wasn't afraid of an opinionated woman.

Still, Franklin had to consider her next career move. It would wrench her away from her soul-country. She was offered a post-doctoral fellowship in the department of biophysics in King's College, London.[20] She started there reluctantly as she faced the same old misogyny of her earlier education, knowing better than ever that it was the culture, not herself, at fault. She joined the lab of the physicist John Randall, who had invented the key to radar during the war. Applying that same Big Science approach, he had now assembled an interdisciplinary team of physicists, chemists and biologists to study living cells.[21] Randall was not alone. He followed the tide of physicists who, over the first half of the decade, had witnessed how their work had manifested in large-scale death. The idea of applying their tools to the science of life appealed.

They joined the biologists in their rapidly accelerating quest to explain the mechanism of heredity. As we saw, Mendel established the link between genes and physical traits in the 1800s. By 1905, geneticists understood that genes are carried by structures called chromosomes, which copy themselves and separate. No one understood, though, how genes on chromosomes transferred between generations. By the 1920s, scientists had identified two elements of chromosomes: protein and deoxyribonucleic acid or DNA. The Swiss scientist Friedrich Miescher had identified DNA back in 1869.[22] But protein appeared more interesting than this subsidiary molecule. There was more of it, and it had twenty different recurring components, while DNA had only four, so protein seemed to have more potential as the code for all life.[23]

A few scientists on both sides of the Atlantic persevered with the DNA molecule. In 1936, the American microbiologist Oswald Avery speculated that the carrier of genetic information between chromosomes might actually be DNA.[24] In Britain, J. D. Bernal (Hodgkin's inspirational tutor) in Cambridge and the physicist William Astbury in Leeds used X-ray crystallography to build a model of the molecule.[25] Using a method that became the backbone to the race to determine the DNA structure, they took hundreds of X-ray photos of DNA fibres and used the diffraction patterns to model, using metal plates and rods, how the elements of DNA (bases, sugars and phosphates) might fit together. In 1938, Astbury published a paper in *Nature* explaining that the bases lay flat, stacked like coins.[26] Meanwhile, Avery showed that DNA transferred between a dead and living strain of bacteria. This proved that it was DNA, not protein, that carried heritable material. He published his results in the *Journal of Experimental Medicine* in 1944.[27]

This work was seismic but still left an unanswered question: how could just four bases carry the specific instructions for all living things? Only the structure of the molecule could explain how the bases fitted together and transmitted their instructions from one generation to the next

Franklin moved into this field, hot with an enthusiasm for the next molecular breakthrough. Her landing was bumpy from the start. Randall assigned Franklin the task of analysing the X-ray photographs that a graduate student, Raymond Gosling, had taken of DNA. The project had been led by Maurice Wilkins, Randall's second in command, but Franklin arrived at the lab while he was on holiday. When Wilkins returned, Franklin was under the impression that she was in charge of the project, while Wilkins assumed that she was a technical assistant (completely unreasonable considering her experience). This made for a fraught start to the relationship. Franklin was in a position of having to defend her territory, while Wilkins resented her presence. Unhelpfully, Randall never clarified the lines of authority and left the two to battle it out.[28]

Besides this awkward beginning, the atmosphere at King's was structurally hostile. With women barely able to attend places like Cambridge, it was unusual for them to take on research positions. At King's, female employees were segregated – excluded from spaces like common rooms where colleagues met to eat their lunch. Franklin wasn't just a woman but also proudly Jewish, while King's had been established as a Church of England college. Its policies were distinctly antisemitic. There were Jewish quotas to prevent 'clever Jews' displacing Christians.[29] Franklin was 'othered' twice over. Her outsider status made it easier to exclude her and barred her from the opportunities for collaboration and publication that, as we'll see with DNA, determined who arrived at conclusions first, who published, and, so, in that competitive arena of Nobel science, whose ideas got recognised. In all of this, Franklin was alone.

Franklin had one ally and friend in Gosling. Gosling describes being in awe of her: 'Her powers of concentration were quite fierce and she could get done in a day what other people might have taken several to achieve.'[30] They had fun together too. When Gosling struggled to visualise the geometry of a set of arcs they were drawing, Franklin sent him to buy some navel oranges, which they used to model the spatial relations of curves: 'By the time we had finished', Gosling delighted, 'there was quite a lot of orange peel on the floor.'[31] The pair worked to attune the X-ray crystallography techniques to DNA. They adjusted the X-ray camera to a needle-fine beam. Then Franklin worked on the DNA sample itself. She drew on the techniques she'd developed with coal to rearrange the fibrous DNA. She used a glass rod to pull out thin strands and bundled them together to create a sufficiently thick material to photograph. She also discovered that humidity helped get a clearer photo: water molecules filled the spaces between the atoms of a crystal to keep it upright and stable.[32] This allowed her to deduce two types of DNA, an A and a B, depending on how much water they absorbed. The B-form photographed like a helix. She also theorised some essential elements of the DNA structure. For example, the location of the phosphate sugars, with the bases tucked inside like a staircase. Franklin spent another eighteen months improving her DNA photographs.

In May 1952, she focused the X-ray beam on extra-wet DNA fibres for an extended time. After sixty-two hours, she had produced vivid photographs of perfect crosses. The cruciform pattern suggested that this was a helix-shaped molecule.[33] She could have used the photograph, at this point, to begin to speculate about the structure of the molecule, using it to build a model, like Astbury and Bernal had. But Franklin still wasn't convinced. She needed more data. She labelled the photos 'Exposure 51' and put them in a drawer. She returned to the dry, A-form photographs that had more detail to analyse.[34] Wilkins didn't appreciate the delay. He repeatedly pushed her to stop with all the calculations and get to building a model.

Wilkins was feeling the heat, it turns out, of a battle Franklin hadn't signed up for.

Watson and Crick had been working on the structure of DNA for a while. And with great urgency. The pair were unabashed about wanting to go down as pioneers in history. As Watson would tell the story in his heroic autobiography, *The Double Helix*, published in 1968, in fact, nothing mattered to the two more than being the first to solve the puzzle of DNA's structure. There was another close competitor, Cal Tech's Linus

Pauling, who they were particularly keen to beat.[35] Crick similarly recalled that their supervisor Lawrence Bragg had been 'quite cast down' when Pauling beat him to the structure of the alpha-helix (an important structural feature of proteins) and that they were determined not to, as they saw it, fail in the same way.[36]

But they were stuck. Watson and Crick were theoreticians, abstract scientists who modelled the structure of DNA, based on the chemical and structural data that was known to date, in a way that made logical sense. Their models depended on the information produced by experimental scientists. For example, the pair were stumped on how the two strands of the helix we know so well today fit together. They were working initially with the wrong base pairs in their models, so they hadn't considered the hydrogen bonding that eventually explained how the structure cohered. A visit from the chemist Jerry Donohue, who spotted this mistake, set them on the correct path.[37] The modelling approach depended on input from different sources like this, in a way expressing the knowledge in the field to date in visual form.

This meant that when Watson popped his head into Wilkins' office in January 1953, he was hoping, as ever, for the next essential clue. The men bonded briefly over their respective experiences of Franklin's angry outbursts (a reaction, let's be clear, to their continual jibes and rummaging through her notebooks in her absence).[38] Nonetheless, Wilkins hadn't produced any better photographs than she had. 'She's got a very good B,' he told Watson and casually showed him Exposure 51. Watson had never seen the image before. He was astonished. Watson said of the moment: 'The instant I saw the picture, my mouth fell open and my pulse began to race… The black cross of reflections which dominated the picture could arise only from a helical structure.'[39]

In their quest to be first, Watson and Crick used Franklin's results without her knowledge or permission. They obtained a non-confidential copy of the report from King's College to Britain's Medical Research Council (MRC), a government funding body for biological research. The pages contained much of Franklin's numerical data and proved helpful as Watson and Crick refined their model.[40] Shortly after, Watson and Crick landed on their insight that the DNA molecule was made up of two chains of nucleotides paired to form a double helix.

The men had taken advantage, it turns out, of Franklin's imminent departure from the research group. Some say Franklin decided to leave King's; some say Randall pushed her to leave. What we know is that, while Franklin was doing her most seminal work at King's, she was undoubtedly the unhappiest she'd ever been. Franklin's brother remembers: 'I think the difficulty for herself was she found it hard to climb out of a blackish mood.' But this only became a problem at King's, he said, when 'things started going wrong for her socially'.[41] In that sense, the final trigger for her departure is less important than the chronic isolation and bullying she endured there.

Franklin arranged to be transferred to Bernal's lab at Birkbeck College (who, as we know, offered a much more welcoming environment for women scientists). She was in a rush to finish up her work on the A-form of DNA (to Wilkins' annoyance, not the helical photos) before she left. The lab was quite adamant that she left her DNA photos behind. Randall had told her, in a tone that gives some indication of the bullishness she dealt with: 'Don't take DNA with you, think about something else, it belongs to our lab.'[42] In the meantime, Wilkins had written to his friends at Cambridge, excited to finally be rid of the stickler for solid data holding him back: 'Our dark lady is leaving us next week,' he wrote.[43] Wilkins had staked their claim to the photos and prepared his alliance. The stage was set to override Franklin entirely and win the DNA race.

The dark lady herself was working on three papers that summarised her work at King's as she found it: that she had discovered two forms of DNA and knew how to transform them from one form to the other. She added what she knew about the structure of DNA. Phosphates were located on the outside, exposed and ready to absorb water, allowing them to hydrate and stretch easily. This protected the DNA from neighbouring molecules. In the first paper, she included the vivid photos she and Gosling had taken of both forms of DNA. The second paper summarised in great detail all the measurements of the X-ray patterns of the A-form. The final paper was much shorter and contained their (as far as she was concerned, inconclusive) findings on the B-form.[44] None of these papers made the leap to the two-strand structure that Watson and Crick did. Only in her notebook, she scribbled, on 10 February (the same week Watson and Crick landed on their double helix idea): 'Structure B: evidence for a 2-chain (or 1-chain) helix?' The same was true when it came to the base pairings, the zipper that binds the two strands of the helix together. Watson and Crick landed on the insight that the equal ratios between the base pairs of DNA meant that they bound together. Franklin only conjectured that the base pairs were *interchangeable* and stopped there.[45]

Franklin was reserved in her conclusions. A true experimentalist, she was focused on gathering the data rather than theorising. She was more of a Hodgkin than a McClintock in that regard; she didn't try to integrate her data into a model. She didn't see the point when the data would inevitably lead her there. Hodgkin herself agreed that Franklin's approach was sound crystallography: 'As Rosalind was necessarily involved in collecting the accurate data on DNA, it was natural of her to postpone model building until her data was complete, and until she had extracted all the information she could that would limit the kind of model she should build.'[46]

Science, Franklin herself wrote, 'is based on fact, experience and experiment'.[47] Nothing more. And her method would have worked; Franklin was on her way to the same insights as Watson and Crick. The Nobel Prize-winning chemist Aaron Klug judged her to be about three months from making the same discovery.[48] In this case, McClintock's metaphor of the computer describes a distribution of labour. Franklin's results fed the computer, in the form of Watson and Crick, that produced the model. Their theory resonated in an age that, as we saw in the McClintock chapter, was one that had sidestepped the question of *how* genes function for the intoxicating idea of a simple code. Their story of a breakthrough, in turn, slotted into that Nobel history of individual eureka moments. All this overlooked the rigorous inputting that the dark lady had done in the shadows. The discovery did not come about in a flash; it was built on the backbone of Franklin's critical data.

But the men worked hard to consolidate their legacy. In his account of the discovery in *The Double Helix*, Watson minimised Franklin's role with spectacularly flagrant sexism. His rendition relied on a caricature of Franklin as a 'difficult woman'. He describes Franklin as a frump, calling her 'Rosy' when he knew that even her friends called her Rosalind.[49] Watson could not understand why a woman from a 'solidly comfortable, erudite banking family' would make a point of prioritising her career over her appearance: 'There was never lipstick to contrast with her straight black hair, while at the age of thirty-one her dresses showed all the imagination of English bluestocking adolescents.'[50] Even the photo he includes of her in the book has her looking buttoned-up and serious, while there are ample of her looking vibrant, lively and relaxed while mountain climbing in Norway or on cycle rides in Tuscany. His unimaginative characterisation allows Watson to string a narrative in which Franklin is uncollaborative; so impossible to talk to that, his implication seems to be, he had no choice but to sidestep her. Later biographies adopt the same stereotype; even those that try to set the record straight seem to agree that Franklin had a personality defect.[51]

It would have been cruel, even if it was true. But Watson gets Franklin wrong. Her fashion sense, while fundamentally irrelevant to her scientific career, was very much *à la mode* in Paris. That aside, though, Franklin's life story gives an impression of a fastidious scientist and a vivacious woman. She made plenty of friends throughout her career. She *had*, repeatedly, unapologetically, posed a difficulty to men who expected deference. But even the men she fought with found her sufficiently collaborative to invite her back. This wasn't about Franklin's personality. This was about Watson and Crick's claim to a discovery.

Biographers are divided over whether or not Wilkins crossed a line in showing the photographs to Watson and Crick; some think the photos belonged to the lab, and, with Franklin gone, this wasn't unusual practice. And perhaps we could all have agreed on that if the men had credited Franklin for her crucial research. They included a footnote referencing the work at King's, possibly after hearing that Franklin planned to publish her findings in the same issue to pre-empt any questions.[52] They also never told Franklin the impact she'd had.

Franklin died very young, at thirty-seven, from ovarian cancer. She wasn't alive when the Nobel Prize in Medicine and Physiology for the DNA work went to Crick, Watson and Wilkins in 1962. The men didn't mention her many papers in their speech; only Wilkins referenced her very vaguely. Her death also gave the award committee an out, since the prize can't be awarded posthumously.

What if Franklin hadn't died? Her photo gave the critical evidence, without which the model would have remained speculation. Would she have been recognised by the Nobel committee in Wilkins' stead? I think we all know the likely depressing, and unscientific, answer to that question.

Many authoritative accounts of the DNA story have diminished, or entirely erased, Franklin's contribution. Franklin had been highly regarded for her ability to produce X-ray photographs with exquisite precision, but she remained unknown to most for decades. It wasn't until the 1970s that a friend of hers, Anne Sayre, whose husband was a crystallographer, demonstrated Franklin's indispensable contribution.[53] In a political climate newly sensitised by the women's movement, people took note.[54] This should never have been necessary. A later *Nature* editor said that the 'Watson and Crick' paper itself gestured to Franklin's unacknowledged presence, when they wrote about being 'stimulated by a general knowledge' of the crystallographer's unpublished findings. Had the paper been subjected to the peer review we have today, where other scientists in the field check over the papers, they would have called out the speculative nature of the model, the editor claims, which meant nothing without the data Franklin provided.[55]

In the end, both Watson and Wilkins felt impelled to modify their respective accounts of the history of the discovery of the structure of DNA, largely in response to Sayre's critique.[56] Whatever your assessment of the protocol involved, it is noticeable just how easily Franklin was dismissed. Franklin's sister referred to the whole saga as 'a warning to girls who might think of a career in science'.[57]

In turn, as tends to happen with controversies, the necessary discussions around Franklin's omission from the story of DNA overshadowed everything else she did. Franklin moved on to Bernal's lab in the crumbling Georgian building in Torrington Square that was part of Birkbeck College; as she put it, 'from a palace to a slum'.[58] There, she was met once again with an egalitarian atmosphere that welcomed her. She found a genuine collaborator in the aforementioned Aaron Klug (who would later win the 1982 Nobel Prize in Chemistry) and built a research group around her. She turned her attention to viruses. She published seventeen papers on the subject over the next five years. First, on the tobacco mosaic virus (TMV) and later on polio. In a beautiful moment of retribution, she left behind a five-foot-high *model* of the TMV virus constructed using 250 polystyrene moulds resembling hatstands. Franklin passed away the day before the model was exhibited at the Brussels World Fair on 17 April 1958.[59] But the symbolism was perfect: an ironic nod to her elegant fashion sense melded into a larger-than-life, bombastic testament to her science, which for most of her life lived in the shadows of men's helical models. Franklin was never afraid to make her point.

When I shouted Franklin's name around The Eagle at the end of the week, I wanted people to hear her story. But clearly, *I* wanted to be heard too. I wanted a platform at an academic institution steeped in the legacy of great men to express my sense of exclusion from their history. I wanted a history I could relate to. Franklin was a woman who struggled socially and emotionally from the consequences of her conviction, a woman who perhaps could have been more diplomatic at times but, equally reasonably, refused. She demanded more: a society that accepts a woman speaking her mind, an academic institution that respects a female research scientist's right to choose her method and a version of science that prioritises rigour over acclaim. The dark lady's story isn't a cautionary tale for women in science; it is a defiant ghost story that will haunt us all until we turn to the shadows of our scientific beacons, to the work and the people they told us not to see.

14

Katherine Johnson (1918–2020)

Mathematician

In February 2017, Katherine Johnson, age ninety-eight, made her way to the stage at the Academy Awards with the cast of *Hidden Figures*, a film that told the story of her career at NASA, to a standing ovation. Dressed in an elegant blue dress, she was escorted by Dr Yvonne Cagle, one of just six Black women to become an astronaut at NASA to date. Johnson had been part of a group of African American women who manually performed the necessary mathematical calculations for the engineers at NASA's predecessor, the National Advisory Committee for Aeronautics (NACA). Johnson had joined these women, known as the West Area Computers, on the cusp of the civil rights movement. The West Area Computers were segregated from their white colleagues. From their separate building, they performed the tests and computations that were essential to what became some of the most iconic moments in US history: the first crewed flight into space, the first orbital space flight, and, of course, the moon landing fix. But despite her three decades of work with the US space programme and the countless calculations and flight paths she analysed, for a long time no one knew Johnson's, or her colleagues', names. Now, as she appeared on Hollywood's most distinguished stage, that had all changed.

In *Hidden Figures* – adapted from Margot Lee Shetterly's 2016 book of the same name – audiences watched in awe as a group of smart, tenacious Black women changed the course of American science and history. Johnson – played by Taraji P. Henson – in particular captivated the public, partly because she was the only member of the cohort still living. And while she was finally recognised for her work, her mythicised status led to misconceptions: the idea, for example, that she was the first Black mathematician to work at NASA, or the only one.[1] The hype glossed over the realities of Johnson's career. Johnson wasn't a superhero but a brilliant Black woman who did the work that is integral to every scientific mission. This grounded view brings us closer to Johnson's experience of NASA's skyrocketing science: how she climbed through the cracks of a segregated society, relying on hard work and talent and sometimes luck.

*

Creola Katherine Coleman was born on 26 August 1918, in the small rural town of White Sulphur Springs, West Virginia, at a time in which, as her biographer, Shetterly, put it, 'the odds were more likely she would die before age thirty-five than even finish high school.'[2] She was the youngest of Joshua and Joylette (née Lowe) Coleman's four children. Her mother was a schoolteacher and her father a farmer and lumberman but also a self-taught mathematician.[3]

Johnson's family, like so many in this book, nurtured her love of learning. Johnson shared her father's penchant for numbers from the get-go. She enjoyed counting dishes, steps on her way to church and, slightly ambitiously, the number of stars in the sky.[4] 'I couldn't wait to get to high school to take algebra and geometry,' she recounted in an interview.[5]

However, receiving the education she was aching for proved almost as impossible as enumerating astronomy. In the segregated schooling system of the pre-civil rights US, there was no local school that would allow so-called 'coloured' students to progress beyond sixth grade.[6] Every autumn, the family moved 125 miles away to the unincorporated community of Institute. Institute was founded by the formerly enslaved Mary Barnes on land 'given' to her by her owner, Samuel I. Cabell, after she'd had thirteen children with him.[7] Here, Johnson attended the high school associated with the West Virginia Collegiate Institute, partly on land sold to the state by Barnes' daughter in 1891 (later, it became West Virginia State College, and today it is known to us as West Virginia State University).[8]

For a while, Johnson's father returned to the farm in White Sulphur Springs, but the Depression eventually forced him out of the pastoral life. He took a job as a bellman at the Greenbrier, a luxury resort for the well-to-do in their hometown. Johnson herself joined him on her holidays, working as a maid to the celebrity guests: presidents and movie stars. All white, of course (with a few exceptions, like the emperor of Japan).[9]

But while her father worked, Johnson was free to attend school. She learned incredibly quickly and skipped several grades. She was just fourteen when she started at the West Virginia State College to study mathematics in 1933.

West Virginia State College was a hopeful place for Black American students. Academics and politics there were very much entwined. The college's president, Dr John W. Davis, was part of a fraternity of Black educators and public intellectuals at the forefront of the debate about the future of Black America.[10] Johnson joined the local chapter of Alpha Kappa Alpha Sorority (AKA), the first intercollegiate historically African American sorority, founded in 1908. Politically and socially minded, AKA members organised to support African American women in and outside the sorority. They established funds to support student scholarships and helped those fleeing discrimination in Southern states find housing during the Great Migration. AKA counted many future West Area Computers among its number.

Davis had recruited a brilliant faculty. In the maths department, the charismatic and irreverent topologist William Waldron Schieffelin Claytor was a veritable celebrity. Just twenty-seven, Claytor looked like a movie star. He drove a sports car and owned a plane, which he once flew so low over the house of the school's president that the machine's wheels purportedly resounded off the roof.[11] Claytor was just the third African American to earn a PhD in maths.[12] His work had been lauded by mathematicians as an important advance in the field. But their enthusiasm didn't seem to translate into a job offer. He waited for the post a graduate of his calibre would usually expect – at one of the country's top maths departments. Only West Virginia State College offered him a position. The civil rights activist W. E. B. Du Bois wrote, in 1939, six years after Claytor received his PhD: 'If young coloured men receive scientific training, almost their only opening lies in the Negro university of the South', while the white universities were 'either completely closed to Negro investigators' or 'only partially opened and on humiliating terms'.[13]

Claytor quickly clocked that Johnson was the only student who followed his pacey proofs.[14] When Johnson completed all the maths courses available in junior year, he designed extra classes for her. 'You would make a good research mathematician,' he told his seventeen-year-old star student after her sophomore year. '*And*, I am going to prepare you for this career.'[15] If Black men were marginalised in academia, Black women were entirely excluded from it. It wasn't clear to anyone how Johnson would forge a research career. Ever conscious of this harsh reality, Claytor prepared her as best he could.[16]

Johnson graduated in 1937 and didn't become a research mathematician. Instead, she took a job as a teacher, which was where most Black Americans who had once aimed for research careers were eventually forced to go. She spent two years teaching at Marion, one of West Virginia's segregated public schools.

And yet, in the background, the National Association for the Advancement of Coloured People (NAACP) was mounting its campaign against legal segregation. They took advantage of the leniency for reform under Franklin D. Roosevelt's New Deal (reforms designed to combat the Depression) to advance their goals. They continued to fight against the abhorrent lynching of Black Americans, but they also homed in on inequalities in public education. In the 1938 case, *Missouri ex rel. Gaines v Canada*, the United States Supreme Court ruled that the state should admit Black graduate students to its white state universities where comparable graduate programs at Black universities in Missouri did not exist.[17] After the ruling, West Virginia's governor desegregated public graduate schools in the state.[18] He asked his good friend Davis to hand-pick three of West Virginia State College's best students to attend the state university and launch this new era, starting in 1940.

Davis picked Johnson. She became the first Black woman to attend the all-white graduate programme in West Virginia University. The principal of the high school she had been working at handed her a stack of maths reference books on the day she left, anticipating that she'd still be barred from the university's library.[19] However, Johnson's main issue was that the course was too *easy* – Claytor had prepared her for a world that exceeded the possibilities.[20] Her research career was eventually cut short for different reasons. Johnson had met her husband, Jimmy Goble, at Marion (he was a chemistry teacher). She fell pregnant after her summer session on the graduate programme.

Johnson spent the next thirteen years raising her children while continuing to teach in public schools.[21] She didn't begrudge this. Her family were a concrete reality: Claytor's vision, a dream.[22] And yet, while Johnson stepped away, that dream for a society that accommodated a Black woman mathematician was taking shape.

The Second World War was the less directly formative of the global conflicts in Johnson's career, and still it created new opportunities for her. Once again, the nation's need for military labour forced leaders to abandon their oppressive policies – not just for women, but for Black people too. The 1941 Executive Order 9902 barred racial discrimination in the defence industry.[23] In aeronautics, things were also changing. The Langley Memorial Aeronautical Laboratory was established by NACA in 1917. They had begun hiring white women with mathematics degrees in 1935, initially to relieve its male engineers of the laborious work of doing calculations by hand.[24] By 1943, Langley was desperate for more and began advertising for Black women with maths degrees.[25] As the job title of computer suggests, these women were considered manual calculators, less than scientists and just as typewriters had been to their employers in the nineteenth century. It was a dead-end job, not a career. Computers were supposed to be grateful for their subprofessional position and not to expect to advance. Still, when Johnson heard in 1952 that the centre was hiring Black women for the role, she jumped at the chance.

Over the course of the war, the Black women recruited at Langley to support the war effort had consolidated their position, turning temporary work into an integral position.[26] Johnson was hired onto the unit of the Black mathematician Dorothy Vaughan in 1953. Like Johnson, Vaughan had found her way through the educational system in West Virginia as a child. After completing a bachelors in maths, she had forgone graduate school to work as a teacher to support her family during the Depression. She worked at the Robert Russa Moton high school (itself a landmark location of the early student strikes against public school segregation) for fourteen years before she was hired by Langley in 1943, at twenty-eight.[27] She was promoted six years later to acting head of the West Area Computing Unit. She had played her part in establishing the unit as an integral part of the US's new space programme. She was a competent manager who knew how to integrate her employees in the organisation's research activities through her understanding of their specific skills and personalities.

Johnson started by filling out data sheets for wind tunnel simulations. Wind tunnels were an important aspect of NACA's research; they allowed them to perform flight tests without pilots having to risk their lives. But the work was menial. Johnson completed the data sheets according to equations that had been laid out by Vaughan; it didn't require much original thinking.[28]

Just two weeks after Johnson arrived at Langley, a white man entered the office. He approached Vaughan and delivered some kind of instruction. After he left, Vaughan called Johnson and a colleague to her desk: 'The Flight Research Division is requesting two new computers,' she said. 'You're going to 1244.'[29]

Vaughan had selected Johnson to fill an opening in the Flight Research Division, in Building 1244, a group that specialised in tests on real aeroplanes. Stationed at her new desk, Johnson had been given a front-row seat to the most cutting-edge work of the whole laboratory.[30] Just before she'd arrived, two of the engineers there had presented their research on fighter aeroplane controls at a conference with the field's top researchers.[31]

On the first day of her new appointment, Johnson left the West Area division and headed over to the second floor, then picked an empty desk. She smiled at the engineer beside her. He got up and left. It was impossible for her to know why – it could have been her Blackness, femaleness or lower professional status that had repelled him. Johnson faced all of these intersecting forms of oppression. She chose not to dwell on the possibilities, focusing instead on putting her best foot forward. Her father had always told her, 'You're as good as anyone in this town, but you're no better.' Johnson carried that aphorism with her.[32] She remained friendly towards her white male colleagues, even when they didn't return the courtesy. Like Claytor, she chose to simulate the world she wanted to live in. In the end, Johnson became good friends with the engineer after he discovered she was a fellow transplant from West Virginia.[33]

Johnson's response to racism frustrated some of her Black colleagues, perhaps because it resembled a kind of complacency in the face of the systemic injustice.[34] But as the civil rights movement slowly dismantled segregation in society at large, this was Johnson's way of navigating Langley's shifting relationships,[35] and it served her well. Vaughan hadn't just picked Johnson for her mathematical talent – NACA's engineers were notoriously blunt (some might say rude and authoritarian). Johnson won them over. Armed with a no-nonsense demeanour and cat-eyed glasses that declared 'don't mess with me', she approached the engineers with endless curiosity, asking them constant questions about the subject they loved.[36] She, in turn, impressed them with her high-level maths – her calculations were impeccable. It didn't take long for her to make herself indispensable.[37] 'The guys all had graduate degrees in mathematics; they had forgotten all the geometry they ever knew,' Johnson later recalled. 'I still remembered mine.'[38]

Johnson and her team were exploring aerodynamic forces on aeroplanes, which helped to improve the safety of military and commercial aviation. Her temporary position became permanent. Slowly, Johnson rose from subprofessional status to closer to equal, not just a 'computer', but a 'math aide'. She also pushed to be admitted into the agency's scientific briefings, previously only open to male staff members, and set the precedent for all women to do so.[39] She worked in the division for five years, continuing even after her first husband died from brain cancer in 1956, powering through intense grief, a demanding job and the tribulations of raising three teenage daughters.

Then, a new war put Langley at the centre of an irresistible project.

'Red-Made Satellite Flashes Across US' and 'Sphere Tracked in 4 Crossings Over US'[40] headlines blared as the Soviet Union launched *Sputnik 1* into space on 5 October 1957. The previous day, Radio Moscow had announced its timetable for the satellite's course over Earth in chilling detail. Anxiety over the Soviet threat had been mounting in the US ever since the USSR detonated its first atomic bomb in 1949, shattering the US monopoly over nuclear weapons.[41] Since then, Americans had read reports of Russian planes 'too fast to be identified' filling Korean skies in the Korean War.[42] But now the threat of Soviet aeronautics loomed directly overhead, beeping across the US. The space race had begun. It was to be a battle of wits unlike anything the world had witnessed before, and mathematics lay at its centre.

Space was the next American frontier. Another war meant, once again, that the nation's leaders loosened their grip on the people they'd long deemed incapable. With technological prowess so central in the conflict, discussions in the press, among policymakers and the public centred around the same question: how can US aeronautics keep up with Russia? Black newspapers answered: 'Who can say that it was not the institution of the Jim Crow School that has deprived this nation of the black scientist who might have solved the technological kinks delaying our satellite launching?' wrote the editor and publisher of the Cleveland *Call and Post,* Charles H. Loeb.[43] The US was lagging behind *technologically* because it was *socially* regressive. During the Second World War, debates had revolved around including white women in technical roles. Now, a nationwide debate guided policymakers to the crucial, geopolitical imperative of including Black people too. Almost every paper, and every review, asked how much brainpower was being wasted by excluding Black people from education.[44]

In 1958, NACA became NASA, the National Aeronautics and Space Administration. The following year, Johnson was faced with a new daunting task – determining the launch conditions required for crewed space flight. The spacecraft for Project Mercury did not have the sleek needle-shape of an aeroplane. Harvey Allen, an engineer at NASA's Lewis Flight Propulsion Laboratory in Cleveland, had shown that the streamlined shape could not withstand the extreme heat caused by the friction as it passed through the atmosphere on its way back to Earth. A more champagne cork-like shape would generate a shock wave to dissolve the heat as it descended. The *Mercury* space capsule took form: 1.83 by 3.35 metres and weighing 1,360 kilograms.[45]

Johnson's task, alongside the engineers, was to calculate the flight trajectory of the newly designed spacecraft. They had to map out the *exact* path that the spacecraft would take from the moment it took off until it hit the Atlantic.[46] The vessel would be fired into space like a bullet from a gun. It would then follow a simple course, as described in the sixteenth century by Galileo Galilei: a parabola. Although this modern trajectory was slightly more complicated. The ultimate goal of the project was orbital flight: a full course around the Earth. This meant that the bullet had to be fired at the correct angle, and with the precise force needed for the shuttle to travel around the planet, so that, when it descended, it landed at a safe distance from the awaiting navy ships to pluck the astronaut from the ocean.[47]

It was an intimidating task. The astronauts' lives depended on absolute mathematical accuracy. But Johnson was confident in her abilities: 'Tell me where you want the man to land, and I'll tell you where to send him up,' she told her bosses.[48]

The work was meticulous and highly complex; there was the Earth's gravity to factor in, as well as its oblateness (Earth isn't perfectly round), and the speed of its rotations. Shooting the bullet in a straight line would see it land in a different spot because the Earth would have moved. Johnson had a solution: 'In the recovery of an artificial Earth satellite it is necessary to bring the satellite over a preselected point above the Earth from which the re-entry is to be initiated.'[49] Slowly, over months, Johnson compiled her equations into a thirty-four-page document. In 1960, she co-authored the paper, 'Determination of azimuth angle at burnout for placing a satellite over a selected earth position' with one of the group's engineers.[50] The title, I admit, isn't catchy, but the paper was seminal. The report set out the equations that carried the first astronaut, Alan Shepard, to space in 1961, and it was fundamental to the first crewed orbital space flight, in 1962.[51]

At the anniversary of the Mercury mission, in 2017, the deputy director at Langley made a speech to mark the opening of the Katherine G. Johnson Computational Research Facility. 'Millions of people around the world watched Shepard's flight… but what they didn't know at the time was that the calculations that got him into space and safely home were done by today's guest of honour, Katherine Johnson.'[52] In the shadows of the iconic images that came streaming in live from outer space (the moon landing still remains the most watched TV programme of all time) were the human hands, making the laborious calculations. Hands reminding us that the human space race was a project of forgetting. A nation doing what they could to deny their vulnerability in the fight for domination; of politicians, bunkering themselves deep in self-serving ideologies, and of ignoring, once again, the segregation that put white men at the pinnacle of this mission, and Black women in the background. The duality of space travel was always that it exposed our vulnerability the further we ventured. Images invoking awe at our achievements along with our minuteness. The tenuous calculations needed just to rise incrementally above the atmosphere. Those hands also remind us of our inescapable vulnerability. Begging us to take responsibility for it.

Hidden Figures did much to weave the Black women doing the maths back into these iconic moments in US history. Like the moment that the astronaut aboard the first crewed orbital mission, John Glenn, requests Johnson specifically to check the calculations for the craft before he agrees to take off. 'Let's get the girl to check the numbers,' he says. 'If she says they're good, I'm ready to go.' The moment is moving for its rarity; that collegiality between the white man the public saw and the Black woman doing the maths behind the scenes. A role reversal, in which Glenn entrusts his life to her. And yet, the scene also reminds us of that double-edged context; how Black people were only granted some degree of their humanity when white lives were at stake. 'Get the *girl*,' Glenn says.

Johnson's role in producing and then checking the calculations for the 1962 orbital space flight established her professional reputation. And yet, Johnson was most proud of the work she did as part of the team on *Apollo 11*, the mission that put the first astronaut on the moon. By the 1960s, the US space programme had set its sights on a new apex.[53] The Soviets had beat them to the first crewed flight – they weren't going to beat them to the moon. Once again, Johnson performed the all-important calculations to determine the exact time at which the lunar lander needed to leave the Moon's surface in order to meet and dock on the orbiting command service module.[54] Between 1963 and 1969, Johnson authored four reports with the engineer Al Hamer. These set out the paths for the lunar orbits. They also addressed the risks: what if, for example, the computers went out? Or there was an electrical failure on board? These questions became more than hypotheticals when, in 1971, the electrical system aboard *Apollo 13* shut down due to an explosion.[55]

The work was gruelling. Fourteen- to sixteen-hour days were standard. Johnson arrived at the office early, checked on her daughters in the late afternoon, then returned in the evening. She fell asleep at the wheel one morning on her drive to work, only waking up when the car hit the side of the road.[56] But to Johnson, this was the acceptable human cost of mankind's giant leap forward. She loved the work.

In the end, though, Johnson watched those first steps on the moon surrounded not by her NASA colleagues but her AKA sisters. On 20 July 1969, she was gathered with a hundred-odd young sorority women, all ambitious Black women of different generations, there to support each other.[57] Surrounded by the Black women who were her family down on Earth, she watched the culmination of her efforts in space.[58]

She watched the nail-biting course, confident that her numbers were on point,[59] though she couldn't quite relax until the astronauts had made their way back to Earth. The moon landing was, of course, just one step on the path she had calculated. There was still the precise timing of getting back to the lunar lander and reconnecting with the mothership. There was the three days back to Earth. There was the task of passing through Earth's blazing atmosphere and dropping into the ocean.[60]

Some of the women watching alongside Johnson may have wondered why, with a wavering civil rights movement, the US was investing its $24 billion on *Apollo* alone.[61] Just days before the launch, Martin Luther King Jr's successor, Reverend Ralph Abernathy, had led a group of two hundred protestors to the launch site at Cape Kennedy. There they had challenged NASA's administrator, Tom Paine, on the national priority of the space programme. Poor Black people in places like Watts and Detroit and rural Appalachia couldn't eat and had nowhere to live. There was a shadow side to the scientific and social progress the space programme brought. The cost of war, cold or otherwise, invariably detracts from building enduring social systems at home.[62]

Johnson, though, had no intention of pivoting from her course. By the time the moon landing was complete, she had already set her sights on her next task: Mars.[63]

The contribution Johnson wanted to make was right there, at the centre of a scientific story of progress. Johnson stayed at NASA for thirty-three years. In that time, she worked on every major crewed spaceflight programme. Besides the solidarity with her sorority members, she fought her battle against discrimination as a Black woman isolated among groups of white colleagues. She worked among white people who had their own circles and discussed work at local barbeques that excluded their Black colleagues. There were no Black administrators or TV commentators and definitely no Black astronauts (Ed Dwight had tried and had been sent packing) to offer her a sense of belonging in her field.[64] Still, Johnson stayed, quietly persistent in the message that her presence sent, until decades later, it was broadcast to the world.

In 2015, President Barack Obama awarded Johnson the Presidential Medal of Freedom, proclaiming, 'Katherine G. Johnson refused to be limited by society's expectations of her gender and race while expanding the boundaries of humanity's reach.'[65] With the release of *Hidden Figures,* the following year, Johnson became an icon. After she retired, in 1986, she turned to the mission that had been at the heart of the civil rights movement and that had facilitated her career: education.[66] She travelled around schools, encouraging the next generation of Black women mathematicians, using her presence in the American narrative to change it.

Three years before the first moon landing, in 1966, a new TV programme aired on American screens. *Star Trek* showed audiences a vision of society that could only have seemed possible miles from the planet they knew: a multinational, mixed-race and mixed-gender crew. Naturally, Johnson was a big fan. As was Martin Luther King Jr. The show, as King put it, imagined Black people 'in the future'.[67] Johnson's presence at NASA did the same for the hundreds of women she mentored and spoke to. She wasn't involved in the political struggle for rights but in the cultural struggle for a collective story that included Black people and women. Working on the cusp of an industry synonymous with technological progress, her presence wove desegregation into the story of progress. Johnson's contributions should be celebrated, not as exceptional, but as a reminder of the hidden figures and dimensions of a history of forgetting. Her story is one of the many vessels we'll need to transport us into a scientific future that supports and recognises the rocket science that happens down on Earth.

15

Anne McLaren (1927–2007)

Embryologist

In 1934, the seven-year-old Anne Laura Dorinthea McLaren stared up at a large screen during a history lesson and watched as mice were flown to the moon. None of it was real, rather the future embryologist was acting in a scene for *Things to Come,* a science fiction film written and directed by H. G. Wells. The film offers a vision of a world that mirrored what scientists in the first half of the twentieth century had seen. One shaped by incessant war and the drive for technological progress. As the film's protagonist watches his world's crewed space flight ascend to the moon, he rouses the father of one of the astronauts:

> And if we're no more than animals, we must snatch each little scrap of happiness, and live, and suffer, and pass, mattering no more than all the other animals do or have done…All the universe or nothing. Which shall it be?

Anne McLaren, whose short-lived acting career soon gave way to working at the forefront of a strange new biology, dedicated her life to negotiating the middle ground between those extremes – to live passively in the world or to claim mastery over it: all the universe or nothing. McLaren is perhaps best remembered for laying the groundwork for IVF. But her work on reproductive biology over the twentieth century morphed into new science- fiction forms, turning from IVF to genetic engineering and stem cell science. She was there, on the cusp of tomorrow, making fictions a reality. But even as she realised the prophetic scene from her childhood role, working with NASA to send mouse embryos skyrocketing into space, she was grounded in a vision of biology that served people and didn't transcend them.

*

McLaren had her mother to thank for the role in Wells' film. Christabel Mary Melville MacNaghten had spotted an advertisement for a part for a little girl in her friend's latest film and decided to surprise him by auditioning her daughter for the part. She coached McLaren beforehand, teaching her daughter how to enunciate and project. By the time the audition came around, McLaren impressed everyone so much that she got the part.[1]

Wells was just one of Christabel's many glitterati friends. She was bona fide aristocracy, and a leading London socialite at the heart of a circle of literary and artistic figures. McLaren's father, Henry Duncan McLaren, 2nd Baron Aberconway, was a businessman and industrialist, often away travelling all over Britain. McLaren grew up in her mother's world. Christabel was tall, beautiful and highly intelligent. From their home in London's West End near Hyde Park, she entertained the likes of the Bloomsbury set of authors, intellectuals and artists: George Bernard Shaw, Virginia Woolf and Sam Courtauld, as well a circle of scientists, socialists and authors including the Sitwells and the Haldanes. Wells was an especially close friend. Of him, Christabel once said: 'Once H. G. and I had settled that I never intended to sleep with him, Wells' company was a delight!'[2]

The late-night dinner parties her mother threw were past her bedtime, but the young Anne was allowed to attend lunch parties, helping her mother write place cards for the guests. She soaked up the conversations of scientists, socialists and writers, whose disciplines blurred as they negotiated their politics. Many were socialist scientists who believed science should serve the public good and who saw imagining and communicating the possible societal consequences of their work as part of their role. Take J. B. S. Haldane (1892–1964), for example, a family friend who later became McLaren's undergraduate supervisor. A lifelong Marxist, Haldane never hid his contempt for capitalism or imperialism. As a scientist, he was known for his early work on the structure of DNA. In his rhetorical writing, he was the first to imagine the possibility of IVF. Communication, social justice, art, imagination, science, human health and the future of the planet – these were the ideas that surrounded McLaren from a young age.

And yet, a darkness was descending on the intellectual utopia that was McLaren's childhood home. McLaren listened to the six o'clock news with her family, registering, at the age of nine, the tension that filled the room at the mention of the Spanish Civil War. She heard about the cession of the Sudetenland to Germany in 1938 and watched her mother shrug off the ominous news. Was this a war averted or the beginning of worse to come?

The bad news kept coming. Sometimes, straight from the source of command, as in 1940, the family had the queen and Winston Churchill over for tea. 'The Queen launched a new battleship', thirteen-year-old Anne wrote in her diary, 'and I presented her with a bouquet…Winston I was rather disappointed in. He was sulky and fat and hardly said anything.' As Churchill reappeared on the pages of McLaren's childhood diary, that initial impression did not wane:

> 10 May 1940. Hitler invaded Holland and Belgium very early this morning. He has also bombed France.

> Churchill is becoming Prime Minister instead. If Chamberlain had to go, which I regret deeply, why Churchill of all people? He is always drunk, and utterly unsuitable for the very responsible post of Prime Minister.[3]

Though these reports affected McLaren, they remained at some remove. In the summer of 1939, McLaren and her family relocated to the eighty-acre family estate in Bodnant, Wales, as they did every summer. Only this time, they stayed for the duration of the war. McLaren had moved from her mother's to her father's world; the estate was very much his. When he came home for the weekends, McLaren followed the energetic man through the grounds. He taught her the botanical names of its rhododendrons and azaleas in animated gestures. She loved how carefree he was on these walks; he even dispensed with formal overcoats, whisking McLaren away into fields that didn't care about propriety.

While industrial powers clashed in the distance, the war continually drove McLaren outside. At fourteen, McLaren found the blueprints for a henhouse in the newspaper. She decided to build it to 'do her bit' for the war effort. Her parents bought her four hens to fill the coup; she called them Alpha, Beta, Gamma and Delta. It was important, for her act of solidarity to mean anything, that the hens produced as many eggs as possible. She fed them Karswood Poultry Spice, a red and peppery supplement that she mixed in with their food, and watched with morbid fascination as they laid their eggs. She was disgusted by the gooey mess they produced. Still, she continued to check in on the hens even as she grew increasingly to dislike them. This was her first experiment in assisted reproduction.

In 1944, McLaren prepared to leave her bucolic childhood behind. She was determined to study at the same university as her father and brothers had and applied to Oxford. She planned to study English. Christabel had to work hard to convince Henry to let McLaren apply – he was more traditional than her mother and still worried that a degree would make her unmarriageable. Christabel won (as she tended to). At sixteen, McLaren triumphantly travelled to a private school at Longstowe Hall, near Cambridge, to study for her School Certificate examinations. One look at the English literature sample papers quickly burst her bubble. She hadn't had the education the boys there had; she hadn't read the right books. Biology, however, looked doable. She recognised the plant names and diagrams of mammals and their reproductive cycles. If switching subjects meant a better shot at going to university, it was worth it. And so, despite her love of the written word and excellent English marks, it turned out to be the long walks in the Conway valley with her father, her informal training in practical horticulture and her experiments with poultry ovulation that determined her future career. The practical skills cultivated in the countryside, far away from the scene of historical events, were McLaren's entry ticket into a man's world.

After two concentrated school terms, she sailed through the Oxford scholarship and entrance exam and entered Lady Margaret Hall in 1945. McLaren felt perfectly at home in Oxford, but it blew open her politically conscious, yet privileged and protected, worldview. Whereas at home, conversation had centred on art and literature, but also land, gardens, business and the topics that preoccupied the upper class, here socialism was a hot topic. This was a period during which the post-war influx of ex-servicemen was breathing fresh life into old-fashioned institutions like Oxford. Everyone had been a bit shaken up in their ideas, and McLaren found herself in a vibrant, open atmosphere of social and intellectual exchange.

McLaren joined the Science Group of Oxford Student Branch Communist Party in 1948, in her second term at Oxford. Her mother was shocked. She became as concerned as her father that McLaren would not return to marry an aristocrat. They were right that McLaren was trying to shed the signs of her privileged upbringing. The process was painful at times. Fellow students noticed incredulously how, on arriving at Oxford, she didn't know how to make her bed or brew a pot of tea. Rather than confessing to her ignorance, she lurked behind doors to watch others doing their chores until she got the hang of it.

McLaren was determined to fit in. While her mother continued to impress on her the importance of dressing well, she opted for increasingly spartan clothing. She favoured what she called 'plimsolls', shoes like flip-flops with toe-holes (ahead of her time, with proto-Crocs). She went on trips to conferences carrying her possessions in a plastic shopping bag. She doubled down on her communism too. On 25 November 1948, she gave a lecture at the Oxford Student Branch. She called it 'The position that science holds in the present conflict between capitalism and socialism'. In it, she argued that science was often under attack by capitalist forces because it was always on the side of social progress. Growing up, McLaren had witnessed the harm technology did in the hands of men (like Churchill) in the Second World War. Now, it was becoming clear to her that science held the potential to solve many of the world's inequalities, but that it relied on politically engaged custodians to guide it.

Science and politics were beginning to meet. McLaren conducted her first research project under the guidance of J. B. S. Haldane. She worked on *Drosophila*, those rapidly reproducing favourites of geneticists. But it took her a while to land on something new or interesting to investigate. It was only when one morning she picked up one of the milk bottles in which the flies had been kept that she had her breakthrough. She noticed that there were small white creatures crawling around in the food. On closer inspection, little pink specks had attached themselves to the flies' legs – a mite infection. McLaren took samples and kept them in separate bottles to observe them over the course of weeks. She eventually noticed that some *Drosophila* strains were more affected, likely because they couldn't fly as well, and so spent longer sitting on the floor where they were more easily parasitised.[4] From there, it took her just a few weeks to write up the project. In 1952, she became the first woman to win a Christopher Welch scholarship, which funded her first year as a graduate student.

McLaren had witnessed how the environmental conditions (presence of parasites) interacted with the flies' own traits (ability to fly) to determine survival. Which side came first in this equation? Broadly, molecular biologists of the 1950s had their answer come down on the side of the innate qualities contained in DNA. But Haldane and other embryologists of his generation had turned to a different question. In 1893, the evolutionary biologist August Weismann asked: 'How is a single germ cell [the cells that go on to form an egg or sperm] capable of reproducing the body in all its different parts?'[5] That question remained unanswered. The structure of DNA explained how a set of instructions are passed on, but it doesn't answer the embryologists' question of how a whole organism emerges from a single cell. McLaren followed in the footsteps of the scientists interested in development itself and continued to ask how organisms interact with their environments, over generations, to become themselves.

McLaren whizzed through her training. She completed her PhD, under Peter Medawar (who was absent for most of it), on neurotropic viruses in just two years. She switched from fruit flies to mice, to study how the viruses infect the nerves and the brain in a similar way that the polio virus paralyses humans.

While working on her doctorate, McLaren met Donald Michie. Michie, known to many as one of the forefathers of artificial intelligence, was a doctoral student in genetics at the time. He had worked at Bletchley Park during the Second World War on cryptography alongside Alan Turing. He'd initially applied to read classics, Greek and Latin at Oxford, but changed his mind; like many people at the end of the war, he wanted to do something more practical.[6] He enrolled as a medical student. After two years, his interest had wandered again, to mouse genetics.

He wasted no time in pursuing his latest fascination. While still in medical school, he bought a few specimens from St Giles' Fair in central Oxford and started breeding mice in the kitchen. The statistics interested him; their coat colours demanded his wartime expertise to process, just like McClintock had done with corn, to deduce the patterns of inheritance. Michie had no experience in animal experimentation. He used names, rather than the standard numbering, to track his research subjects – one mouse was called Marmaduke Smith, another David Parry. He eventually contacted the statistician and geneticist Ronald Fisher at Cambridge, who invited him and his mice to visit his department. Michie took his mice on the 'brain train' that ran between the two university cities at the time. When he arrived in the genetics department, students and professors gathered around his cages and giggled at his amateurish nomenclature. Michie was swiftly put out of his misery. Fisher strode out into the hall, took one look at the names on the cages, and said, 'Mr Michie, what an excellent method of record-keeping.' They became instant friends.[7]

Michie went on to complete his doctorate under the supervision of the ecological geneticist Edmund Brisco 'Henry' Ford, at Oxford. McLaren was Ford's first ever female tutee. She valued their tutorials. She wrote to her mother: 'It is not often one has the chance of having tutorials on a particular subject with one of the greatest experts on that subject, and my admiration for his mind goes up and up.'[8] And still, Ford could be a little stiff. McLaren's friend, Avrion Mitchison, J. B. S. Haldane's nephew, decided to organise a small dinner party to bond with their teacher. He invited a few other students, including Francis Huxley (son of the evolutionary biologist and eugenicist Julian Huxley) and Philip Shepherd (Ford's 'pet' researcher). He also asked McLaren along, who, with a sense of mischief, wrote to her mother: 'I thought he might be more at his ease if it was an all-male party, but that my presence would certainly make it an interesting scientific experiment.' In the end, though, the 'comic dinner party,' as McLaren describes it, went well. 'After dinner', she tells her mum, '[Ford] even unbent to the extent of asking me whether I had ever shared my bedroom with a cow.'[9] She also mentions one of the other guests: 'a charming medical student who has established himself in the Zoology Lab and does research on mice.' Donald Michie.[10]

McLaren and Michie shared a verve for mouse genetics and a growing commitment to socialist politics. They were also both inclined to a pragmatic, sometimes makeshift approach to experimentation. Using a tiny grant from the Royal Society, they bought baking tins from Woolworths that they used to make mouse cages at home. They married the same year that McLaren finished her doctorate.[11]

The pair eventually received more substantial funding from the Agricultural Research Council (ARC). Between 1952 and 1959, McLaren worked with Michie at University College

London (UCL) and simultaneously published four papers on polio; the first in *Nature* in 1953 showed that aged pertussis (whooping cough) and diphtheria vaccines protected mice from neurotropic virus infections.[12] But the most interesting work she produced at this time, as far as McLaren was concerned, was the final 1959 paper that explained age differences in the resistance to infections in mice. McLaren saw how this work could help tackle the epidemic of polio-induced infantile paralysis in humans.[13] She was a scientist who worked on mice, but always with an interest in how her findings applied to the human.

Human applications were increasingly prescient in McLaren's embryo transfer experiments with Michie. Together, they transplanted mouse embryos from the uterus of one mother to another, a surrogate – a technique called embryo transfer. There were several reasons for her wanting to do this. The central aim was vision. Embryologists study the earliest stages of development, from cells to organism. The problem for embryologists working with mammals is, of course, that they are interested in studying what happens inside the womb, which is hidden from a scientist's sight. McLaren realised that she could use embryo transfer to visualise the interactions taking place between the developing organism and the uterus. By moving the embryo between maternal environments and observing the effects this had on its physical development, she would be able to 'see' what was going on behind the scenes.

This was progress but it remained an indirect method. McLaren could still only infer and couldn't directly observe development. Embryologists still faced the same obstacle they had throughout human history. At the same time, however, McLaren was developing another, very promising technique that could provide the solution: embryo culture. This would involve transferring embryos not to another womb, but to a dish, in a so-called culture, a solution that would provide the necessary nutrients for the embryo to grow. This promised to be a real window onto development, simulated under the microscope.

The techniques of embryo transfer and culture in mice didn't just solve an age-old problem for embryologists; they also laid the foundation for IVF in humans. IVF had been floating on the horizon for a while. In 1924, Haldane suggested the idea in a series of lectures on the future of science, later published as *Daedalus; Or, Science and the Future,* before the Heretics Society of the University of Cambridge. Haldane referred to the first reported case of embryo transplantation by the scientist Walter Heape, in 1890, to suggest the same was possible for humans. Like McLaren, Heape had designed his experiments to visualise the process of reproduction hidden inside the wombs of rabbits. Haldane extrapolated a future in which the first 'ectogenetic' mammal, or an animal grown outside the body, would be born in 1940, the first child in 1951.

Haldane was right that, as the techniques of embryo transfer developed, the potential to transform human reproduction was quick to follow. In 1930, the American biologist Gregory Pincus (co-creator of the combined contraceptive pill) reported his first attempts in Cambridge to transfer rabbit embryos that had been fertilised outside the body to another rabbit's womb. In 1937, five years after Aldous Huxley published his vision of a future of genetically modified citizens in *Brave New World,* an editorial in the *New England Journal of Medicine* announced that Pincus and his colleague Ernst Vinzenz Enzmann at Harvard had isolated an egg, fertilised it in a watch glass and re-implanted it in a surrogate doe to produce a successful pregnancy. By 1948, the research scientist Miriam Menkin and doctor John Rock began exploring the same possibility in humans in Boston. They managed to cultivate the initial stages of embryo growth in a dish but got no further.[14]

A series of small but crucial discoveries eventually led to the birth of the first IVF baby, only a few decades later than Haldane had predicted in 1978. Scientists in Australia and the US, for example, showed that sperm had to mature past certain stages before it could fertilise an ovum. In 1951, the Chinese–American biologist Min Chueh Chang drew on these insights to impregnate a rabbit using IVF. Also, as Haldane actually identified in his 1924 speech, the culture medium proved pivotal. McLaren and her colleague John Biggers developed the culture medium, and in 1958 showed that, with it, a mouse embryo could survive outside the womb for over twenty-four hours, then be placed in the uterus of a mouse mother and develop into a normal healthy mouse. In 1968, the gynaecologist and pioneer in laparoscopy Patrick Steptoe, at Oldham General Hospital, teamed up with professor of human reproduction at Cambridge Robert Edwards and the often-forgotten nurse-turned-research scientist Jean Purdy. They embarked on a decade-long struggle to use IVF to help infertile couples conceive. They drew on interdisciplinary techniques and interspecies findings along the way. After several excruciating near-successes, one couple, Lesley and John Brown, became proud parents of the first IVF baby, Louise, on 25 July 1978.

Despite the dystopian visions of genetic engineering contained in *Brave New World*, the general public were enthusiastic when Louise Brown was born; the press celebrated the new 'miracle babies'. Scientists were more sceptical. They had been shocked by how quickly IVF had gone from experimental technique in animals to clinical treatment. They wanted more experiments to understand the impact of IVF on human development. And yet, when they started to discuss the technique with members of the public, they found, much to their surprise, that, although people were supportive of IVF itself, they were generally against the experiments needed to make the technique safe.

McLaren was involved in the extensive public debates in Britain that determined the legislation on human embryo research. She made the same argument that other scientists did: that without laboratory research IVF might have adverse effects, and so, instead of experimenting on animal or human embryos, they'd effectively be experimenting on members of the population. But she didn't dismiss people's concerns either. She understood that the public needed to understand what research on human embryos could do for them.

As the only scientist on the government committee tasked with developing guidelines for the legislation on IVF and embryo research in 1982, McLaren did a lot of 'non-scientific' work. She spoke to expectant mothers, for example, about the kind of research they thought was needed, discovering that they were generally in favour of work with human embryos, provided it meant that scientists could prevent their children from being born with genetic disease. With three adult children, McLaren could speak to these women as both a mother and a scientist and truly empathise with their concerns. Ever a socialist scientist, McLaren believed that the public should have a choice in the direction science takes, and that scientists had a duty to explain their work. She didn't believe in imposing her views. This stance ran counter to a culture that had long considered scientists as experts who shouldn't have to answer to anyone. Many of her contemporaries still resisted oversight. But McLaren believed that scientific progress should reflect a social consensus. Her work on the committee shifted public opinion from vehemently against human embryo research to largely in favour. The legislation still in place today remains some of the most permissive in the world. It is built on the laws enacted in 1990.

McLaren's career in a field centred on motherhood shows us another dimension of how 'maternal' skills are integral to making science happen. McLaren didn't talk only to mothers about reproduction. She did this work of communicating wherever she went: in policy, at home, between fields and scientists and, we'll see, political regimes. Her colleagues called her 'the great connector'. That aspect of her work was not just essential to knowledge exchange and research collaboration among scientists, but also to the multifaceted work of ensuring that scientific discoveries contribute meaningfully to society. Conversations determine whether science will be allowed to happen, who hears about it and whose priorities drive research; it determines what research becomes application and who has access. The undervalued skills of communicating, connecting and relating determine which science reaches people in the ways that they need and which science fails to, and so becomes dystopian. McLaren gives us an answer to Wells' question: we don't need to choose between submitting to reality or imposing our will on nature. Scientists' power should always be tempered by the relationships that are the basis of shared decision making.

Of course, part of McLaren's effectiveness as a communicator is explained by the access she had. She was, palpably, almost as upper class as it gets. That social status meant that she talked as easily to members of the public as decision makers. She is remembered for wearing plain, casual clothes to important meetings and carrying her notes around in plastic carrier bags. During the debates on human embryo research in the 1980s, at the height of 'power-dressing', she visited the likes of Margaret Thatcher wearing a dirndl skirt and a tennis shirt (what would Watson have made of that!). It didn't matter; her intellect and self-assurance put her on a par with anyone she spoke to.[15] For all her abilities, a large part of this story is once again privilege, not genius. There are other women who, given these opportunities, could have done something similar, albeit in their own, distinctive way.

McLaren's skills of connecting became all the more important as a new conflict seized the scientific world. A split had emerged in the field of embryology. The polarisation between the East and West in the Cold War now shot though her discipline too. With the discovery of the molecular structure of DNA in the West, we have seen, scientists became increasingly attached to the idea that the gene alone drove the development of an embryo. In the Soviet Union, there was a different view. They'd designed an agricultural policy based around the ideas of Trofim Lysenko, a notorious friend of Stalin's, who said that the environment of the gene, not the gene itself, produced the heritable traits in plants (drawing on the earlier ideas of the eighteenth-century French naturalist Jean-Baptiste Lamarck).

McLaren didn't align completely with this 'Lysenkoist' view of inheritance; she subscribed to the idea of the gene as a carrier of heritable traits. But she did feel that the role of the environment had been neglected in those theories. Her career, and everything that has happened in genetics since, has proved her right.

Despite the total rejection of Russian ideas by most in the West, McLaren continued to travel to Russia and exchange ideas with Lysenko and others. These scientists helped her develop her understanding of epigenetic controls: the mechanisms that turn genes 'on' and 'off' *without* changing the DNA sequence. These processes explain how environmental triggers, including diet, stress, toxins, prenatal conditions and also a cell's position in the body, regulate gene expression through different, reversible mechanisms, like adding chemical tags to DNA or changing how the DNA is packaged. The concept of epigenetics today acknowledges that genes alone cannot explain entirely how an organism develops.[16] Epigenetics explains how a cell specialises into a particular type, becoming a nerve or muscle cell, for example. And so, it is key, among many other therapeutic applications, to the process of reprogramming cells to become induced pluripotent stem (IPS) cells. IPS cells can be guided down developmental pathways to specialise into a variety of different cell types. They are used in transplants to treat many diseases, from heart disease to spinal cord injuries. The stem cell field has opened up all these therapeutic possibilities, beside the often-hyped prospects of creating ova, sperm and embryos.

McLaren was a scientific mother who took a pragmatic approach to building connections to form a shared perspective. By working broadly and flexibly, she contributed to a paradigm shift in biology. Unlike McClintock, who sequestered herself with a heretical idea for years, McLaren talked to people on their terms, finding points of resonance, and ignited research that would reveal more of the problem she was interested in. While other scientists stick to their niche until they make that Nobel Prize-winning discovery, McLaren hopped around to explore her underrepresented perspective: that self-same environmental impact on development of the embryo transfer experiments. Over the course of her career, from the 1950s and well into the 2000s, McLaren moved between the fields of genetics, reproductive science and embryology, from questions of foetal implantation to embryo cultures, and how the body's immune system can be harnessed in contraception, gonad development and germ cells (the predecessors to sex cells), and how genes are activated or deactivated by environmental triggers. She built a body of findings but also connected researchers across fields, leaving a prolific web of techniques and collaborations behind her.

McLaren never ceased to be ambitious in her scientific work of connecting people, resources and ideas. After the collapse of the Soviet Union, during her tenure as foreign secretary of the Royal Society (as the first woman, in fact, to hold office in its 330-year history), she continued to travel to Russia, strengthening ties with its science academies, sometimes carrying money hidden in her clothing to support their underfunded research. She was also involved in scientific projects that brought together Russian and Western scientists, including her collaboration with NASA in which she turned her childhood appearance in Wells' sci-fi film into another scientific reality. From 1979 well into the 1990s, she investigated the effects of microgravity on the development of embryos. Space was never the final frontier for McLaren; science was just a vessel for progress down on Earth.

McLaren worked as a public advocate for research on human embryos from the late 1970s well into the 2000s. This was more than necessary, because her career overlapped with rapid advancements in the field. Scientists were repeatedly as amazed as the public by the possibilities IVF spawned in a matter of decades. In quick succession, remote science fiction possibilities became realities: prenatal screening for genetic disease in the 1980s, human therapeutic cloning in the 1990s (what we today call somatic cell nuclear transfer, where the nucleus from one cell is transferred into an egg cell to produce patient-specific stem cells), mitochondrial donation, stem cells and genome editing in the 2000s. At each juncture, McLaren worked on the scientific techniques themselves as well as their regulation. She was at the forefront, ensuring techniques tackled the problems that mattered to people.

Her science was all about giving people a voice. She began almost every public talk with the following George Eliot quote: 'The strongest principle of growth lies in human choice.' In a 1989 keynote for the World Health Organization in Geneva, McLaren elaborated: 'Every animal can reproduce itself; only humankind can choose when and how and whether to reproduce.' Those decisions are a human, social imperative.[17] I can't help but hear an echo of Wells' line at the end of *Things to Come*. 'And if we're no more than animals.' McLaren gives us a different version of what it means to be human and to take responsibility for our humanity.

Despite her seismic impact on the science of embryology, developmental biology and reproductive biology, McLaren has not received the same degree of scientific or public recognition as scientists like Hodgkin, who moved in similar circles but worked on a specific area. In interviews with me, her former, male colleagues often told me that McLaren was more of a connector than a contributor. 'Not a Nobel-quality scientist,' some of them say. I think this tells us more about the kinds of scientific contributions we value than about the quality of McLaren's science. McLaren did not make a single, break-through discovery. She did something much more difficult: she worked for over half a century, building a sprawling web of knowledge and relationships that allowed scientific research to flourish and birthed legislation to ensure science *mattered* to people. McLaren is the mother of many breakthroughs and a whole infrastructure of embryonic research. Where, I ask you, is the Nobel Prize for that? How do we even begin to grasp the value of contributions on that scale? It wouldn't make for a flashy headline. Or a concise epitaph. But we have to find a way.

McLaren died, tragically, in a car crash with her by then ex-husband, Michie, in 2007. They'd divorced in 1959, but they remained lifelong friends. They spent family holidays together and were in constant communication. They continued to share the same politics. They collaborated on left-wing causes and let their commitment to the social good shape their science; neither of them specialised in a single field, and both their careers reflected the interconnectedness of people, systems and ideas. The author of their obituary in the *Morning Star* attributed their lasting bond to 'an affection for – indeed, a fascination with – each other'.[18] In fact, I spent much of my PhD wading through the impressive wad of their printed email exchanges. Michie shared his thoughts with McLaren on an almost daily basis. She responded, nurturing the connection with all the difficulties and joys it brought, and with an eye to all the other relationships theirs enabled. She was good at that.

16

Jane Goodall (1934–2025)

Primatologist

Who hasn't heard of the primatologist Jane Goodall? Since her 1963 article 'My life among wild chimpanzees' was published in *National Geographic*, she had been an icon of conservation. Anyone reading that article would have seen Goodall's photo before they read her words; on the cover page sits the mysterious silhouette of a woman, writing field notes against the 'sun's fading glow on Lake Tanganyika', strands of hair falling loose around her face. A sensual image of natural femininity in a primordial landscape. The article that follows continues to take us back to our evolutionary roots, led by this embodiment of all the contradictions of femininity – angelic, yet sexualised and also maternal – as we see her in intimate contact with her chimpanzee children, feeding them and watching them play with books. This was Goodall's introduction to the world.

We have just encountered a scientist who proved the value of the skills associated with motherhood to research. But we have also seen examples, throughout history, of how the image of motherhood can be used to discredit or displace the work of the scientist. The message of motherhood changes depending on the perspective we take. *Who* is telling the story? Since Goodall's first studies on the social behaviour of chimpanzees in their natural habitat, she has become one of the most iconic scientists of our time. Over the course of her life, Goodall claimed her own narrative, turning the camera outwards, like McLaren, helping people to *see* what she saw. Only, if McLaren's career reflected the view that our responsibility to the world sets us apart from animals, Goodall's question has always been one of what connects us to nature.

*

Goodall always saw the world differently. As a child, she was diagnosed with the neurological condition prosopagnosia, also known as face blindness, that affected her memory for faces and patterns.[1] Goodall struggled to recognise people's faces, but she had no trouble recognising animals. She grew up in a world of women and nature. Even before her parents divorced, her father, Mortimer Herbert Morris-Goodall, was often travelling to compete as a race car driver; that world did not interest Goodall in the slightest. In his absence, her mother, Margaret Myfanwe Joseph, or 'Vanne', nurtured her daughter's affinity for the natural world.

As a young child, Goodall played in the garden in the suburbs of Weybridge, and then later in the idyllic Kent countryside, where the family lived for a while during the war. After her father went to war, they finally settled outside Bournemouth. 'The Birches' was a red-brick Victorian house where Jane's grandmother, two aunts and an influx of women left homeless by German bombardment of London lived together.[2] This female household, nestled in rolling Dorset hills, remained Goodall's home after her parents divorced when she was twelve.

Nature defined Goodall's childhood. She had a favourite tree, a beech, that became a kind of emotional barometer. When she was happy, she climbed high into its branches and listened to the birds sing. But Vanne also taught her daughter about the solace nature brought. She taught Goodall to lay her cheek against the bark and feel the sap, the lifeblood of the tree, coursing through its limbs.[3]

Here, Goodall was free to prioritise the relationships she loved. Her dog, Rusty, became a touchstone in her later work with chimps. Watching Rusty convinced Goodall that animals had personalities, minds and feelings.[4] She observed how he reasoned: if she threw a ball from an upstairs window, he understood that it had landed outside, barked for the door to open and went to fetch it. He could plan: when it was very hot, he took a trip to the beach, had a swim and returned refreshed.[5] These are behaviours many of us have observed in our pets, and yet, Goodall took note. Later, she became determined to bring these observations, so indubitably real to her, into the realms of serious science. Putting data to an intuitive experience.[6]

This was partly how Jane made sense of the world. She experienced nature fully, she let herself be affected by it. Then, she went looking for the words to describe what she'd felt. Her mother helped her. When they walked together, she taught her daughter to greet the natural world, as a friend: 'Hello stream,' or 'Oh, wind, for heaven's sake, calm down.'[7] Vanne continued to gift her daughter the language to describe her world. She brought Goodall a copy of Hugh Lofting's *Doctor Dolittle* from the library that Goodall read with a flashlight under the covers at night. Next came *The Jungle Book,* then *Tarzan.* She read stories about wolves, bears and wolverines in North America and Canada, jaguars, anacondas and sloths in South America, orangutans, Indian elephants and tapirs in Asia.[8] Nature and stories became interconnected, one always guiding her to the other. Nature implored her to name what she saw, words inspiring her to look further, deeper, again. It was a self-perpetuating cycle that would propel her all the way to Africa.

Goodall wanted to go and see the animals in person. It wasn't a vague desire; she recalled in an interview that she 'decided' she would go.[9] In the many accounts of Goodall's life, including her own, there is a sense of inevitability in her travelling to Kenya at age twenty-three. Part of this is the product of the elaborate publicity that came to surround her.[10] But Goodall really did seem focused on her goal. After finishing school, she had no ambitions to enrol for a degree or pursue a research career. She just wanted to get to Africa.

First though, her mother had other plans. Vanne had guided and facilitated Goodall's connection with nature. Those relationships were based on a felt understanding: empathy. In the wake of a war, when, as Goodall put it, just 'the sound of a German voice sent chills down one's spine',[11] Vanne sent her daughter to spend four months with a family in Germany. Her mother considered it important that Goodall visited the country that had been their 'enemy'. That, as Goodall writes, her daughter should realise that there were 'ordinary Germans who did not like Hitler'.[12] Goodall may not have gone to university, but she received a thorough education in empathy.

Goodall returned to England, still unwavering in her intentions. Her mum had told her that secretaries could work anywhere in the world, so Goodall commenced secretarial training in London, strolling around the Natural History Museum in her spare time, or reading books about African wildlife.[13] The jobs she worked after graduating weren't what she'd hoped for. She'd just landed herself a relatively more interesting gig, at a documentary production company, when (the heroine's story goes) she received her call to action. A letter, mailed with stamps depicting elephants and giraffes, from an old school friend, Marie-Claude Mange. Marie had moved to a farm in Kenya with her parents. She invited Goodall to visit.[14]

This was the opportunity Goodall had been waiting for. She quit her job, cut off all flirtations and moved home to save for the trip. Eventually, she had enough to pay for her spot on the *Kenya Castle,* a passenger liner that would carry her to the port of Mombasa. It was suddenly very real, but in the face of the unknown Goodall wasn't even a little scared. Looking back, she sees how that confidence stemmed from her own carefully cultivated sense of the world: 'I could move into this new era without fear', she writes, 'for I was equipped, by my family and by my education, with sound moral values and an independent, free-thinking mind.'[15]

Goodall arrived in Kenya and was awe-struck by the wildlife. On her way to the farm, the car stopped to let a giraffe cross the road. She took in the animal, towering above the car, its beautiful dark eyes and surprisingly black tongue, and couldn't believe her luck: 'I had actually gotten to the Africa of my dreams – the Africa of *Doctor Dolittle* and *Tarzan*.'[16]

Goodall's story, in her many retellings, is infused with a sense of inevitability (as good stories are). Goodall herself attributed the sequence of events in her life to her sense of inner calling. But she was also travelling to Africa at a very particular point in history. The British Empire was losing its foothold. Even on her way to Africa, in 1956, a crisis had erupted as the British tried to regain control over Egypt's Suez Canal; it almost blocked her passage through to Africa.[17] This was a time of colonial reckoning. As Goodall searched for work in Nairobi, Kenyans were pushing back against British rule. Colonial authorities responded by detaining more than 10,000 Kenyans in camps and ordered troops to hunt down and capture members of the Mau Mau group, the rebel group at the forefront of the struggle.[18] It wasn't until 1963 that Kenya formally gained independence.

This context gives another perspective on the inevitability in Goodall's story. Her trip to Kenya was supported by the historical conditions of colonialism. She joined a lineage of Western, white naturalists and explorers who had travelled there over decades, entrenching the colonisation of people with the knowledge they claimed over their culture and natural habitat. Goodall was enamoured with the 'roar of a lion' under the 'hot tropical sun' as many foreigners like herself had been. 'I had never been so happy,' she writes in her memoir about her first expeditions.[19] And yet, as she entered the Serengeti, she was staking her claim to land previously inhabited only by the Masai people. An American hunter, suitably named Stewart Edward White, had stumbled upon the luscious land in 1913 and claimed its 'discovery', turning it into an untapped paradise for visitors.[20] This context fell outside Goodall's view. Goodall, in her early twenties, seemed uninterested in the human footsteps she traced. Her empathy was rooted in her immediate connection with the world. It was, in the end, though, that intimacy with her experience that would lead her to a more encompassing view.[21]

Then again, Goodall wasn't exactly in control of her narrative. Her mother had guided her in finding the words for her own experience of nature, but men's stories were impossible to escape.

Goodall took a secretarial job in Nairobi so as not to overstay her welcome on her friend's farm. Soon, she caught wind that for anyone hoping to work with animals there was only one man to talk to: a certain renowned paleoanthropologist. In 1957, Goodall visited Louis Leakey in his office in the Coryndon National Museum in Nairobi. The man sat there, with a sparkle in his eye, amidst strewn papers, fossils and teeth.[22] At this point in his career, Leakey was an eminent pioneer in the field of palaeoanthropology, the study of early human beings and their antecedents. He worked not with live animals but fossils and cultural artefacts, like stone tools, to flesh out the theory of human evolution. His work was edgy, and he was always controversial. In his doctoral dissertation, he'd tried to prove that the earliest human origins lay in Africa, not Asia, as was widely accepted at the time. He had discovered early human remains, but his work was initially rejected by some of his most eminent contemporaries. This, and an affair with the geologist and archaeologist Mary Douglas Nicol, had cast him into disrepute.[23] Leakey had only started to regain his credibility with a series of finds, made with his team and Douglas Nicol, that supported the existence of a stage of evolution that connected ape to man.

When Goodall entered his office, they found they had a lot to talk about. Leakey was impressed by Goodall's knowledge of African wildlife (and her beauty too; he had a strong record of workplace harassment).[24] He offered her a job as his assistant on the spot.[25]

Though he had bigger plans for her. Leakey wanted to take advantage of Goodall's passion for wildlife. He had a groundbreaking study in mind. He had not given up on his old, controversial darling that the earliest humans emerged from chimpanzees and bonobos in Africa. He thought that perhaps studying live chimpanzees would offer another lens onto the behaviour and appearance of early humans. He needed actual, extensive observation to supplement the fossil record he was building.[26] And yet, who would be willing to disappear into the wilderness for months on end?

Leakey gave Goodall an opportunity to do extensive academic research in the only laboratory that would have suited her. He had a talent, it turns out, for picking competent women to study primates in the wild – he recruited Dian Fossey and Biruté Galdikas after Goodall, whose work together laid the foundation for modern primatology. Leakey liked to refer to them as 'the Trimates' (make of that what you will). As far as Goodall was concerned, this was another once-in-a-lifetime opportunity. She'd strayed far beyond the boundaries of the domestic life set out for women. Not just that, but she had no formal education in research. To Leakey, though, both her gender and lack of expertise were what qualified her for the role. He believed that Goodall would be open and unencumbered by academic prejudices when observing the animals. He also believed that, being a woman, she was naturally more patient and more sensitive than a man would be, which made her a better observer.[27] This is a more complicated point that we'll come back to.

At first, Goodall's gender gave Leakey a small education in the realities women faced. Leakey had to obtain permission for Goodall to work in the intended region in the Gombe Stream chimpanzee reserve in Tanzania. The British colonial authorities still presided over the country. And they would not allow a woman to travel there alone. She chose to bring her mother.[28]

Vanne had, after all, nurtured Goodall's naturalist instincts into being. Armed with empathy and a shared experiential education, mother and daughter arrived in Gombe on 16 July 1960, in what was then still one continuous wilderness habitat. They had permission to go, but most authorities and locals remained sceptical of sending two unarmed women into the bush alone. Many doubted they would ever return.[29] They were accompanied by their cook and a game ranger for the first few days. The ranger irritated Goodall when he enlisted various locals to help. She'd waited a whole year for permission; she was wary of being controlled now that she was finally there. Her irritation was understandable but also short-sighted. Time with the environment and people of Gombe would eventually reveal the value of local expertise.

To the animals, though, Goodall was a humble student. She did what her mother had taught her. She named the animals she watched and got to know them individually. But it wasn't easy to get close. In the early months, she grew dejected as the chimps kept running away. Her mother was there to remind her of the value of description. Watching from a peak, using binoculars, Goodall was still finding a language for phenomena that no one had ever observed before. 'You're learning how the chimps make beds at night, bending the branches over,' she said. 'You're learning how they sometimes travel alone, sometimes in small groups, and sometimes in big, excited gatherings. You're learning the foods that they eat and the calls that they make. *You're learning more than you think.*'[30]

Vanne was right. Goodall's detailed notes in themselves gave unprecedented detail on chimp behaviour. No one had studied chimps in the wild before. Other ethologists who studied animal behaviour were wedded to so-called objectivist methods. To them, that meant experimentation, not observation. Goodall's work used the entirely different method of putting language to her experience of the natural world. As Goodall later explained: 'An anecdote can be a very carefully recorded observation. It's an anecdote, because you only see it once, but those anecdotes are sometimes the key to unlocking a puzzle. They're terribly important. And a collection of anecdotes, stories, has been very, very important in my research.'[31] Goodall was gathering data-rich stories about the chimpanzees' world even before she understood that as science.

Vanne left Gombe just two weeks before Goodall made her breakthrough. Leakey had asked Goodall to determine whether chimps used tools; this was the crucial piece of information he needed to complement his fossil studies and make the case for a hominin split in Africa. Goodall found her answer after just one day. One of the chimps, 'darling Greybeard', who had 'just begun to lose his fear', allowed Goodall to come close enough to watch him push small blades of grass into the termite hill he was sitting on.[32] Goodall watched him extract the stems, then pick termites from it – to eat. Greybeard had devised his own form of cutlery.

Goodall was thrilled. This was the first recorded instance of a non-human making tools. After confirming her findings, she telegraphed Leakey the news. He responded, the story goes, with a phrase that would be recited as a pithy summary of the significance of the finding: 'Ah! We must now redefine man, redefine tools, or accept the chimpanzee as human!'[33]

The statement rings very much like Watson and Crick's 'We've discovered the secret of life.' Sweeping proclamations that punctuate the history of science. The words commemorate people but stifle our understanding of the evolution of ideas. I'm still not really sure what it would mean to 'define' humanity or life; if we could do so in a sentence, I'm not sure there'd be much point to living.

Of course, the words were chosen for their impact, not their accuracy. The phrase was part of Leakey's broader efforts to reframe Goodall's discovery as a revision to the evolutionary narrative. Leakey had recruited Goodall to *his* project after all. He was supportive of her work, and continued to be, but he was also determined to claim the discovery as evidence for his own theories. That still proved a stretch. An article since has traced his famous phrase roughly (not verbatim) to a memo Leakey had written to *National Geographic*, intended as a supportive statement on the significance of Goodall's work. His conclusion was rejected, after Sherwood Washburn, another leader in the field, argued it was a step too far.[34] Scientists were far from agreed that tool use was proof of our closer evolutionary relationship with chimps. Nor was Goodall convinced. In her descriptions, she drew an analogy: chimps making what we as humans would describe as a tool. But she never argued, as Leakey does, that this proves chimps' evolutionary relationship to humans.[35] Goodall's observations were always centred around affinities, not causality. Her work shoved humans from their pedestal in other ways.

If anything, her observations complicated the very story of evolution. The toolmaking Goodall observed is an example of social learning. This is the process of learning from others, not only from their actions but also from consequences of their actions.[36] It's different, in that sense, from individual learning based on personal discovery. Social learning explains how knowledge is maintained in a population over time. Goodall was the first to show that animals, like humans, have cultural traditions. Animals impart knowledge about behaviour, or how to use artefacts, or make nests. For humans, these traditions might extend to imparting attitudes and beliefs. In either case, their learning and sense of social cohesion extends beyond the present.

Social learning described a whole new inheritance system, separate from and in dialogue with evolution as it was understood. Evolutionary theory at the time was understood primarily through a genetic lens. In this view, a species evolved based on the process of natural selection: in Charles Darwin's words, the 'principle by which each slight variation [of a heritable trait], if useful, is preserved.'[37] This explained how, in a population with a variety of characteristics, those whose traits were best adapted to their environments would be more likely to survive and reproduce, thus, conserving their genetics in the next generation. But if animals could learn collectively over time, there was another mechanism driving evolution, determined not by genetics, but social intelligence.[38] While Goodall's research showed a connection between humans and apes, it was an affinity, not a direct link. She complicated the frameworks for understanding but didn't corroborate them.

National Geographic rejected Leakey's pronouncement but embraced Goodall wholeheartedly. They funded Goodall's research and did a lot to shape her public image. 'My life among wild chimpanzees' got such an unusually strong response that the senior executives arranged for a second article, and then a broadcast on the CBS television network of a National Geographic Society special, *Miss Goodall and the Wild Chimpanzee.*[39]

The images of Goodall that accompanied her writing were affectionate, taken by the photographer Hugo van Lawick, who she later married.[40] As a public story, though, these media representations foregrounded Goodall's femininity – whether as a sexualised object or a maternal figure. 'Miss Goodall lathers her blond hair with water pure enough to drink'[41] reads the text, drawing attention to her appearance. Another sequence of photos shows her crouched down, affectionately grooming and tending to Greybeard, then, in a similarly maternal posture, feeding cough syrup to a local fisherman's son; 'open wide' reads the caption that frames this visual story of Goodall as a mother, competing for a readership's attention with her observations as a scientist.[42]

Goodall didn't have control of how she or her work were represented. Like Hodgkin, the media presented her first as a woman, not a scientist. In the end, though, Goodall argued that the media campaigns were pivotal to the acceptance she found as a *scientist.*

When, in 1962, she entered the University of Cambridge, two years after making her ground-breaking observations in the field, many scientists were still sceptical of her and her methods. Leakey had secured her the spot; she entered the programme as the eighth person in the history of the university to be admitted to work for a PhD without so much as a bachelors. And yet, her lack of an academic background made it easier for the bigwigs there to dismiss her approach. They ridiculed Goodall for her method of naming the chimps she watched.[43] These scientists, who numbered rather than named their research subjects, told her she'd undermined the most fundamental law of the field: she'd *anthropomorphised*. She had projected human qualities and emotions onto animals. She was told by the established ethologists there that her results were invalid.[44] In this climate, the *National Geographic* article and documentaries 'forced science to believe what I was saying,' Goodall reflected. The reporting accelerated a process that could have taken years if it had happened inside the ivory tower. But the story gained momentum, drawing the scientists' gazes to what Goodall herself had witnessed. 'When they saw Hugo's film', Goodall said in a 2020 interview, 'they had to believe.'[45]

Goodall had always understood the importance of seeing. Any child who has played with their pet dog or watched birds flutter among the trees appreciates that animals have a kind of intelligence. Goodall understood that actually seeing what she had seen in Gombe allowed people to recognise this same intelligence, even in an unfamiliar animal.

Once people were open to Goodall's view of animals, they could embrace her method. They could understand how little you would learn by hovering objectively over a lab animal; how much more you would gain by getting close to them in the wild. As Goodall put it, her empathic reasoning about animal behaviour ('if they were human, they'd do that because of—') gave her a 'platform'. You can 'stand on that platform and then try to analyse what you've seen in a scientific way'. But it's that initial intuitive insight, she says, that delivers the 'aha moment' that so many scientists speak of.[46] This is reminiscent of McClintock's descriptions of her method – of the computer integrating her data based on an intimate knowledge of the corn she studied. In this view, intuition isn't the unreliable (and feminised) opposite to reason but a storehouse of knowledge based on experience. Intuition is the ever-evolving model through which we all make sense of the world.

Goodall believed that the media campaigns helped to promote her approach to science. Perhaps, despite themselves, they allowed her to show the world what a 'feminine' approach revealed about primates that 'male' science couldn't. Over the course of her career, Goodall wrote close to thirty books. In 1986, she published *The Chimpanzees of Gombe*, which summarised twenty-five years of research. She contradicted and elaborated on numerous widely held beliefs about these animals. She proved that chimpanzees wage wars over territories, eat meat (she watched them devour a piglet early in her time at Gombe), exhibit emotions like love, jealousy, anger and sadness, and that they embrace, play and care for each other. Goodall's work has contributed to a growing body of research that demonstrates that organisms on all branches of the animal kingdom, not just humans, have 'culture'. Not just apes – even bumblebees learn collectively, collecting nectar from flowers that are popular with other bees in the hive.[47] In Japan, crows use pedestrian crossings as nutcrackers by dropping hard nuts onto the road and waiting for cars to crush them. They even wait for the traffic light to turn red to safely collect the cracked kernels.[48]

Goodall's approach, maintaining a relationship with the animals she studied, pioneered a new way of relating not only to the animals but the people of Gombe too. She built a research centre to train local scientists while letting their own priorities direct the work: handing knowledge of their environment back to them.[49] In this, Goodall claimed mothering, once again, not as a trope but as a verb – that *how*, rather than the *what* of science. Seen this way, the image of the mother-scientist shows us new ways of practising science with a sense of interconnectedness and empathy for our world.

In later life, Goodall turned her lens more determinedly to the human world: from science to conservation and reform. With it, she underwent a transformation in the public eye: from mother and sex object to wise sage. She travelled around the world speaking about conservation for a generation disillusioned and afraid in the face of the climate crisis. No longer the leggy blonde, she had found a way to inhabit yet another stereotype. Rooted in her years of experience as a scientist and an intimacy with the natural world, she spoke and wrote books in a new language, one that rings as quasi-spiritual and somewhat out of step with our hyper-scientific, hyper-materialist age. Her favourite word was hope. Despite the monumental environmental disaster subsuming us all, she inspired people with what *was* possible and argued for this innately human capacity as a necessary catalyst for action.

The spiritual lens is important and sorely missing from discussions of science. Goodall herself evoked the great scientists we know, who, like McClintock, were always inspired by some manifestation of the unknown. And yet, the language of hope can lull us into complacency. This was not Goodall's intention. Hope for her was the prerequisite to action. In our age of environmental crisis, Goodall offers a vision of a scientific method based on humility. An approach driven by curiosity and by acquaintance. Of using your intuition, and recognising your limits, to move closer to another being, not to dominate. The unknown in that intimate encounter gave Goodall confidence that change is always possible.

Goodall never stopped spreading her gospel. She died, at age ninety-one, on a speaking tour in the US in October 2025. There was an international outpouring of grief, as was expected, but perhaps even bigger. Everyone from colleagues and naturalists, animal and human rights organisations to presidents and celebrities paid tribute. People were moved by her death. Goodall's final words to us help us understand why. In the month of her passing, Netflix released the first episode of *Famous Last Words*. In the episode, the executive producer of the show, Brad Falchuk, interviews Goodall. Ever playful, she had brought a stuffed monkey to the interview, but she unflinchingly acknowledged the horrors we all faced at the hands of world leaders like Trump or Netanyahu. At the end of the episode, Falchuk leaves the set. Goodall speaks directly to the camera. 'I want to make sure that you all understand that each and every one of you has a role to play,' she says. 'Your life *does* matter, and every single day you live, you make a difference in the world. And you get to *choose* the difference that you make.' She reassures us, and she implores us, 'Even today, when the planet is dark, there still is hope.'[50]

I think we all needed to hear that.

17

Virginia Johnson (1925–2013)

Sexologist

In 1968, the obstetrician and gynaecologist William Masters and his co-researcher Virginia Johnson prepared for their next interview on human sexuality. It was part of their groundbreaking medical study, approaching a topic long considered utterly taboo. However, the location of the interview was... unusual, perhaps as far removed from the clinical world of the laboratory as you could get – the Playboy Mansion.

Founded in 1959 by self-appointed 'playboy' Hugh Hefner, the Chicago-based, seventy-room red-brick townhouse was ironically Victorian. Its existence, at the end of the decade of the newly domesticated housewife, in some ways represented a radical shift in American attitudes to sex: hyper-visible and openly commercialised. In other ways, it was simply a reinvention of the patriarchal home. In his TV shows, Hef invited the public into his pleasure dome, where they could watch uninhibited as adolescent 'bunnies' (otherwise known as women) strutted around at all hours, sliding down poles in near nudity (all this while Hef, much their senior, remained fully clothed, surprise, surprise...). This male fantasy was the new mainstream of liberated sex in 1960s America.

Masters and Johnson arrived punctually and knocked on the door, ignoring the sign that read, in Latin, *si non oscillas, noli* *tintinnare* ('If you don't swing, don't ring').* What ensued was a titillating conversation.

They bonded over science, discovering their shared admiration for the work of the entomologist-turned-sexology-pioneer Alfred Kinsey. Kinsey had conducted over 10,000 sociological-style interviews with men and women in the 1940s about their sexual habits. Published in two books that became known as the *Kinsey Reports* (on men in 1948 and women in 1953), these medical books became popular bestsellers. They challenged some of the most commonly held ideas about sexuality: homosexuality, affairs, sadomasochism, all previously viewed as the aberrations of small subpopulations were in fact, so widespread, they would almost have to be considered normal.

Before he founded *Playboy* in 1953, Hefner had studied at Northwestern University, where he wrote a paper assessing the legality of men's sexual behaviour according to Kinsey statistics on male sexuality. He reported that most American men, according to law, would be classified as criminals. Hefner delighted in this non-judgemental, amoral conversation about sexuality with this pair of scientists. He was especially impressed by the female half of the research team.[1]

Virginia Johnson was a far cry from a bunny. She was forty-three. Her hair was always pinned back and she wore the same doctor's coat for every media appearance. She presented as a scientist: Masters' equal. She talked frankly about women's experience of sex, not as objectified pin-up dolls but as people. Their research surpassed Kinsey's in Hef's eyes, because it was informed by a woman's perspective.[2] Their first interview led to the debut of a new piece of anatomy to the pages of *Playboy*, the

* Why Latin, you ask? Your guess is as good as mine.

clitoris, as well as $300,000 in funding towards their research programme.[3]

Johnson was the female half of the research duo Masters and Johnson, who took the world by storm with their book *Human Sexual Response* in 1966, which described their live observations of people having sex. Subsequently, they founded the discipline of sex therapy. They did this at the dawn of a sexual revolution, and I could have started my story there. I chose to start, instead, with Johnson talking about clitorises with two highly visible men. I wanted to start with her in conversation with a man who mistook his own pornographic fantasies for the sexual liberation of women, because it shows us the world she worked in. It is also the context we need to make sense of her seismic contributions and her ambivalent legacy.

*

Johnson was born in Springfield, Missouri, on 11 February 1925, to a gentleman farmer, Harry, and his wife, Edna Eshelman. Harry had hoped for a simple life when he'd met Edna, a schoolteacher, at twenty-nine. After witnessing his share of bloodshed in the First World War, he was content to settle down and farm his plot of land. Edna had different plans. She'd married up, found a gentleman and she was determined to take full advantage of the opportunity for a better life.[4]

When Johnson was five, it was Edna's idea to move to Palo Alto, California, to escape the economic Depression. For a while, her father worked as a hospital groundskeeper there. Edna had been right to suggest the move. Their surroundings were a testament to her sound judgement. The family lived on the lush green hospital grounds: a veritable paradise compared to the arid plains they'd left behind. The education was better too.

Johnson was enrolled in a progressive school. She excelled in her studies and was even able to finish eighth grade two years early, at age twelve.[5]

Johnson's mother wore the trousers in the relationship. She had traditional values, rating marriage over a career for women, but she was ambitious in her domestic role. She made sure Johnson had piano and singing lessons and taught her to cook: everything a refined lady might need to know.[6] And, as the Palo Alto move exemplified, she was uncompromising in asserting her will over her husband. Johnson copied her mother's pose of carefully feminised authority. She later said in an interview: 'I always lived the façade of mother's little lady but I always did exactly what I wanted to do. I just never let it be known.'[7]

It had been Edna's idea to leave Springfield, but now she wanted to go back. She missed home, and her husband's job had lost its glamour. She was pushing against the tide of people escaping the ever-more derelict Midwest. The family disagreed, but they knew better than to resist Edna's will.[8] They moved to a tiny town five miles outside a small, dusty town called Golden City, leaving Johnson with two clear options in life: get married or get out.

Ambitious mother that she was, it was clear to Edna too that Johnson needed to move on to bigger things. She sent her daughter to Drury College in Springfield to study music. Johnson loved music and had an excellent voice. She hoped to sing at the Metropolitan Opera or become an international classical singer.[9] Meanwhile, Edna had found relief from the boredom of life in Golden City by getting involved in politics. Her status as a respected committee member of the Republican Party helped Johnson secure a patronage job with the insurance department in Jefferson City, and she later advanced to the legislature, supporting the state senator. She hoped that her position might help her daughter pin down a successful husband.[10]

Johnson left home at sixteen. Her mother's education had prepared her for her new-found, high-society surroundings. She knew how to befriend powerful people, secretaries and rank-and-file admins alike.[11] She knew how to maintain her social acceptability as a woman, while advancing her own ambitions. Just like her mother had at home.

Johnson entered the workforce at the start of the Second World War. The war changed women's understanding of what was possible: it complicated ideas about acceptable womanhood that had governed the lives of Edna's generation. Women took traditionally male jobs in factories and businesses, while men were fighting overseas. Wartime propaganda played a role in conserving traditional ideas about femininity while imploring women to take on roles outside the home. Subversive posters featured the voluptuous, carefully made-up 'Rosie the Riveter' brandishing a drill or flashing her biceps: 'It's a tradition with us, mister!' The power-woman on the poster could do it all. In real life, this new twist of duplicitous binds became a minefield for women to navigate. Johnson was irritated by the behaviour of her female colleagues, who shapeshifted to fit the new ambiguous mould. She watched how women acted virtuous, then dropped the act shamelessly when Saturday night came around, only to resume the act the following day.[12] 'There were a fair number of women who were sexually active', she explained in an interview, 'but the "nice" girls – the ones who were so labelled, of which I was too, for that matter – a lot of them were not.'[13] It was a complicated charade that exhausted Johnson. She happily mingled with the soldiers stationed at a local military facility, where she performed with a women's quartet in the Army-post shows.[14] But she was unapologetic about her relationship to sex: 'I literally didn't go out with anyone, to any degree, that I didn't have a sexual relationship with. I enjoyed sex.'[15]

In a climate of shifting gender roles, women found their own way and decided which risks to their reputation they were willing to take. They made decisions. Birth control expanded their options. Campaigners had been fighting restrictive contraception laws for over two decades. In 1937, the American Medical Association officially endorsed contraception as a medical responsibility.[16] But the medical community was slow to take up its role. In some states, contraception remained illegal; everywhere, provision was inconsistent. Women struggled to access the methods available in the pre-pill era, which included cervical caps and diaphragms. Where provision was lacking, social stigma still reigned. Johnson didn't seem to feel these risks so acutely. Throughout her life, respectability never outweighed Johnson's need for freedom.

Nonetheless, after the war, Johnson found herself grappling with social convention, particularly the economic necessity of marriage. Women in the US still didn't have the right to open a credit card or apply for a bank loan without a man to sign for them. It could still be difficult to start a business or buy a house as a single woman. You could still get fired for being pregnant, and equal pay was unheard of. Johnson would marry three times (although in some interviews, Johnson names a first husband, a political figure whom she married at nineteen. The legal papers have never been found).[17] Each marriage was, in her recollections, motivated by a sense of pragmatism reminiscent of her mother's. Necessity always subjugated love. In a social system that made life difficult for unmarried women, she retained some autonomy by making the marriages work for her. Her first marriage, at age twenty-two, to a 43-year-old lawyer whom she met in the state insurance department in Jefferson, offered a welcome economic severance from her parents. Still, she was immediately ambivalent about the marriage. When the minister asked her if she wanted pictures taken of the ceremony, she said, 'No, I don't want to record this.' She had an inkling that the marriage wouldn't last. As soon as Johnson realised her husband didn't want to have children, she filed for a divorce.[18]

Johnson moved to St Louis and found a job as a secretary at the *St Louis Daily Record*, a law and business journal, where she met her second husband, the leader of a nightclub band. Their shared interest in music bound them for a while, with Johnson performing alongside him. They had two children. But the late-night schedule eventually lost its allure for Johnson, and her husband didn't seem up to the responsibilities of fatherhood. She felt stuck and frustrated; she had left home for college and opportunity. She may have opted into another marriage, but she hadn't planned on a life as a housewife. She left her second husband too.[19]

At age thirty-two, Johnson, an unemployed, twice-divorced mother with two small children, wanted a fresh start. Divorce was still rare in the US; stigma and judgement surrounded women who broke from convention. Johnson struggled with feelings of personal failure, but still she prioritised her own desires. She no longer wanted to be a singer; she wanted to return to university. She'd hoped to pursue social anthropology, to investigate the cultural differences between nature and nurture in human development. But there wasn't a major in this field, so a school advisor had directed her to sociology. Only she needed a campus job to cover tuition. There weren't any in sociology. She looked around and found that the medical school offered a working assistantship. Her first appointment with an eminent psychiatrist was cancelled. Undeterred, she trotted on to her next meeting in the obstetrics and gynaecology department, with one William Masters.[20]

It was an entirely chance meeting. She hadn't chosen the job, and yet, with a commitment to independence, a thick skin in the face of judgements and a pragmatic attitude to her career, Johnson was well suited to the project Masters had in mind.

Johnson thought she'd been hired to work as a secretary on a project to help infertile couples conceive. She didn't know that Masters had been observing sex workers in local brothels. Like Kinsey, he was interested in human sexual behaviour, though Masters made a point of distinguishing his highly scientific, medical method from Kinsey's sociological speculation.[21] Masters wasn't a people person. He didn't dabble in the nuances of human utterances, nor their interpretation. Interviews, as far as he was concerned, were wholly unreliable. From his own experience, he knew that people didn't always tell the truth about sex. Rather than finding ways to establish trust, he saw a more direct method. He would measure the body's functioning during sex: every heartbeat and labial quiver.[22] He would reduce this activity riddled with sentimentality to verifiable fact. He did what all scientists do; he found an approach that suited his personality.

Masters was determined to learn from Kinsey's mistakes in more ways than one. Kinsey's work became unexpectedly popular, and it exposed him to widespread criticism. Even Margaret Mead, who, as we saw, conducted her own studies that challenged accepted gender roles and sexual norms, found the language inappropriately clinical. Kinsley, Mead aptly concluded, 'confused sex with excretion.'[23] Masters would avoid such unnecessary discussion around a matter by keeping it a secret.

There were other criticisms of Kinsey's work though that were more difficult to avoid. Kinsey had mainly interviewed prisoners, who weren't representative of the general population (though I think this objection begs verification; it's intuitive only if we rely on prejudice). Masters was bumping against a similar issue with the population of research participants available to him. Sex workers, he was finding, commonly suffer from chronic pelvic congestion, making them atypical in this regard to the rest of the population.[24]

Masters wanted a more representative research sample. But here, he hit the limits of what his approach to science could do. He wasn't a personable man; that put him at a disadvantage in an area as intimate and taboo as sex. When it came to recruiting women for his study, he didn't stand a chance. He realised that he needed a female research partner if he wanted to convince members of the public to let him observe their physical reactions while they had sex or masturbated.

Masters had approached female doctors, but they'd all been understandably cautious of risking their professional reputations. Johnson was slightly baffled when she got the job; she'd answered Masters' few perfunctory questions ignorantly while he'd stared at her as if inspecting a subject through a microscope.[25] She didn't know that her ignorance qualified her for the job. Like Goodall and Leakey, the power dynamic between the dynamic duo was prewritten. Masters hired Johnson as his secretary and told her nothing about the project. He held the power; she had the illusion of a choice.

For a while, Johnson didn't notice anything suspicious. She had been compiling case histories that sometimes included intimate personal details, but this was consistent with the work she thought she was doing on infertility. And yet, everyone around her knew. The truth came out over lunch in the canteen. When Masters walked in and realised what had happened, he asked her the question he should have in their initial interview: 'Is that going to bother you?'[26] Johnson, surrounded by a group of highly amused male medical residents, answered, completely collected: 'I can't imagine why.' It wasn't the subject matter that bothered her, but that she couldn't understand what was so interesting about the research. She thought for a moment, then asked, 'But why does anyone need it?'[27]

That second question convinced Masters that Johnson was the research partner he'd been looking for.[28] Johnson was matter of fact about sex, as she was about most things: it didn't shock her. Sex was a fact of life to her: 'It was important but I never categorised it that way. To me, it always was a natural requirement, *a need*.'[29]

Masters was aware of the power imbalance between himself and Johnson that persisted when he started his career-defining work with Johnson in 1957. It suited him perfectly. As he put it, when describing his decision to make her his research assistant: 'The nonmarried female inevitably is a professional virgin,' (which showed how little he knew about the sex lives of ordinary women). He'd cast himself, *Pygmalion*-like, as a Professor Higgins, sculpting his Eliza into his image of a respectable scientist.[30] It is true that Masters provided Johnson with a medical education that surpassed anything a standard medical education would have taught her. Besides medical terminology and compiling data, Johnson mastered the most unconventional lab techniques. In Netflix's *Masters of Sex*, we get a visual of how the pair struggle with the lack of suitable equipment available to them in their novel field, fashioning an enormous dildo-like Perspex camera called 'Ulysses' to explore the inside of their subjects' vaginas.

While Masters congratulated himself on creating a prodigy, he forgot (again, *Pygmalion*-like) to recognise Johnson's unique and independent contribution to their studies. Masters's quest for a female partner proved how stuck he was before he met her. Johnson recruited all the research participants they needed. She convinced students, nurses, secretaries and even doctors' wives at the University of Washington to participate for next to no money, just for the good of science, all while the project remained shrouded in secrecy and taboo. Two-thirds of the women she approached agreed.[31] Her skills of relating were as crucial to this work as building the correct equipment or designing an experiment. Johnson learned the skills conventionally associated with research from Masters and even ran the study while Masters continued to run his ob-gyn practice. But when it came to the pivotal human skills that facilitated their work, her contribution was singular.[32]

Masters and Johnson studied 382 women and 312 men between eighteen and eighty-nine for a decade, wiring them up and filming them to measure a range of reactions.[33] The study was scientific, but as it progressed, like Kinsey, they began to undo many of the accepted social scripts that had long shaped people's ideas about what 'natural' sex looked like. They described the mechanism of vaginal lubrication as the result of increased blood flow to the vaginal walls,[34] thus debunking the earlier, widely held notion that this originated from the cervix or glands. They studied female orgasms and proved, among other things, that some women are capable of having multiple orgasms. They obliterated the many long-standing misconceptions, like Freud's idea that female orgasm was either clitoral (potentially achieved independently, and so, according to Freud, 'immature') or vaginal (given by men and thus 'mature'). The polygraph-like instruments Masters and Johnson designed and created showed that female orgasm, whether stimulated directly or indirectly, was always clitoral. They identified the four stages of sexual response – excitement, plateau, orgasmic and resolution – and showed that these occurred in both sexes. They demonstrated that the size of the male member is not linked to virility and that elderly people still have sex.[35] The list of myths they busted goes on and on.

They finally published their findings in *Human Sexual Response* in 1966. The book contained their findings as well as a framework for a clinical approach to sex. In their classification of sexual dysfunctions, they described problems that presented obstacles to orgasm in sex, such as premature ejaculation, impotence, vaginismus or the still-Freudian 'female frigidity'. But they also offered something different from Freud. Using what they had learned about the sexual response cycle, Masters and Johnson designed a therapy to help couples affected by their newly diagnosed sexual dysfunction. While psychoanalysis saw unconscious psychological conflicts as the cause of sexual problems, Masters and Johnson reduced the problem to something simpler. They argued that fear of sexual inadequacy was generally the block to orgasm. That anxiety put people in a 'spectator role' and took them out of the moment while they monitored themselves. Instead of long-term psychoanalytical therapy, they offered a short, focused course to help people recognise their own triggers, blocks and preferences. Their programme, they reported, was much more effective than psychoanalysis: eighty percent of their clients 'recovered'.[36] Their method offered a way for couples to reconnect and for individuals to develop sexual self-awareness that is still fundamental to sexual therapy today.[37] In 1970, Masters and Johnson published their findings on this sex and relationship work in *Human Sexual Inadequacy*.[38]

Their books were written in the same dry, academic language that, as with Kinsey, was the subject of some ridicule. For Masters and Johnson, though, it ultimately hit a sweet spot that won over medicine, feminists and industry. Their work was academically accredited; it became standard reading for the medical field. But it also resonated across society.

The book took the world by storm. The publisher, Little, Brown, shipped it to doctors wrapped in plain brown paper.

They sold 300,000 copies in the first months. 'We feel for the first time that we are working with the support of public opinion, not against it,' Johnson told *Newsweek*. The magazine referred to their work as 'the most daring and explicit experiments ever conducted in the scientific studies of sex'.[39]

The medical language in the book allowed the media to quote the book unabashedly. The words found their way into the mainstream and opened the floodgates to a new conversation. Words like 'clitoris', 'orgasm' and 'masturbation' were suddenly public parlance. Sex could be discussed and worked on for the benefit of both partners. As one biographer put it: 'Masters and Johnson's mechanical approach…made their book palatable to a tongue-tied nation.'[40]

The public appetite and receptiveness to sex talk had been growing for a while. The ground had been primed by a wave of books and films. Even pent-up suburban wives were freed from their sexless existence with the figure of Mrs Robinson in *The Graduate*.[41] At the same time, books exploded the bedroom doors to the sex lives of ordinary women. Even before Friedan's *The Feminine Mystique*, women found unlikely manifestos in books like *Sex and the Single Girl* (1962). In this classic, advertising executive Helen Gurley Brown took aim at the sexual double standard that demanded celibacy from women out of wedlock while men roamed free. Written in the style of a chatty advice book, her book combined political manifesto with salacious entertainment.[42] After Masters and Johnson's studies gained traction, 'the floodgates', as Johnson remembers, really 'opened'. Now, 'every single magazine … was being sold on the basis of sex. A little food, a little fashion, a little parenting, and all the rest is sex, so the media actually created the concept of a revolution.'[43]

Of course, this mattered more to women. Friedan, as we saw, had launched her own critique against the neo-Freudian diagnoses of housewife's hysteria. At the same time, prompted by new reproductive possibilities like the contraceptive pill, a second wave of feminism was mobilising against the control of women's bodies. Masters and Johnson's book became a touchstone to the movement, offering a scientific rationale for a new social reality feminists were creating.[44] It spoke to the need for practical choices that helped women claim control of their own bodies. That wasn't just about the choice to reproduce. As the media publicised the sexual escapades of rebellious social movements, of anti-war hippies and housewifely rebellion, orgasms became another universal human right.[45]

Unlike Freud and even Kinsey, Masters and Johnson gave unprecedented space to the female experience of sex. Together, they appeared on the 1970 cover of *Time* magazine, as the face of sexual liberation, passed around by feminists along with copies of Germaine Greer, offering a version of women's sexuality so different from the female frigidity instilled in people's minds in the previous decades. Johnson, in particular, symbolised the new progressive woman. Her talent at connecting with people allowed her to reach a wide audience. She appeared on chat shows, was interviewed by magazines and wrote columns. In all of these, she was presented as an equal to Masters and a spokesperson for women's sexual freedom.[46]

And yet, Johnson didn't consider herself a feminist.[47] Through the 1970s and 1980s, famous feminists and public figures repeatedly invited her to rallies. She always refused. 'I would never march with the ladies,' she said. She even declined an invitation to appear with First Lady Betty Ford in support of the Equal Rights Amendment.[48] This is somewhat frustrating; like McLaren, Hodgkin or Goodall, Johnson could really have added clout to the social causes her science influenced. Her individualistic outlook reflected the life she'd had: one of scrambling independently for opportunities. Johnson hadn't benefited from a social movement to build a career. She'd wound her way through the post-war confusion on her own. Women needed to 'step up', she believed, to claim their own destiny.[49] That counted for their careers as much as it did for their sex lives. 'If a woman doesn't have an orgasm, it's her own damn fault!'[50]

Masters and Johnson's conversation with *Playboy* was indicative. Their book may have been radical but not enough to alienate the mainstream. Their book existed as feminist fuel *and* marketable product. 'You must remember that in publishing this book we were concerned primarily with acceptance – that is the reason that it wasn't in English to start with,' Masters quipped (though it is doubtful he knew a language other than 'science').[51] Public appeal was essential to funding their work. Because, despite their public popularity, at universities their work was still controversial. Washington University ceased to back Masters in the 1960s. Public support, on the other hand, proved lucrative. Talking to *Playboy* funded their Reproductive Biology Research Foundation (which would later become the Masters and Johnson Institute), a non-profit organization dedicated to sex research. Wide support produced funding and, as Goodall observed, it also accelerated public acceptance that would have taken far longer if kept behind university doors.

Masters and Johnson were not scientists with a social vision. The pair had little interest in the social ramifications of their work. In fact, they fell sadly short when it came to interrogating the social constructs and gendered ideas underpinning their research. Their classification of sexual dysfunctions, for example, relied on a cultural stereotype in which women are the recipients of male action and desire, and in which women's desires are only ever secondary. They also assumed that normal sexual response was the body's involuntary preparation for heterosexual, penetrative sex. This implied that the only 'healthy' form of sexual intercourse happens between male and female partners.[52] In their 1979 book, *Homosexuality in Perspective*, they championed the right of gay men and women to get treatment for their 'problem'. Many readers were, appropriately, appalled that the pair suggested 'cures' for being gay.[53]

Johnson's attitudes to feminism or homosexuality are just some of the inconvenient facts omitted from her image as an empowered and emancipated research scientist. Even the more recent Netflix show skims over the inequality between her and Masters established from day one. In the show, Johnson is a sexually liberated proto-second-wave feminist who separates sex from love, and so is unfazed when, in the early days of their studies, Masters asks her to 'participate' in the study *herself*; not just as a researcher but also as its subject, by having sex with him.

In the biography on which the Netflix show is based, however, we read that Johnson was reluctant at best. 'No, I was not comfortable with it, particularly… I didn't want him at all and had no interest in him.' Her job was on the line; it was clear to her that sex with Masters was a condition for their ongoing collaboration: 'I had a job and I wanted it,' Johnson says, and conceded that in more modern times 'it might have been sexual harassment'.[54] The pair may have been the face of a new kind of equality, but this was not Johnson's personal reality.

Johnson took her characteristically pragmatic approach to the relationship and married Masters in 1971. Stuck in another sexist bind, she turned marriage, once again, into a choice. The highly publicised wedding was a business arrangement. It helped consolidate Johnson's expertise as a sexologist and therapist, as a married woman herself, and their shared brand as sex and relationship experts. Masters left his wife of twenty-nine years to secure their image.[55] In time, though, both halves of the research team publicised their lack of love for each other. Though only one could make the decision to leave. Masters eventually fell in love with someone else and asked for a divorce. They agreed to continue to run their clinic, which was Johnson's primary source of income, and arrived at a practical arrangement that was somewhat agreeable to both.[56] But Johnson's solo ventures failed to take off without Masters' credentials to substantiate her expertise.[57]

Johnson's life, her relationships and her career all reflect the ambivalences of a changing public consciousness around sex. Masters and Johnson won new-found acceptance for academic investigations into sex. They turned their lens to a human experience previously hidden from science and discovered a whole new set of experiences. But they paid little attention to *how* they were looking. They only interrogated their social beliefs when the science forced them to. Unlike McLaren or Goodall, Johnson did not actively try to steer her innovations for the social good. She straddled that line, honed from childhood and through the Second World War, between respectability and rebelliousness. She worked under the authority of men, finding agency where she could. In the end, she was met with hard evidence of what she had always felt: that society would only accept her as one half of a pair. Being a mistress of sex, it turns out, isn't as easy as being its master.

18

Youyou Tu (1930–)

Malariologist and Pharmaceutical Chemist

In 2015, in an era of Western pronouncements and apologies over their histories of colonialism, Youyou Tu became the first scientist from mainland China to be awarded the Nobel in Physiology or Medicine. She shared the prize for her work on a new malaria drug with the Japanese scientist Satoshi Ōmura, who worked on a treatment for roundworm parasites and river blindness. The selection was supposed to draw attention to non-Western scientists tackling diseases that disproportionately affected people in developing countries. Tropical medicine has long flown under the radar, underfunded and deemed lower priority by pharmaceutical companies, so the Nobel committee had hit upon an important research agenda. However, the framing missed the far less worthy context of Tu's work.

The truth is that the revved-up scientific machine of the second half of the twentieth century zoned in on malaria only when it began to affect military populations. From 1967, Tu joined the Chinese effort to find a cure for malaria as it waged a proxy war with the US in Vietnam. She was included, nonetheless, in the story of Western science. She stayed there as a strange and rare anomaly in a system that strained to include her. Her presence draws our attention to the contradictions that are forcefully assimilated in the progress-driven history of science: like the war that drives global powers to save lives in the process of taking them, or the authoritative objectivity hiding a myriad of decisions about what to value and who to ignore. The story of Tu and her world-class drug for malaria is not a story about what should have happened, but of what did.

*

Youyou Tu was born on 30 December 1930 in Ningbo, a port city just to the south of Shanghai. It was a turbulent time; China was on the brink of a period of civil wars (1930–37), a war with Japan (1937–45), and a later three-year civil war between the communist and nationalist armies (1945–49). Tu was lucky enough to be born in a relatively stable region of the country to a wealthy and well-educated family. Her father worked in a bank, while her mother stayed home to tend to Tu and her four brothers. Her family sent her to the best schools in the region. Tu was part of a privileged elite whose daughters were encouraged to study.[1]

Her parents also gave her a prophecy. Her father had named her Youyou, which was adapted from a sentence from the Chinese *Book of Odes* that describes how 'Deer bleat, *youyou!*' while they eat a substance, called 'hao' or 'qinghao' – the very plant from which Tu would later extract the breakthrough malaria treatment that would make her name in the wider world.[2] It's hard not to read fate into this (although we won't, because we are scientists!).

At sixteen, Tu caught tuberculosis, which took her out of school for two years. The experience made her want to pursue medicine.[3] On graduating from high school, she took the university entrance exam and was admitted into the department of pharmacy in the medical school at Peking University. Tu attended the university at a time when new investments were being poured into university's infrastructure and curriculum.[4]

This included appointing professors who had received educations and advanced degrees in Western countries. Western education and science connoted the most advanced science, while, at the same time, China was working hard to assert itself as a scientific power. This ambivalent relationship with Western science, both emulated and rejected, is still palpable in the nickname the media gave her when she won the Nobel Prize. In China, she was dubbed the 'three noes'. She had no medical degree, no doctorate and had never worked overseas.[5]

Tu's career would come to encompass traditional practices that had been excluded from the 'modern' science associated with the West. Despite her westernised training, Tu's interest in Chinese traditional medicine grew. She had been treated with folk recipes as a child, but this was her first in-depth introduction to the topic. One professor in particular, Lin Qi-Shou, one of China's most respected phytochemists, wanted to impart traditional practices to the next generation. He instructed Tu and her fellow students on how to select the appropriate extraction solvents to determine the structures of the chemicals within folk remedies. Another professor, Lou Zhicen, taught them about the origins of medicinal plants and showed them how to classify, distinguish and identify these plants based on their botanical descriptions.[6] These courses applied the tools and language of modern science to show how ancient herbal medicines worked. This is a tried and tested method. A Western example is acetylsalicylic acid, or aspirin, derived from the precursor molecule salicylic acid, found in the willow tree. Willow was used by the Assyrians and Babylonians to treat pain and inflammation.[7]

After graduating from university in 1955, Tu was sent to work at the Academy of Traditional Medicine. This was a programme run by the China Ministry of Health to compensate for the dearth of traditional Chinese medical practitioners compared to physicians.

The government set up initiatives aimed at improving healthcare services by combining Western and traditional Chinese medicine. The goal at the academy was to search for medication to treat disease, particularly those that remained uncured by Western medicine. It was an interdisciplinary affair, where traditional Chinese practitioners and historians worked alongside chemists, pharmacologists and physicians.[8] Tu ended up on a full-time training course in the theory and practice of traditional Chinese medicine. Here, she learned the skill of 'processing' natural remedies. This traditional chemistry was a knowledge base of methods developed over thousands of years to harness the properties of naturally occurring substances. Tu learned how to use washing, cutting, drying or stir-frying with adjuvants, like wine, vinegar or honey to increase the potency of compounds or reduce harmful side effects.[9]

But academics had difficult times ahead. By 1966, Mao Zedong had launched his Cultural Revolution. This was a government-enforced regime geared towards expelling capitalism and forging a socialist society. Many (especially Western-trained) scientists and intellectuals were publicly humiliated, persecuted, sent to 're-education' farms in the countryside or even driven to suicide or 'disappeared'.[10] Scientific publications were banned.[11] However, in the midst of this terror, one project was protected because the government had an urgent scientific problem to solve. North Vietnam, a key ally at war with the US, had appealed to China to help reduce the toll of soldiers dying from malaria, a lethal infectious disease spread by mosquitos. Not only was malaria killing more soldiers than the Americans, it was also devastating the population of southern China. The government needed a solution. Mao set up Project 523, named in utilitarian communist style after its starting date, 23 May 1967. Over the next fourteen years, the project would recruit over 500 scientists from sixty military and civilian institutes.[12]

Mao needed someone to lead the project.[13] The government turned to the Academy of Traditional Chinese Medicine in Beijing for a capable scientist.[14] In 1969, two directors and another member from the National 523 Office visited the Academy of Traditional Chinese Medicine and the Institute of Chinese Materia Medica, seeking help with their hunt for novel Chinese remedies. The academy's leadership team put Tu in charge of the Project 523 research group.[15]

Tu was lucky that her research expertise aligned with government aims. In propaganda, Mao was pushing for greater gender equality, pronouncing, 'Women hold up half the sky.' But the real impact of Mao's vision was less decisive. Women were still less likely to gain high-ranking positions, were generally regarded as weaker and less capable, and continued to do most of the housework and childcare.[16] Amidst the persecution of academics and the ongoing oppression of women, Tu's appointment was remarkable.

Tu, by this point, was a mid-career scientist who knew that finding a cure for malaria would prove exceptionally difficult. In the race for strategic advantage, scientists in the US and China had already screened 240,000 compounds without any positive results.[17] And this war-fuelled quest was far from new.

The first treatment for malaria in the West was quinine, a substance in the bark of the cinchona tree. The substance was brought to Europe from South America by Spanish colonisers in the 1600s, where people in the tropical parts of the Andes had long known its medicinal properties.[18] In 1820, the French chemists Pierre-Joseph Pelletier and Joseph Bienaimé Caventou extracted the active ingredient, quinine, from the cinchona.[19] British and Dutch botanical explorers scoured the Andean forests for more of the trees, competing to establish plantations in India, Ceylon and the Dutch East Indies.[20] Spain held a monopoly over production at first, and later the Dutch secured their hold over plantations in Indonesia in the early 1900s.[21] But it was the First World War that animated the next stage in research. The Allies controlled quinine stores in Java. The German army in East Africa suffered heavy malaria casualties as a result. After the Armistice, the German government wasted no time in commissioning a search for a quinine substitute. Chemists at the newly established headquarters of IG Farben in Elberfeld tested thousands of compounds.[22]

In 1934, the chemist Hans Andersag at IG Farben found a workaround, when he synthesised chloroquine, which worked in the same way as quinine did, in the lab.[23] Andersag collaborated with the French firm Specia in North Africa to conduct the clinical trials. When the Allied forces landed in Tunis in May 1943, the Americans discovered the reports on the research and sent it on to the US to conduct its own trials.[24] Chloroquine proved ten times more effective than quinacrine at the same dosage and had fewer side effects.[25] It was introduced into clinical practice in 1947. However, chloroquine soon lost its effectiveness. Various malarial parasites were fast becoming resistant. The same was true for the only other line of attack, the insecticide dichlorodiphenyltrichloroethane (DDT), developed in 1939. DDT's widespread use as an agricultural pesticide after the Second World War in particular quickly diminished its effect on mosquitos. When Tu began her work, there was once again an urgency for alternatives, in another martial incubator.[26]

Tu, like these Western malaria scientists before her, was working in the context of war, only from the 'other' side of the divide, drawing on a tradition of medicinal practices that only she, as a Chinese scientist, had access to. She began by visiting traditional medical practitioners across China. She took notes of her conversations in a notebook, 'A collection of single practical prescriptions for anti-malaria'.[27] At the same time, she began searching ancient texts for remedies for the fever that was a symptom of malaria. She and three assistants scoured every archive they could find for recipes. Their search took them away from the lab and into the archives. As Tu later described: 'Most ancient texts were readily available from the institute, some from the university libraries and even the museums. Traditional Chinese medicine is a real "treasure house".'[28]

Tu and her team catalogued more than 2,000 traditional Chinese recipes and then compiled a list of the ingredients that appeared most frequently in fever treatments. After two years, they had reduced their list to 380 herbal extracts that they planned to test on mice (including some commonplace condiments like pepper). One of the recipes contained *Artemisia annua* L. (qinghaosu in Mandarin),[29] derived from the common Chinese plant sweet wormwood. The earliest written record of a qinghao treatment was discovered on a piece of silk unearthed from the Mawangdui Han Dynasty tombs (168 BCE) and prescribed the herb as a treatment for haemorrhoids. Another *Compendium of Materia Medica* by sixteenth-century herbalist Li Shizhen of the Ming Dynasty, however, prescribed it for intermittent fevers in humans.[30] In the 1950s, officials in rural parts of China had brewed qinghao tea to treat infections. Still, it hadn't been studied scientifically.[31]

That meant that, when Tu and her team tried to apply the wormwood to mice, they couldn't explain why its powers seemed to evaporate.[32] They tested *Artemisia annua* to see if it could suppress the parasite that carries malaria in rodents, *Plasmodium berghei*. Tu recalls, 'It was a very laborious and tedious job, in particular when you faced one failure and another. This was the most challenging stage of the project.'[33]

Tu prepared her extracts using a range of methods, including boiling and ethanol (in both cases, to separate the active compound from the rest of the plant material). By 1971, she started to see that some of her formulas did reduce the parasite count in the mice – by about sixty-eight percent.[34] This was positive, but the results were unreliable, ranging from twelve to forty percent when the team repeated the experiments.

Tu and her team could not explain these fluctuating success rates. Tu turned to an ancient text for the explanation science couldn't provide. The recipe, by the fourth-century physician Ge Hong, was included in the *Handbook for Prescriptions for Emergencies*. The directions were to soak one bunch of wormwood in water and drink the juice.[35] Tu had been boiling the wormwood. Perhaps that damaged the active ingredient. She made another preparation using an ether-based solvent, which boils at 35°C.[36]

Tu had discovered a new method of extraction, but there were other questions the Ge Hong text left unanswered. Qinghao is a general name of a category of herbs in traditional Chinese medicine, but there are six types of herbs within this category. Ge Hong's handbook did not specify which type of qinghao should be used to treat fever. Tu showed that just one type, *Artemisia annua*, worked against malaria. Her team also specified the geographical locations where *Artemisia annua* grows and discovered that only the fresh leaves of the herb contained the active compound they needed. They even determined that it was best to pick the flowers in spring, before the plant flowered. 'All these uncertainties', Tu said, 'caused confusion and inconsistent results in our early stages of research.'[37] These nuances were excruciating but pivotal. With them in place, the team found an extract that, when tested on mice and monkeys, proved 100% effective at reducing the presence of malaria in the blood, as well as reducing the symptoms.[38]

But would it work in humans? The Cultural Revolution had erased the facilities required for clinical trials. Tu and her colleagues decided to test the drug on themselves. Tu gives a sober retelling of the risky decision: 'There was no formal procedure at that time in China regarding safety and clinical evaluation.'[39] No one was going to stop them. They figured out a method that would allow them to progress safely, but also, she admits, quickly: 'Truly, the real motivation was our eagerness to prove the clinical efficacy of this rarely found remedy as soon as possible.'[40]

Luckily, they experienced no side effects. The team felt ready to progress the clinical trials. They tested the drug on twenty-one labourers who had contracted malaria in Hainan Province, an island off the south coast of China.[41] Their fevers subsided; in about a day, the parasite disappeared from their blood.[42] The drug had worked. Tu reported her team's success to the academy in March 1972.[43]

The team had created an effective drug, but they still hadn't isolated *Artemisia annua*'s active substance in its pure form. Tu and her team identified the molecule and deduced its formula, following the same laborious processes of the other crystallographers we have met in this book. The chemical structure opened the way to creating synthetic forms of the natural extract. Tu created an artificial version of the extract in 1973 to confirm the structure and prove it had a hydroxyl group. Her artificial version turned out to be even more effective.[44]

It took until 1978 for the team to publicise their results in China and another year for their work to appear in an English-language journal.[45] They had to wait for the ban on publication to lift – for the Cultural Revolution to end. Even then, the authors remained anonymous to abide by the collectivist ethos in China, though the practice also accurately reflected the collaborative effort.[46]

The paper's anonymity, however, made it difficult for the work to register in the international scientific community. For a long time, it was not widely known who was behind the discovery of the drug that became known as artemisinin. In a scientific system that was shaped by individualistic, Western approaches to valuation, it was difficult to know how to recognise this contribution to global, not just Chinese, science and health. International awards went to individuals, sometimes two or three collaborators, but not teams.

The importance of the discovery of artemisinin was officially recognised in the West in 2011. Louis Miller and Xinzhuan Su, two malaria researchers at the US National Institutes of Health (NIH), had started digging into the drug's history in 2005.[47] It wasn't clear to them who they should credit for the discovery. In lieu of a publication record, they reviewed letters, researchers' original notebooks and transcripts from once-secret meetings. The credit, they decided, should go to Tu.[48] Tu had been the principal investigator at the Academy of Chinese Medical Sciences. She was the researcher who had led the team in their literature reviews and experiments. She had been the voice of the project when she presented the findings in 1981 to a World Health Organization visiting study group on chemotherapy for malaria in Beijing.[49] In 2011, Tu received America's top medical prize, the Lasker Award (sometimes called the 'American Nobel').[50]

Not everyone agreed with the decision. Some malariologists, both Chinese and Western, were outraged. Many argued that the discovery of artemisinin had not been an individual effort, and that the prize should have recognised the clinical trial leader, Professor Li Guo-Qiao, at the very least. Li had worked with the malaria researcher Dr Keith Arnold at the Roche Far East Research Foundation in Hong Kong to test artemisinin and published the first high-profile clinical trial of the treatment.[51] Li and Arnold also conducted important studies showing that mixing artemisinin with slower, more persistent drugs improved success rates. This so-called artemisinin combination therapy is now the most effective malaria treatment.[52] I spoke to Dr Nicholas J. White, a malaria researcher at the University of Oxford and Mahidol University in Thailand, who has been one of the voices arguing the discovery was collaborative and should have been credited as such. He argued that the accounts of Tu's rise to acclaim omit the contributions of her collaborators in and beyond her immediate research team. He added that Professor Li Guo-Qiao was offered the Lasker in a joint award but declined, because he couldn't share the prize with everyone who contributed.[53] Even Dr Xinzhuan Su said to me about his decision to nominate Tu for the Lasker that, 'If possible, I would nominate the team for the award.'[54] It is clear that the artemisinin research was a team effort. It is also true that, while other scientists felt it inappropriate to accept the award, Tu did accept it.

The argument that the malaria research was a collective effort is, of course, perfectly reasonable. But this is not unique: we've seen many examples of very collaborative science in this book. I am also wary of admonishing a non-Western woman for claiming a rare moment of institutional recognition. Tu did no more or less than a vast history of other scientists did by claiming her prize.

The debates around Tu's prize could have prompted a broader discussion around the conventions for awarding scientific accolades in the West. Instead, gatekeepers funnelled her into the same old self-justifying Western narrative of science. The same was true when it came to the traditional Chinese aspects of her science. In 2015, Tu became the first Chinese woman to win the Nobel Prize. Tu didn't even have a medical degree, and yet the committee made a point of diminishing the role of Chinese medicine in its press announcements. The award was not going to a discovery in 'traditional' practices, they clarified, but to a scientist who, inspired by them, went on to use 'sophisticated' (read: modern and Western) research methods to develop a 'new' therapy for malaria.[55] Tu could not be recognised in the context of the intellectual tradition that had shaped her approach.

Tu sits on the margins of a Nobel history, somewhat uncomfortably, as a scientist who worked for the opposing camp in a history of 'productive' divides. We cannot separate her work from the military machine. Nevertheless, it has also saved millions of lives. While malaria is still a global health challenge, artemisinin is a cornerstone in modern treatment. Together with its derivatives, the drug has saved approximately twenty-two percent of the 663 million averted clinical cases.[56] While other scientists had their sights fixed on the next frontier, Tu proved the benefit of looking backwards. Today, around forty percent of common pharmaceutical products have been developed using traditional knowledge. Some of our most celebrated modern scientific developments draw on long-standing community-based health practices. The smallpox vaccine enacts the principle of variolation, where samples from smallpox sores are transferred to healthy people, producing milder forms of the illness with lower mortality rates. Variolation was used in Asia and Africa as early as 200 BCE. The practice of exposing babies with jaundice to sunlight, today the recognised medical treatment of phototherapy, is an Ayurvedic practice and so belongs to a system of medicine that emerged in ancient India. Modern scientists have unearthed these methods and explained their mechanisms.[57] As we excavate the remedies of our ancestors, maybe we'll find other patterns too: the cycles of conquests and defeats, of cures and resistances, of the knowledge lost between the cracks of broken parts as we delineate victors, giants and eras. Perhaps we'll ask ourselves what it would take to cultivate a science wise with history, as well as progress.

19

Flossie Wong-Staal (1947–2020)

Virologist and Molecular Biologist

'Rare cancer seen in 41 homosexuals,' reported the *New York Times* on 3 July 1981. 'Outbreak occurs in New York and California – 8 died inside 2 years.'[1] Kaposi's sarcoma, a rare, soft tissue cancer characterised by purplish lesions, was clustering in the US's gay communities. Doctors were baffled. Kaposi's sarcoma predominantly affected older men; these patients were young and healthy. Initial reports used the term gay-related immune deficiency (GRID), or 'gay cancer', to name the unknown entity driving a rapidly escalating death toll in the US,[2] but doctors soon saw the same symptoms in other communities: haemophiliacs, recipients of blood transfusions, intravenous drug users and Haitian immigrants. Nevertheless, it was the gay community who bore the brunt of the epidemic – and the stigma. By 1982, almost 300 Americans had died, and it was fast becoming a worldwide epidemic. In September of that year the Centers for Disease Control and Prevention (CDC) rechristened the disease: acquired immunodeficiency syndrome or AIDS. Nobody understood the cause.

However, as Dr Wong-Staal and her employer, Dr Robert Gallo, at the NIH took in the events, they were struck by how closely the virus seemed to resemble the one they had just identified a few years previously. In 1980, Gallo and his colleagues had discovered the first human retrovirus, human T-cell lymphotropic virus 1 (HTLV-1), a type of virus that invades the nucleus in a cell and then inserts its genes into the DNA of the host. Until then, many scientists had thought retroviruses existed only in animals. At thirty-three, Wong-Staal had worked alongside Gallo for almost a decade, providing the definitive molecular evidence that the retrovirus causes cancer. The virus driving the new epidemic seemed to have a lot in common with HTLV-1.[3] They set out to investigate.

*

Born as Yee Ching Wong on 27 August 1946 in Guangzhou, China, to the businessman Sueh-Fung Wong and housewife Wei-Chung Chor, she was the third child of four. In 1952, when Wong-Staal was just five, her family fled to Hong Kong in the wake of the Chinese communist revolution. Once there, she attended a Roman Catholic girls' school run by American nuns, who taught in English and followed the British schooling curriculum. Seeing her academic potential, the nuns asked Wong-Staal to choose an English name to prepare her for international opportunities. Wong-Staal wanted something that would set her apart. She asked her father for suggestions. He came up with Flossie – a little disconcertingly, after a hurricane that had recently hit the region.[4] Of course, it wasn't destructiveness that her father wanted to impart to his daughter. In the end, though, the name aptly predicted the storms Wong-Staal would endure throughout her career, as well as the surprising ways she found to embody this force of nature.

While she had two brothers, Wong-Staal was the only one of the children to go to college. As unusual as this was, her parents 'never had the concept that girls should not have higher education'.[5] Wong-Staal believed her mother was frustrated at not having had the opportunity herself and was determined to rectify this for her daughter.[6]

Wong-Staal was an academically gifted child and loved reading as much as she did science. But high schoolers in Hong Kong had to choose their path early on. Science was generally considered the more ambitious choice. Wong-Staal explained: ‘Part of the mentality is that if you’re smart, you should go into science. So it’s almost like you’re told that you should go into science.’ People accepted the advice, she added, ‘because they feel it’s an honour and a privilege’ to be regarded as sufficiently gifted.[7]

Wong-Staal enrolled at the University of California Los Angeles at eighteen. She applied for universities in the US, rather than the UK, in part because she’d been exposed more to American than British culture while in Hong Kong and especially, of course, the West Coast America showcased in popular entertainment.[8] She hadn’t just chosen a cultural hub but a scientific one too. She was going to study molecular biology.

By the 1960s, the molecular revolution was in full swing. Franklin, Watson and Crick had established the structure of DNA, which also explained the mechanism by which DNA replicated itself to transport its instructions between generations. That said, the molecule remained long and chemically monotonous, making it difficult for scientists to handle and identify individual genes. Scientists couldn’t study DNA directly.[9] But over the course of the next decade, a string of tools and insights from across the fields of chemistry, physics and embryology transformed DNA from one of the hardest macromolecules to analyse to one of the most workable. Starting in the 1950s, bacteriologists had identified so-called restriction enzymes that could cut DNA into manageable fragments. By the 1970s, they had found ways to replicate the strands of DNA, so-called molecular cloning, and to ‘amplify’ the DNA they were studying, reproducing the fragments in sufficiently large quantities to analyse. DNA sequencing followed. Scientists could now determine the order of the base pairs – those four letters binding the strands of the helix that allow DNA to replicate itself and carry its instructions to the next generation. DNA sequencing was the gateway to understanding the role of specific genes, tracking its function and, in the case of harmful or dysfunctional genes, developing treatments to target it. DNA sequencing was at the heart of the Human Genome Project, launched in 1990, with its lofty aim to sequence and map the entire human genome (i.e. to determine the sequence of the base pairs of human DNA and identify all the human genes). The enthusiasm around the project rested, once again, on the potential therapeutic applications that defining the code would bring. The work would allow scientists to study how genes functioned and also to target the genetic basis of the biggest health threats of the century, like cancer, and to develop genetic tests and treatments for inherited disorders.[10] Together, these techniques opened up a new era in which scientists were working very differently, not just observing or studying life, but altering it in order to understand it. *Using* biology as a tool in the lab and in medicine.

Wong-Staal was entering into a field that had become, in just a few decades, a playground for biologists. She completed her undergraduate degree in bacteriology in 1968, graduating cum laude, and moved straight onto a PhD programme in molecular biology, which she finished in 1972. The mode of study reflected the interdisciplinary nature of the field. 'At UCLA', Wong-Staal explained, 'molecular biology was not a department. It was an intradepartmental institute.' She interacted with chemists and microbiologists, exchanging ideas in the same way working scientists in the field did.[11] She also met her first husband, Stephen Staal, there: a fellow scientist. You'll have noticed how scientists often find love through their work. Given the hours spent working and shut off from normal life, it's no surprise that connections are forged. The couple spent a short time at the University of California San Diego, after which they both found opportunities at the NIH in 1973.[12]

Wong-Staal was interested in several of the appointments at NIH, but one stood out. Dr Gallo's lab was working to identify a pathogenic, that is, disease-causing, human retrovirus. Retroviruses, like other viruses, reproduce by infecting a host cell and hijacking it. Retroviruses differ, in that they insert their own genes into the host cell, permanently altering its DNA.[13] Studies on animal retroviruses were in full swing, and there was much excitement about the insight they could give into carcinogenic mutations, or as a model (another molecular tool: a kind of test ground for human biology). They were not, however, considered to be *human* pathogens. Wong-Staal started off working with retroviruses in primates as models for humans. What could these show about the way retroviruses might work in human bodies?[14]

The NIH, at this time, was another exciting place to be. Researchers weren't required to teach or write grant applications as researchers at universities were; they got to dedicate all their time to research. It was, as Wong-Staal herself describes it, a veritable Mecca for biomedical research, especially as a young researcher.[15] She thrived in the lab, quickly establishing a reputation for being a bit of a 'whiz kid', very skilled at manipulating the components of living things like DNA and proteins. This was a skillset that Gallo desperately needed in his team.[16]

Both Gallo and Wong-Staal believed, against convention, that retroviruses *did* infect humans, just as they did other animals. There were so many variations of the virus: why wouldn't there be a human form too? Other scientists generally didn't share their sense of possibility. If there were human retroviruses, they argued, they would have been identified by now.[17] Gallo's lab was in the minority.[18] Wong-Staal remembered: 'Even the really well-respected scientists –particularly the well-respected scientists – were very strongly against the idea that human retroviruses existed.'[19]

But by 1979, Gallo had found evidence of the first human retrovirus, HTLV-1, in the cells of a leukaemia patient. Once he had isolated it, Wong-Staal (by now the head of her own group in the lab) set to cloning the virus and sequencing its genome.[20] This would allow scientists to identify the virus in infected patients, track how it evolved and was transmitted, and to develop treatments and vaccines. Wong-Staal studied various parts of the virus and showed how it interfered with human DNA to activate cancer-causing genes called oncogenes. She had identified the mechanism by which the virus caused cancer in humans.[21]

In 1982, the same year Wong-Staal was made chief of the molecular genetics of hematopoietic cells section, Dr James Curran, the doctor leading the task force on human immunodeficiency virus (HIV)/AIDS at the CDC, visited the NIH to present on the new epidemic recently labelled AIDS. He urged Gallo to launch a research programme to identify the virus behind the disease.[22]

Wong-Staal, as chief of molecular biology, was intimately involved. Given the similarities between HTLV-1 and AIDS, Gallo was pretty sure that they were looking for a retrovirus. Like HTLV-1, AIDS infected T-cells, a type of white blood cell involved in the body's immune response. They noticed, as Wong-Staal put it, a 'yin and yang phenomenon', where T-cells were being depleted as the virus proliferated, just like HTLV-1. In short: 'AIDS "smelled" just like a retrovirus.'[23]

And it does make you wonder what would have happened if the lab hadn't already found the first human retrovirus a year before. Without that discovery, maybe their line of inquiry wouldn't have been deemed credible. As Wong-Staal herself said in a 1997 interview:

> The discovery of the first human retrovirus, HTLV-1…broke the barrier of credibility. Right? I mean, before, people would not have even entertained the idea that a retrovirus could be involved. But the fact that a human retrovirus that targets particular T-cells had been identified made a retrovirus as the possible cause of AIDS more plausible.[24]

Had they not taken this step, treatment would have been out of the question. Epidemiologists may have figured out how the virus was being transmitted, through population-level studies, and would probably have taken preventative action, but the mechanism underlying transmission would have taken much longer to figure out.[25]

As devastating as the effect of the virus was, this was another case in which widespread human disaster facilitated an electrifying time for science. Wong-Staal afterwards referred to this time as the 'highlight' of her career and as a 'period of intense discovery'.[26] 'Working with this virus', she said, '[was] like putting your hand in a treasure chest. Every time you put your hand in, you pull out a gem.'[27] We have seen repeatedly throughout this book how the connection between public crisis and scientific advancement is not a mysterious but a material one. More resources are pumped into research programmes and, surprise surprise, they make discoveries. In this case, it moved the marginal field of retroviruses to the centre of scientific interest. Funding made the work more attractive to scientists. As it did so, widespread dismissiveness towards human retroviruses turned to fierce competition.

But it took a while. While Wong-Staal's niche research interests garnered new-found public relevance, the US government waited a long time to fund AIDS research adequately. This sense of inertia is exemplified in AIDS activist Larry Kramer's 1985 play *The Normal Heart* (adapted for film in 2014), where Dr Emma Brookner – inspired by Manhattan physician Linda Laubenstein, who encountered the first cases of AIDS in 1981 – fiercely admonishes a government board for failing to fund necessary medical research on the virus:

> Your National Institutes of Health received my first request for research money two years ago. It took you one year just to print up application forms. It's taken you two and a half years from my first reported case just to show up here to take a look…it is an unconscionable delay and has never, never existed in any other health emergency during this entire century. While something is being passed around that causes death.

The virus did not discriminate, but, of course, it was known in those early years as a 'gay cancer'. Under Ronald Reagan's administration, right-wing pressure groups seized on AIDS to spread homophobia. The televangelist Jerry Falwell, who led the Moral Majority, popularised the view that AIDS was 'the wrath of God' enacted on gay people. Pat Buchanan, the White House communications director from February 1985 to March 1987, described the crisis as nature 'exacting an awful retribution against gay men'.[28] In the face of this callous bigotry, the playwright William H. Hoffman compared the epidemic to the Holocaust, writing that 'as with the present situation, the general population by and large viewed the victims with a mixture of fear, hatred and indifference'.[29] The National Association of People with AIDS (NAPWA) were forced to spell out the rights people with AIDS deserved at the National Lesbian and Gay Health Conference in Denver, Colorado, including the basic rights 'to privacy, to confidentiality of medical records, to human respect'.[30] In the midst of a mounting atrocity, these patients had to fight to affirm their humanity.

Reagan did not publicly acknowledge the AIDS epidemic until his friend, the actor Rock Hudson, died in 1985. That year, Congress allocated $70 million to AIDS research. Four years previously, researchers at the CDC had proposed allocating $40 million to pre-empt the disaster; they received less than $1 million. In 1981, the death toll at the end of that year was 74. In 1985, it was 12,529.[31]

Reagan now claimed that AIDS was one of his administration's top priorities.[32] By the final year of his presidency in 1989, his government was investing $2.3 billion into HIV/AIDS research and prevention.[33] But his delayed response cost thousands of preventable deaths. His biographer, Lou Cannon, called his response to the epidemic 'halting and ineffective'. The stats are hard to argue with. 'There were only 199 reported cases of AIDS in 1981,' Cannon writes. 'Eight years later more than 55,000 persons had died from this new scourge, exceeding the total of U.S. combat deaths in either the Vietnam War or the Korean War.'[34]

Within the scientific community too, politics seemed to distract AIDS researchers from the lives at stake. There was, in Wong-Staal's own words, 'a lot of collaboration, but also a lot of competition'.[35] One of the first things Wong-Staal herself did in 1984 was to clone and characterise the AIDS virus. The results were published in *Nature* in November 1984.[36] A month later, researchers at a lab in France reported the cloning of a different virus they were working on.[37] The next milestone was the sequencing of the genome of the virus. Wong-Staal and her team published in January 1985.[38] The same month, another lab published their sequencing results.[39] It was a confusing time, as various groups across the world were coming up with genes and calling them different things, making it difficult, as we'll see, to know where the research stood, and who had made any given discovery. We've seen throughout this book how the discovery is, problematically, the dimension of science we tend to recognise – the breakthrough, the first, the event – rather than the painstaking, meticulous and, often, collaborative work over time. We credit the Watson and Cricks; not the Franklins or the McLarens or the Kelseys of the scientific world. Everyone in the AIDS field knew that. With a sense that an important discovery was about to be made, every lab put its hat in the ring, trying to be the first. Scientific journals too, rushed to accept papers so that they could be the first to publish breakthrough results on a type of virus that, just a few years prior, nobody even believed existed.[40]

It was one of those scramble moments in science that came with the stealthy competitiveness that Venki Ramakrishnan, in his description of the race to determine the molecular structure of the ribosome (the gene-reading machine that processes the DNA in a cell), compares to the premise of the 1963 film *It's a Mad, Mad, Mad, Mad World*. In the film, a group of motorists all feign disbelief on hearing a dying man's confession hinting at hidden treasure, while secretly turning against each other to find it.[41] And yet, to me, the logic exemplified by the comedy film made at the height of the Cold War is only a continuation of that same competitive mode of scientific production that we have watched emerge from the Second World War: an approach born from the goal of dominance, and so with the potential for destruction written into its DNA.

Wong-Staal herself later described the AIDS research as 'like the Manhattan Project',[42] with different labs working towards a common aim – and vast rewards. Wong-Staal found herself increasingly at the heart of another man-made nuclear storm cloud. This time, it was scientists fighting for acclaim. During the race, Wong-Staal tried to remain collaborative. AIDS, as she saw it, was an interdisciplinary problem. Some institutes, she said, were 'more involved in the immunological aspects of AIDS, and certainly we were not immunologists'.[43] Each lab approached the same problem from their own area of expertise. She joined the scientists who chose to continue to share samples, information and publications in the spirit of improving understanding and finding the best possible solutions. She consolidated that nexus at the centre of an accelerating field.

And yet, she did defend the importance of her own contribution. Wong-Staal maintained, for example, that her lab was the first to produce a molecular clone of the virus. They then showed that these copies functioned in the same way as the viruses isolated from patient DNA; the clones depleted T-cells in culture in the same way. This was strong proof that the virus acted as a retrovirus. Unlike a patient sample, the clone in culture had no other genetic information associated with it. In this isolated context, this molecular clone did exactly what Wong-Staal thought it would. This was decisive evidence that it was this particular virus behind the AIDS epidemic.[44]

But as much as cloning helped to prove the existence of what came to be known as HIV, it also revealed the obstacles to doing anything to help the patients who were dying: namely, its genetic diversity. It had become clear that HIV was a slippery beast: not a single genetic entity. Every sample the team sequenced had slightly different genes. This meant that there was genetic variation of the virus between individuals. But what was especially important, and what Wong-Staal showed, was that there were variations in HIV *within* individuals too; different clones from the same patient would have different genes. HIV was a so-called quasi-species. Not only did it have different strains, but the species could drift; its composition could change over time, under different immune pressures.[45] Wong-Staal and her team demonstrated this variation in HIV at the molecular level, whereas other groups had shown it using immunological data.[46]

Wong-Staal was utterly absorbed by the virus: not only was it a novel virus, it was a very interesting and complicated virus. That meant that there were a lot of discoveries to be made.[47] That very same complexity made it difficult to treat. Wong-Staal's scientific fascination was, after all, a real-world problem; HIV's genetic diversity was causing thousands of deaths. In that sense, Wong-Staal's enthusiasm could strike us as detached, as insensitive to the atrocity around her. But she exemplifies a tension that exists for most scientists. The real-world problems they work on coexist with their wonder at the minutiae they explore. Without a sense of fascination, scientists would not be animated to pursue experiments or make sense of their findings – McClintock wouldn't be inspired by her corn; Curie wouldn't be enraptured with her radium. And yet, discoveries find their way, in some form, into the world. We've seen repeatedly how the outside world is sometimes eclipsed from a scientist's immersive view. The back-and-forth between inside and out is inherent to any scientist's job; those working on the cusp of a societal crisis register the tension most acutely.

In the end, Wong-Staal's molecular biology was crucial to developing the first medication. She knew how to map out the roles of individual genes and proteins, their organisation and how they replicated. This created the possibility of targeting various aspects of the virus' mechanism. The first FDA-approved drug, azidothymidine (AZT), released in 1987, targeted the protein HIV reverse transcriptase, an enzyme involved in HIV DNA replication.[48] Later, a class of drugs known as protease inhibitors targeted HIV protease, a key enzyme that breaks down protein that combines with DNA to form a functional virus.[49] When administered with other inhibitors, this drug suppressed the replication of the virus, often to barely detectable levels.[50] Wong-Staal also applied her tools to HIV prevention. She helped to develop a second generation of blood tests that identified the HIV genome rather than antibodies in the bloodstream, allowing medical professionals to catch the virus much earlier.[51]

Wong-Staal was doing this meticulous and important work in the midst of the storms that her name prophesied: the destruction that HIV caused in the population and the increasingly heated environment giving form to AIDS research. In this new era of hyper-productive science, productivity also meant financial reward. Some researchers abused the system, drawing tenuous connections to AIDS research to access this pocket of government funding. 'AIDS money' itself became very competitive.[52]

Not just this, but the whole research area was premised on the need for real-world applications. AIDS research was, in that sense, a production line, meaning there would be products to sell. Patenting became an issue. Prior to the AIDS crisis, Wong-Staal reflected, she and her colleagues 'hardly thought about patenting'. Their 'interest was in scientific discovery and gaining new knowledge…not the commercial implications'.[53] But the team's work on HIV now offered solutions to a public health problem. They were told to file for a patent to attract pharmaceutical companies to manufacture a diagnostic test.[54]

It is here that scientists' attention began to divert more jarringly from the human crisis driving their research; while gay people were still denied their right to healthcare, and with escalating death tolls across the population, news reports showed a scientific community more preoccupied with who received credit.[55] Inside the lab, economic competition increasingly quelled the spirit of open collaboration that Wong-Staal had relished at NIH. Gallo himself had become locked in a fight that would extend over years with the head of the Pasteur Institute in Paris, Luc Montagnier, over who would get credit for discovering HIV. Gallo was adamant in his claim to the win. He had published four papers in *Science* in May 1984 in which he proved that the retrovirus he called 'HTLV-IIIB' (for human T-cell leukaemia/lymphoma virus type IIIB) caused AIDS.[56] Only exactly a year earlier, in the same journal, the French lab had reported their discovery of a retrovirus they called LAV (for 'lymphadenopathy associated virus') that they had isolated from the lymph nodes of an AIDS patient.[57] After Gallo's papers were published, DNA analyses showed that the two viruses were the same. Montagnier's lab accused Gallo of stealing samples of their virus from them that became the basis of his findings.[58] With national patents and personal pride at stake, the matter escalated into a fully fledged federal investigation.[59]

The argument escalated in 1985 when patents for the first HIV blood test were issued. If the decisive sample of the virus had come from France, did the Pasteur team have the right to exclusivity?[60] The leaders of the scientists' respective countries eventually intervened. Jacques Chirac and Ronald Reagan agreed to list the two researchers as 'co-discoverers' of the virus and to split the patent royalties equally. Gallo and Montagnier accepted the agreement (publicly, at least) and co-authored a chronology of AIDS research published in *Nature* that year.[61] For a while, the history seemed to have been settled, the credit and prize money agreed upon. Though the debate was reopened once more in 1990, when the *Chicago Tribune* published a long investigative story by the journalist John Crewdson followed by a lengthy book on the matter.[62]

Today it is generally agreed that Montagnier's group was actually the first to isolate HIV, but Gallo's lab showed the virus *caused* AIDS and produced many of the insights and techniques that made the discovery possible. Gallo was nonetheless excluded from the 2008 Nobel Prize that Montagnier shared with his colleagues Françoise Barré-Sinoussi and Harald zur Hausen, who discovered the link between HPV and cervical cancer (itself a seismic development for women's health).[63] Wong-Staal was reticent to comment on the saga but attributed the negative press Gallo received to his alleged polarizing personality.

Wong-Staal was never going to divert a field long bound to its course. That path built on an old story of scientific breakthrough, carried forward by new scientific heroes over the decades, and compounded by military goals and national interests in modern history. Of course, Wong-Staal herself didn't have a stake in the game. She was not one of the individuals who would be recognised in the story of 'the discovery', nor would she be the one to determine the research culture scientists worked in.

But she could weather the hurricane. Gallo himself described the men at each other's throats as 'bulldogs' and Wong-Staal as 'resilient' in the thick of it all.[64] In one way, she had no control over the forces swirling around her. But she made her quiet centre into a node. She continued to support researchers with the crucial tools they needed. She emphasised collaboration across the board. She was known for supporting younger researchers in her lab.[65] And like Katherine Johnson, Wong-Staal held steady, inserting her presence, offering us a hidden figure and different view onto the science that made our modern world.

Though Wong-Staal did important work in Gallo's lab, by 1990 she'd decided to move on.[66] She joined the University of California San Diego, where she continued her work on HIV treatments, mainly gene therapy and ribozymes (a kind of catalyst inside the ribosome). She also co-founded a biopharmaceutical company with her then husband, Jeffrey McKelvy, where her research interest eventually turned from HIV to finding better drugs for hepatitis C.[67]

Wong-Staal had chosen molecular biology at university because it was a burgeoning new field. Little did she know the extent to which she would live out her name's prophecy. She couldn't have anticipated that her interest in the very basics of molecules would put her at the centre of a new whirl of scientific interests; she couldn't have anticipated that she would be the most cited female scientist of the 1980s. Nor would she have foreseen a historic epidemic or the heated spats between scientists. And yet, like the crystallographers in this book, she did her meticulous and quiet work. She studied the details that filled out the explosive discoveries. Details that carried the work forward, firmed up findings, tethered themselves to other fields and expanded into real therapies for patients. She held her place in the eye of the storm.

20

Jennifer Doudna (1964–)

Structural Biologist and Biochemist

For much of the last century, genome editing was the stuff of science fiction, not scientific journals. From Huxley's *Brave New World* in the 1930s to *Blade Runner*, *Jurassic Park* and *Dune*, artists have created innumerable visions of the possible misuses of genetic engineering. But alongside these dystopian scenarios was something else: hope. What if we could tackle seemingly uncurable genetic diseases, like Huntingdon's, sickle cell anaemia or cystic fibrosis? Could we one day eliminate genetically linked early-onset Alzheimer's and cancers? The possibilities were endless and terrifying.

Then, in 2012, came a crucial breakthrough. Jennifer Doudna and her colleague, the microbiologist, geneticist and biochemist Emmanuelle Charpentier, developed a method for editing sections of DNA. They discovered that the enzyme CRISPR-Cas9 could be used as a genome editing tool. The pair made history when, in 2020, they were awarded the Nobel Prize in Chemistry for the discovery; this was the first time that the Nobel Prize had been shared by two women and no men. Through it all, Doudna has shot to fame. Unlike many of the women we have met, she worked in a context and an era where she could claim credit for her findings. Reporters dubbed her the 'Beyoncé of science'.[1] Today, Doudna is one of the most culturally significant scientists of the twenty-first century.

CRISPR can be used to gene edit plants, animals and people – sometimes to eliminate disease, optimise crop yields or modify traits in people. The possibilities for advancement and corruption coexist in news reports on the technique: the possibility to prevent disease, grow organs or reprogramme cancer patients' cells to fight their own tumours exists, alongside the kinds of scenarios Huxley predicted, where gene-editing is used to reinforce social divides.[2]

Doudna has become a spokesperson and a symbol for the hope of genome editing for human health. As a scientist at the forefront of discovery, she also registers the challenges biological tools could bring. Doudna's story is no longer just her own: it is a story about us all.

*

Doudna was born on 19 February 1964 in Washington DC. Her father, Martin, was a speechwriter for the Department of Defense, and her mother, Dorothy, taught history at a community college. The whole family moved to Ann Arbor so her father could complete a PhD in literature at the University of Michigan. When Doudna was seven, Martin was offered a job teaching American literature a little further afield – at the University of Hawaii.[3]

Doudna felt like an outsider in Hawaii, so she used books to learn about the world and figure out how she fitted into it.[4] She carried the stories with her when she explored the island – the volcanoes, rainforest and beaches. She wanted to understand how such natural diversity was possible, though she still lacked the tools to answer these questions.[5] Like Goodall, words and nature became the touchpoints of Doudna's life. She returned to both shores: for retreat, hope, discovery and sometimes to make sense of the life she had made.

At twelve, Doudna discovered a story that particularly inspired her. Her father brought her a copy of *The Double Helix*. He left the book on her bed, and she read it in a single rainy afternoon. Doudna says that she was amazed and inspired by the way in which these scientists had been able to solve one of biology's greatest mysteries.[6] Of course, it's hard not to be enthralled by a nail-biting race to a breakthrough. Maybe Doudna wanted her own hero's journey. It was a story that she would come to replicate and then outgrow.

Doudna, after all, was reading the tale of mastery over life from a different perspective. She was especially interested, she reflected, in Franklin's role in the discovery. She realised for the first time that women could be scientists too.[7] Watson's recollections of Franklin in the book may have diminished her contributions, but they gave Doudna the faint outline of that hidden figure, a faint suggestion of what was possible for a woman in science. At high school, Doudna discovered cell biology. She recalls a particular class in which a working scientist presented her research on how normal cells become cancerous.[8] The scientist, another woman, offered a tantalising glimpse of what was possible.

Doudna applied to study chemistry at Pomona College in California. She completed a summer placement at the laboratory of Don Hemmes, a biology professor and family friend who lived in Hilo. She joined the team studying the fungus *Phytophthora palmivora*, which was infecting papayas, presenting a problem for local growers. The team discovered that calcium ions signal fungal cells to grow when nutrients are available. Doudna enjoyed the work, the nitty-gritty of experimentation as well as the euphoria of the breakthrough.[9] 'It was my first taste of the thrill of scientific discovery', Doudna recounts, 'an experience that I had read so much about, and that left me hungering for more.'[10]

Doudna applied to the graduate programme at Harvard Medical School, joining the lab of the Nobel laureate-to-be, Jack Szostak. Szostak was working on ribonucleic acid (RNA), the single-stranded molecule that carries copies of DNA to the ribosomes in the cell, where it directs the process of protein synthesis; proteins are the building blocks of the human body.[11] The history of the discovery of RNA, like the molecule itself, runs parallel to DNA as an essential dimension to the process of inheritance. The biologists David Davies and Alex Rich identified the structure of RNA in 1956, just three years after DNA.[12]

RNA is obviously part of the story of how DNA becomes a complete living organism. And yet, scientific attention at the time was more on DNA than it was on RNA. When Doudna joined Szostak's lab in 1986, she was making a risky career choice. As Doudna put it: 'When I was taught biology, we learned about the structure and code of DNA, and we learned about how proteins do all the heavy lifting in cells, and RNA was treated as this dull intermediary, sort of a middle manager.'[13] As the world vested its hope in the Human Genome Project to explain everything we would want to know about how life took shape, she took a different turn.[14]

Szostak was more interested in the ontological questions left unanswered by the hunt for biology's 'Holy Grail'.[15] RNA, he believed, held a lot of those answers. He was interested in the hypothesis, proposed by Francis Crick himself among others, that RNA was the egg that had birthed the other components of life, DNA and protein.[16]

A cohort of molecular biologists in 1961 had already shown that RNA was a multi-functional molecule: it could store genetic information, and directed protein synthesis in the ribosome.[17] But to prove the primacy of RNA in evolution, Doudna and Szostak would have to show that RNA could self-replicate. If

RNA could store information and copy itself, it could act as the proverbial 'chicken and the egg' in its own creation, adding a lot of plausibility to the idea that it might be the molecule that sparked all other life.[18] The question, as Szostak posed it to Doudna, was: 'Did this piece of RNA have the chemical chops to copy itself?' He suggested she focused her PhD on answering it.[19]

Their approach was distinctly modern. The operative word in biology was no longer observation, but engineering, following an assumption that if you can make something work, you understand it better.[20] Using the tools of molecular biology, Doudna worked with Szostak to engineer a non-protein coding segment of RNA into an RNA enzyme that could self-replicate. By altering the structure and function of the molecule, they would be able to see if the phenomenon was biologically possible. This was highly advanced, technical work. The biochemist Richard Lifton called it a 'technical tour de force'.[21] The team published the results in the 1989 *Nature* paper 'RNA-catalyzed synthesis of complementary-strand RNA'. '[The work] was exciting', Doudna later explained, 'because it suggested that RNA could function as a polymerase.'[22] A polymerase is an enzyme that catalyses the production of new molecules – including DNA. RNA could plausibly have preceded the current star of the show in the story of evolution.[23] The work earned Doudna her doctorate.

Doudna then joined the lab of Tom Cech. In the 1980s, Cech had noticed, accidentally, how the RNA of the single-celled organism *Tetrahymena* cut and reassembled itself in a test tube, without an enzyme. He called the enzymes ribozymes and opened the possibility, taken up by Szostak, that RNA had sparked life on Earth.[24] The discovery that RNA could function as a catalyst won Cech the Nobel Prize in Chemistry in 1989.

Now Doudna spliced herself into this exciting lineage of ideas. This is a moment in which Doudna's story, and her retelling of it, starts to parallel her beloved *Double Helix*.

Just as the missing piece for Watson and Crick had been the experimental evidence of the structure that came from Franklin's photographs, Doudna now set to work at a lab at the University of Colorado, to use X-ray crystallography to obtain pictures of RNA enzymes. Just like Franklin had.

Only Doudna didn't have formal training in X-ray crystallography. She worked with a graduate student, Jamie Cate, who was an expert in using the technique to photograph proteins. For a while, the research stalled. They got as far as crystallising the enzymes, but then the X-rays destroyed the structures.[25]

Doudna relied heavily on Cate and other researchers to develop the techniques she knew she needed. Like Franklin, she was grasping in the dark; it wasn't even clear that RNA molecules would have a defined structure like proteins do.[26] Unlike Franklin, she was well connected at the university. Not just that, but she moved freely there and so she also benefited from the casual conversations that led men like Watson and Crick to their crucial insights.

One conversation with the fellow RNA scientist, Tom Steitz, proved pivotal. Over coffee in the lab's lunchroom, Doudna shared her troubles with her crystals with Steitz. He told her about a new technique his team had been testing at Yale of 'cryocooling' crystals (aka plunging them into liquid nitrogen). He flew Doudna to Connecticut to learn the technique. It worked. The frozen crystals held their structure under the force of the X-rays. The technique was akin to the breakthrough. 'At that point we knew that we had crystals that were ordered enough that we would eventually be able to solve the structure,' she later said.[27]

Doudna wrote up her paper containing the structural model of the enzyme and got the credit she deserved for the discovery (though she ensured that Cate, who pioneered the crucial techniques, was the lead author). She explicitly compares the 1996 publication to Watson and Crick's paper on DNA. They, too, had located every atom of their structure. And here too, the structure revealed the mechanism – just as DNA's double helix explained how the molecule stores and transmits the genetic code, their model showed how RNA could replicate itself.[28] They showed how the single strand of the RNA folds into complex three-dimensional shapes through base pairing – the same mechanism that joins the two strands of the DNA helix. These three-dimensional structures allowed the molecule to perform its function as an enzyme.[29]

Doudna had replicated the Watson and Crick story – with a difference. Her style of working was entirely different. This was a collaboration, with a woman at the heart of it, giving credit where it was due. She looked at biology differently. Doudna was working, after all, in an era of synthetic biology.[30] She was thinking from the get-go about the ways in which the biological entity she was studying could itself be re-engineered to work differently, on itself, and on other molecules in the body. That thinking about the functions of biology in the lab also had potential real-world applications. Specifically, Doudna and her team hoped that they might be able to modify RNA enzymes to act on and repair defective or harmful genes.[31] She wasn't just writing a story about an individual discovery or an individual scientist; she was oriented, by her very approach, to think about how her lab-made origin stories might unfold in the world.

By now, Doudna had received multiple awards – and still found the time to get married to, surprise surprise, Jamie Cate and have a child. She'd started a new job at Berkeley. Her prominent position led to another career-defining collaboration that, again, in stark contrast to the women we have seen, simply fell in her lap.

Jillian Banfield was an environmental researcher who had been sequencing the DNA of microbes that lived in the salty lakes of her native Australia, as well as, less idyllically, the acidic waste drained from a Californian copper mine into a salt marsh.[32] Banfield had landed on Doudna's lab using another pivotal modern invention – the Google search. She thought they might be able to help each other.[33]

In her work, Banfield kept bumping into the same section of DNA code, a sequence known as CRISPR, or clustered regularly interspaced short palindromic repeats (I'm sure you appreciate why the acronym is necessary). This length of DNA plays a role in the mechanisms that allow microbes like bacteria to protect themselves from viruses. Banfield needed the help of a biochemist to understand how CRISPR worked in combination with a complex of adjacent enzymes, or CRISPR-associated proteins (Cas; hence we get the now familiar CRISPR-Cas, accompanied by a number denoting the specific enzyme).[34]

At the time, it wasn't clear what CRISPR did. Microbiologists had come up with theories, but no one had tested them in a controlled setting. Doudna and her colleagues dissected the CRISPR system into its chemical subparts to analyse how each part worked.[35]

In 2008, Doudna had helped organise the first international conference on CRISPR research at Berkeley. The field was small, but the presentations proved that new findings were proliferating. That same year, the microbiologist Luciano Marraffini and his advisor Erik Sontheimer at Northwestern University had shown that CRISPR-Cas did not work by silencing genes, as Doudna had thought, but actively targeted and cut the invading DNA.[36]

This was more exciting than Doudna could have imagined. Sontheimer spelled out the implications: 'If it could target and cut DNA, it would allow you to fix the cause of a genetic problem.'[37] This was a direct path to Doudna's dream of turning RNA into a DNA editing tool.[38]

CRISPR quickly became less niche.

The race to be the first to claim profits began even before there was a product to speak of. Marraffini and Sontheimer filed a patent for CRISPR as a gene editing tool the year of their discovery. The patent was denied; they only had an idea, not an invention. Doudna herself was not shy about throwing her hat in the ring. In 2011, she founded her company, Caribou, to commercialise the findings emerging from her research lab. As the competition accelerated, she was sometimes self-admittedly aggressive in pushing her team: 'I knew we had to speed up, and it wasn't clear that our collaborators had the same commitment… That was a frustration to me. I'm competitive.'[39] Though Doudna had never worked with human cells, she was determined to win.

But to put the idea to work, researchers would have to understand exactly *how* the CRISPR enzyme cut DNA.

*

Emmanuelle Charpentier had been studying CRISPR too. She was less competitive than Doudna. Born in Juvisy-sur-Orge, just outside Paris, on 11 December 1968, she had completed her graduate studies at the Pasteur Institute and started her career in antibiotics research.[40] Since then, she'd opted for a nomadic career. Like McLaren, she'd refused to commit herself to a niche; she was willing to forgo Nobel acclaim (though, unlike McLaren, she got it) if it meant staying interested. In 2009, she left her position as head of the microbiology and genetics lab at the University of Vienna to travel to the far-flung University of Umeå in Sweden, best known for its research on trees.[41]

That same year, the CRISPR buzz landed on a specific enzyme. The research on Cas9 had cast the system in satisfying clarity. A key piece was understanding the role of CRISPR RNAs (crRNAs). These segments of RNA contained the genes of viruses that had previously attacked the bacteria. They acted as the system's memory and guided the Cas enzymes to cut the same virus if it returned. RNA that guides the enzyme; enzyme that acts as 'scissors' to excise the virus. These were the two basic elements of the CRISPR system.[42]

Charpentier had taken up Cas9 as her latest research subject. But her nomadic past left her with no obvious collaborators. With the help of just two graduate students at Umeå, she homed in on another minute but crucial aspect of the CRISPR system. She and her tiny team showed that trans-activating CRISPR RNA (tracrRNA) helps produce the crRNA and hooks onto the invading virus to indicate where crRNA should mark the spot for Cas9 to cut. They published their results in *Nature* in 2011.[43]

The work showed the CRISPR world the whole system in three simple components: just two forms of RNA and an enzyme. But Charpentier still wasn't satisfied. She wanted to know what happened to the tracrRNA after it had produced crRNA. Did it stick around to help the crRNA? Or was its mission complete? Charpentier didn't feel she'd get her answer by studying living cells.[44] This level of precision needed a biochemist to isolate each chemical component in vitro. She set off for the March 2011 conference of the American Society for Microbiology in Puerto Rico with the intention to speak to a scientist she knew could help. 'I put in my mind that I would find a chance to talk to her.'[45]

Doudna didn't take much convincing. One stroll with Charpentier through St Juan's cobblestoned streets had her writing another breakthrough story in her head.[46] Working with two postdoc researchers, the pair joined forces to determine exactly how Cas9 worked with RNA in CRISPR.[47] It turned out that once the tracrRNA created the crRNA, it also became the scaffold that held the other components in place to allow Cas9 to cut. Now that they had the elements of the system, the team took another leap: they tried to reprogramme it. This was how the fundamental science became a tool. They published their results in 2012.[48]

Doudna and Charpentier hadn't just found the secret of life; they had found a way to recode it.[49] This was not just an insight, not even just a tool, but also a potential application. The stakes in scientific discovery today extend far beyond the acclaim Watson and Crick coveted. Around a billion dollars were funnelled into CRISPR after Doudna and Charpentier proved the biological system worked as an applied technology. Venture capital firms, pharmaceutical companies and public stock offerings all hedged their bets on a new technology that promised to be useful for everything from improved crops and livestock to better animal models and industrial chemicals.[50] Much of that money – tens of millions of dollars – went to lawyers as academic licence holders scrabbled with each other at the US patent office.[51]

A battle began over who was the first to invent the genome editing technique. One of the other groups of CRISPR researchers, led by Professor Feng Zhang of the Broad Institute, had filed a fast-track patent application for the use of the technology in plant and animal cells. Their application was approved first. Doudna's team appealed, and a decade-long legal battle ensued.[52]

Doudna herself seemed to register her story expanding around her. In some accounts, she describes the time as reminiscent of the joy of discovery she'd experienced in the Hawaiian lab at the start of her career.[53] Then, too, her knowledge had had direct applications for farmers struggling with papaya pests. Now, though, her discoveries might have a far bigger impact. And as the possibilities dawned on her, her mind increasingly travelled back to the shores of her childhood and to the kind of overwhelm that young people experience most acutely on encountering the world for the first time. She put it into words.

Doudna's *Double Helix*, in the end, was not the straightforward tale of a breakthrough. In the more ambiguously titled *A Crack in Creation*, Doudna lays out the path to the CRISPR-Cas9 discovery – and the ethical questions it raised. The book opens with a dream in which Doudna stands on a beach in the Hawaii of her childhood. She faces a colossal tidal wave. Like everyone on the island, she grew up with the latent threat of tsunamis. In this dream, that deep-rooted fear of nature's force has coalesced with a strangely subverted version of it: a fear of the natural power that she, using laboratory tools, felt to have unleashed. In the dream, Doudna refuses to submit to her fear as the wave approaches. She decides to paddle out to meet it, slicing through it on a surfboard. She emerges, not triumphant, but in awe, soaking up the view of the island's volcanoes that, she now sees, reach 'protectively' into the sky.[54]

The dream expresses Doudna's relationship to her invention. The wave represents CRISPR-Cas9's awesome potential; frightening, she admits, but rendered harmless when faced head-on. The message is clear: we must all meet the force the CRISPR scientists have unleashed. When we meet the ethical challenges head-on, we'll find a safer world on the other side.

And yet, the wave proved temperamental.

In 2018, just a year after she published her book, Doudna received an email from the Chinese biophysicist Dr He Jiankui, with the news that he'd gone ahead and edited, using CRISPR-Cas9, the genomes of twin girls to make them immune to HIV.[55] He announced the discovery at a conference later that week, confronting the scientific community with their failure to regulate the research in their field.[56] Jiankui has since been sentenced to three years in prison. The incident mobilised scientists to decide the direction they wanted research to take and to enforce regulation. But this all came a little too late – the twins had, after all, already been born.

That brings me back to Doudna's wave and the threat it poses. No one person should be experiencing this degree of responsibility, not even, I would argue, in their dreams. The modern *Double Helix* story is about invention. In the era of genome editing, every scientist is an experimentalist like Franklin. Inventions reveal structures at the same time that they connect scientists to the world. Like the atomic bomb, this is a moment for collective decisions about how we put inventions to work. What are our research priorities as a society? How do we determine access? How do we protect scientists and their work's integrity while working towards this goal? Where and how do we draw lines in the sand as to how far we want the research to go? These choices exist at the heart of scientific progress even in the midst of rapid commercialisation and advancement. In the age of genetic engineering, scientific discovery is, needs to be, socially programmable.

Times have changed since the era of Franklin. Doudna was free to express her competitive spirit and vied for credit and profit the same way male scientists did. She didn't win the patent wars, but she did win a Nobel Prize for CRISPR in 2020. She has raked in millions in awards for her research and her proliferating CRISPR companies, and in 2015 *Time* magazine named her among the 100 most influential people.[57] In that sense, Doudna could be said to be a traditional scientific success story. And yet, from the beginning, that story was straining at the seams. As a woman working on an application that connects her unavoidably to the world around her, Doudna sits on the cusp of a tidal wave. I am watching Doudna's story evolve and science with it. Like the immune systems of the bacteria now used in humans, plants and livestock, I want to see this tale of individual discovery become a dialogue about a human-made application that affects us all.

21

Timnit Gebru (1983–)

Computer Scientist

In 1993, the science fiction author and mathematician Vernor Vinge published his essay 'The coming technological singularity'. In the essay, Vinge predicts the end of the 'human era'. Technological progress, he says, has accelerated over the course of the century and will culminate, within thirty years, in the ability to create 'superhuman intelligence'. The consequences for society, he says, will be 'comparable to the rise of human life on Earth'. The 'intelligence of man would be left far behind' in an intelligence explosion that would soon outstrip humans and then human control. His concept of this new 'technological singularity' became a vision that framed people's debates about the ethical and, indeed, existential implications of artificial intelligence. Our conversations became centred around a singularity that split into a duality: equal parts disbelief at these outlandish possibilities and fear nagging us that it was all too possible.[1]

Timnit Gebru has a more grounded vision, and yet she is no less committed to the discussions on the human impact of artificial intelligence. Gebru makes it her business to put ethics at the forefront of technological development. She is one of the most influential computer scientists in the still relatively new field of ethical artificial intelligence (AI).[2] She produces research on and sometimes using AI that exposes the existing racial, class and gender disparities in the tech industry, from worker mistreatment to discriminatory product design.[3] With singularity-like visions dominating people's imaginations, Gebru shows how the consequences of AI aren't futuristic but historic. Turn away from this latest frontier of scientific progress, Gebru says. The ethical impact of the technology will always be in how it is used in the present, in the conditions for ongoing oppression that we create *today*. Gebru, as a scientist, sees it as her role to alert people to these dangers. In the previous chapter, Doudna raised the question of what a scientist's role can be in ensuring that scientific progress serves humanity. Working from the forefront of AI, Timnit Gebru offers an answer.

*

Timnit Gebru was born in Ethiopia, but her parents were from Eritrea. When Gebru was a teenager, she was forced to flee Addis Ababa, the Ethiopian capital, after war broke out between her parents' homeland and her own. She applied for a US visa, but it was declined, forcing her to stop off in Ireland, before she endured what she describes as a 'miserable' experience with the US asylum system. Eventually, she made it to Boston as a refugee.[4]

Gebru got to know a new country – and a whole new world of discrimination. She'd heard about the racism in the US growing up but, she repeats across multiple interviews, she was 'not prepared at all for the way it was going to happen'.[5] She was shocked by how unapologetic the hatred was. At high school, teachers would make their racism explicit: 'I've met so many people like you who think that they can just come here from other countries and take the hardest classes,' one said to her. Other times, people's views hid behind systems, blocking her from taking advanced classes in physics, despite her exceptional grades.[6] It was an experience after high school, though, that really shook her to the core. Gebru had called the police after her friend, a Black woman, had been assaulted at a bar. When the police arrived, they arrested her friend and never filed the incident. Everywhere she went, Gebru saw American racism at work.[7]

At the same time, Gebru had made it to a centre of power.[8] Stanford University in the early 2010s was the epicentre of AI. Since the 1950s, almost every decade had seen massive government investments in the dream of intelligent machines. Every wave was followed by an era of disappointment when the task proved too great for scientists. But a decade after another lull in research and deflated morale in the field of AI, the second of the so-called 'AI Winters', intelligent machines were an electrifying possibly once again. More data was available due to people using smartphones. Technology could now also handle these vast quantities of date to identify patterns in it. Silicon Valley tech hoped to unlock the new, hugely profitable possibilities. New products abounded. Everything from facial and speech recognition to targeted advertising to warehouse logistics. In 2011, Apple launched Siri, an individualised assistant and operating system that adapted to users' preferences and voice patterns.[9]

At the centre of the industry buzz, Gebru was still facing the same old sobering realities. White peers would ask her if she'd made it into Stanford through affirmative action. Senior administrators talked a lot about diversity, but by the time she reached her PhD in Stanford's Artificial Intelligence Laboratory she was struck by how alone she was among a network of white men. As she put it in one interview: 'There was nobody who looked like me at all. It was just *shocking*.'[10]

At a conference in 2016, Gebru noted that she was the only Black woman out of 8,500 delegates. She posted on Facebook afterwards, with a statement that would come to define her work: 'I'm not worried about machines taking over the world; I'm worried about groupthink, insularity and arrogance in the AI community.'[11] Gebru co-founded the organisation Black in AI with fellow computer scientist Rediet Abebe the following year.[12] Their aim was to improve inclusivity in AI by advocating for Black representation in its policies and practices.[13]

Gebru had studied electrical engineering – specifically, analogue circuit design – in her undergrad and masters. Though she explored other areas like device physics (designing and constructing devices that can sense and detect, and process information), while also studying the fundamental properties of matter.[14] Gebru was developing the material basis of her science, but she was ever aware of the world technology lived in. In her PhD, she ventured into image and signal processing (all this while working a job engineering the first iPad).[15] She became interested in computer vision.[16] As a field in AI, computer vision trains computers to interpret and analyse visual data, like images and videos, to extract meaningful information; in that sense, it allows machines to 'see' and make sense of the visual world. It is part of technologies like object recognition, self-driving cars, medical image analysis and security systems.[17] Gebru was developing algorithms to recognise and classify objects.[18] As we have seen, these technologies fuelled a hot industry. But Gebru was less interested in the technology as a product than as a lens. She turned it to the outside world. In one piece of research, she worked with colleagues on a system that deduced income, race, education and voting patterns from patterns in car ownership detected in Google Street View images. For example, higher numbers of sedans compared to pickup trucks make that city more likely to vote for a Democrat in the next presidential election, and the reverse indicates the vote is likely to be for a Republican.[19] This study confirmed an intuitive sense using verifiable data. This became very characteristic of the research Gebru and members of her teams did: using technology to clarify the social divisions that make our world.

That thinking also led Gebru to interrogate how the social divisions she had experienced were reflected in the datasets that had reanimated AI research. In computer science, discrimination had a deceptively innocuous name: bias. Gebru joined Microsoft Research's Fairness, Accountability, Transparency, and Ethics in AI (FATE) group in 2017. There, she studied the algorithmic bias and the ethical implications of data analytics.[20] She started to get more explicit about her views on how racism functioned *within* research. In the paper 'Data sheets for datasets', Gebru and her colleagues proposed that all research datasets should be accompanied by datasheets that state how and why any given dataset was created, what information it contains and what it should and shouldn't be used for, along with any ethical or legal concerns.[21]

Gebru wasn't the only one now asking how racism shaped the research and applications in the field. She'd been inspired by a 2016 investigation conducted by the non-profit ProPublica that showed how the AI being used in US courtrooms to predict the likelihood of defendants reoffending – designed by Northpointe – was starkly biased.[22] The predictions informed decisions about who can be released at every stage of the criminal justice system: from assigning bond amount to release and sentencing.[23] The journalists compared the actual reoffending rates with the software's predictions. They found that Black defendants who did not reoffend were still more likely to be classified as 'high risk', while white defendants who actually reoffended were more likely to be marked as 'low risk'.[24]

The algorithm-generated risk scores, many of them produced by for-profit companies, were widely used, without much research into how reliable they were. Jurisdictions across the US adopted Northpointe's software before rigorously testing whether they worked. The companies also did not publicly disclose the calculations they used to arrive at defendant risk scores, making it impossible to hold them accountable for their process.[25] Cases like these offered stark examples of the ways in which the hype around AI sneaked in forms of discrimination that anyone would be forced to recognise as discriminatory in any other context. In the case of sentencing, race, nationality and skin colour had become a politically unacceptable basis for risk assessment since the 1970s.[26] And yet here was the same carceral logic, exercised obliquely under the guise of cutting-edge technology.

AI systems in the 2010s relied on advanced computer science. So-called artificial neural networks, composed of a series of nested mathematical functions, were 'trained' using a process that had been honed since the 1960s, based on iteratively feeding it data, evaluating its performance and refining the algorithm until it produces reliable and accurate results.[27] And yet, these highly advanced machines still operated on the same datasets, collected by humans with the same old prejudices. So-called biases did not emerge miraculously through the often-mysticised processes of computation; they mirrored the same institutionalised racism that meant Black people were more likely to be sent to prison. AI was complicit in reinforcing these existing inequalities.[28]

Gebru's work turned into a call for transparency. There is no such thing as unbiased data, she argued. It is always collected with particular social values and priorities in mind. Gebru was clear: this is a human issue. Taking responsibility starts, she said, with transparency about biases in the data. As she later put it, in a 2023 talk:

> People think that just because they're pulling data from the Internet, that everybody's point of view is represented. And that's just not true… So what we need to do is make those values explicit, and those values are often decided by society.[29]

What world were the tech gods creating? Computer science has always existed alongside the possibility of creating a human-like intelligence. In 1950, Alan Turing, father of modern computer science, posed his rephrasing of the sticky question 'Can machines think?' in his famous paper 'Computing machinery and intelligence'. In it, Turing proposed his 'imitation game', a thought experiment based on the more tractable 'Can a machine exhibit intelligent behaviour indistinguishable from that of a human?'[30] Turing reduced the ineffable question of human intelligence to a behavioural experiment. This language has proven useful in the context of variable experiments, but, for a field always in dialogue with consumers and a social world, it has excluded the ethical questions that define our collective humanity.

Not just this, but AI, like CRISPR, is a scientific field always on the cusp of a commodification. Its goals are equally Promethean. Anthropomorphic phrases express the scientific ambition and then travel into a market eager for hype. Terms like 'intelligence', 'learning' or 'expertise', or references to human biology, like 'neural networks', promise, as the history shows, *over*promise, what science will soon be able to do. But they also perpetuate, in flashy bursts of dopamine, a hype that has always sidestepped the question (that defines our humanity) of what their so-called intelligence means for the human world. That's where Gebru turned her attention.

As Gebru's career progressed, the study of racism and computer science became increasingly entwined for her. In an intersectional study published with the computer scientist Joy Buolamwini at MIT, she showed that the facial recognition software developed by tech companies IBM, Microsoft and Face++ did not accurately recognise the gender of people who are not white men: error rates for lighter-skinned males were under 1% compared to a staggering 34.7% for darker-skinned females.[31]

The study left no doubt that algorithmic systems encoded social inequalities. Gebru presented quantifiable, peer-reviewed proof, along with a framework that researchers and businesses should use to measure, report and ultimately mitigate bias. In this uncompromising call for accountability, the researchers also boldly named specific companies and products.[32]

The paper helped usher in a new era of AI governance. It turned a faint whisper about the ethics of AI from a few researchers on the fringes of Silicon Valley into central public concern. The study hit the mainstream news: 'Facial recognition is accurate, if you're a white guy,' reported the *New York Times*.[33] The American Civil Liberties Union, inspired by the research, magnificently input members of Congress to Amazon's facial recognition product, Rekognition, and it mismatched twenty-eight of them to mugshots of people who had been arrested.[34] Gebru's data gave new force to existing public concerns about digital surveillance. Newly under scrutiny, companies were forced to respond. Two years after the paper was published, IBM exited the facial recognition business. Their reason? 'The technology's potential for mass surveillance, racial profiling' and 'violations of basic human rights'.[35] The same year, Amazon (who hadn't even been named in the study) conceded to a rare moment of external oversight and announced a one-year moratorium on police use of Rekognition. Microsoft announced that it would not sell their technology to the police without federal regulation. There were hearings in the US and EU on biometric oversight. The paper set a new standard and reframed the discussion around the deceptively technical 'algorithmic bias' as a public issue.[36]

Gebru was fast becoming an expert in the long-overdue field of AI ethics.

In 2018, Gebru joined Google as a research scientist. With her colleague Margaret Mitchell, an algorithmic-bias specialist, she built the company's first ethical AI team. She and Mitchell became known for fostering a supportive environment for non-white researchers at the company, where still only 1.6% of researchers were Black women.[37] Simultaneously, they supported Google product teams to consider the social impact of their applications.[38] It was exciting; Gebru had new-found institutional support for putting ethics at the heart of AI research. Her team even hired Google's first *social* scientist, to integrate ethical questions into the technical process.[39]

At the same time, Gebru co-led work on a paper about the ethical risks of large language models. This AI software generates fluent prose based on vast amounts of data and is used in systems like ChatGPT or Google to perform search queries.[40] Together with Mitchell and two other colleagues, Gebru published the paper 'On the dangers of stochastic parrots: can language models be too big? 🦜' in 2020.[41] In it, they made several ethical arguments. They pointed to the environmental impact of running AI models that consume huge amounts of electricity (to train one language model in a particular task produces the equivalent of about the lifetime output of five average American cars' worth of carbon dioxide).[42] This means that only wealthy organisations can afford to train and benefit from AI, while, we know, climate change continues to ravage marginalised communities the most. They also argued that language models work indiscriminately on online content, replicating the dominant societal views reflected within it. When a system learns the complexity of language from an ever-growing tide of hate speech, the technology could reproduce it.[43] The systems will also not register changes in language use that are so important to social change, for example, according to the revisions spurred on by social movements. Nor will it reflect the languages of communities with less internet access. Add to this the difficulty of checking large datasets for embedded biases.[44] The study called into question once again the historic, and unequal, foundations on which AI technology was built.[45]

The paper ruffled some feathers. This time, Gebru had singled out one particular organisation. The fallout led to Gebru's contested resignation. Feel free to Google it.

Gebru went on to found her own research institute, the Distributed AI Research Institute (DAIR), in 2021. Through her work there, she continues to argue that the biggest challenge we face with AI is not its capacity to outpace humans, but the social impact it is having in the world. In that vein, she works not just with computer scientists but former Amazon delivery drivers and people with experience of the gruelling work of manually labelling online content to train AI systems, often processing disturbing and also illegal material.[46] Workers are frequently paid less than $2 per hour. Many are traumatised by the conditions.[47]

Much of this work happens in developing countries, and so Gebru collaborates with people in Africa as much as the US or EU. She continues to employ social scientists, as well as refugee advocates, labour organisers and activists. These diverse perspectives and forms of expertise are required to design research and applications that, at the least, don't exploit, and at best benefit a marginalised majority of people. 'AI is not magic,' Gebru says. 'There are a lot of people involved – *humans*.'[48] The most pressing issues around AI, like data theft, labour exploitation, environmental damage or copyright, are not new. They are continuations of social problems. That also means, Gebru points out, that in many cases we already have the agencies designed to regulate these kinds of practices, like the Federal Trade Commission.[49] We need to govern AI like every other scientific field or industry.

AI *can* be used ethically. At DAIR, Gebru works with her multi-disciplinary and diverse team to design research that responds to people's needs, rather than industry hype. At DAIR, research is targeted, not to building generalised, machine gods but to technological solutions to people's specific problems. One piece of research, for example, came from a computer vision researcher who grew up in a township in South Africa. Drawing on her experience, they asked if people's lives have improved in South Africa since the end of apartheid. They got very practical about the question. Because the South African government doesn't distinguish the townships created during apartheid from the other suburbs, the team turned to satellite images of the provinces. They used this data to create an AI system to identify the townships, alongside the suburbs and vacant land.[50] This allowed them to compare characteristics affecting quality of life, such as the greenery, health services or education centres available.[51] People already knew their quality of life had worsened since the end of legal apartheid. Gebru said in an interview: 'People know that their lives are actually worse, but they don't have the data to back it up.' Gebru empowers people with the scientific evidence they need to advocate for real-world change.[52]

Gebru and her institute work independently of the big tech companies to show how researchers can put ethical questions at the heart of science. As a scientist, she believes she is uniquely positioned to act as an 'early warning system' to potential issues.[53] This is more than she could do in government, where her work on policy and legislation would respond to the realities already enacted by tech giants. That is not where she believes she can make her most important contribution. As a scientist, she wants to be free to influence technology in the making: 'I want to be able to institute my imagination of whatever technological future I want to live in.'[54]

Freedom of imagination and the ability to influence scientific developments for the social good – that is how Gebru sees her role as a scientist. She works collaboratively and prioritises human needs over profit or acclaim. That approach reflects the experiences of racism and sexism she has experienced in her life. Gebru does science differently because she has had a different kind of life. As she puts it:

> I'm literally a scientist and an engineer, whose default is something different. It means that I am centring the communities that I come from… It doesn't mean my science is different. My default is *doing* it differently.[*55]

Gebru shows us why it is important that we improve diversity in science. Without the vision of scientists with different life experience, the questions we ask and the knowledge we consequently build will always reflect the priorities of an exclusive elite. Marginalised scientists will 'do' science differently, which, as we

---

* Emphasis my own.

have seen throughout this book, leads to a different view of reality.

Gebru continues to share her thinking on the impact of AI on the world today. Nonetheless, in her upcoming memoir, *The View from Somewhere*, she does turn her gaze to the future. Here she shares her vision for a positive technological future in which AI serves communities rather than centralised power.[56] The tone is different from Vinge's essay. Gebru's book isn't a dystopia or a utopia; it isn't a vision but a set of questions. A new set of conversation starters for students, researchers and the public as they shape a hopeful future around AI.[57]

Gebru is a scientist whose work is intimately tied up with social justice, as well as the kind of care needed to foster an inclusive environment; a scientist who shows us a new vision of technological progress and how science can be done. This is where I want to end this collection of women working in science. With this case study in what happens when women bring their life experience, their perspective, their skills and their priorities to the deeply human business of illuminating and building our world.

Conclusion

The beauty and agony of writing a book is that you grow with it. A book is never complete, least of all a selection of just twenty-one women from the breadth of the history of science.

There are scientists I would have included – women I discovered or discovered more about in the shadows of those I already knew (Emmanuelle Charpentier and Lise Meitner are just two). Maybe, if I had my time again, I would have written a collective biography of scientific teams spanning the globe, the history of a non-Western science or a lineage of the lab technicians never credited with so much as a mention in an academic paper.

The seventeenth-century sociologist Thomas Hobbes drew the analogy between a human body and society to explain how humans organised into a state. In science, too, the shoulders of giants are groups of people, inextricably connected to other limbs. Writing this book has impressed on me powerfully that we cannot escape our interconnectedness. And so, it has also shown me the limits of any history of individuals in realistically describing how science is done.

What I have shown, though, is how at the heart of the male-dominated Western scientific narrative, so often driven by war and profit, there *were* women, offering us glimmers of a different vision of progress.

Though the path wasn't always revolutionary. These scientists, as women, were caught in a double bind: sometimes remembered only as paragons of womanhood, 'too feminine' to be taken seriously (Nightingale or Goodall); at other times dismissed by male colleagues and contemporaries for not being feminine enough (say, McClintock or Cavendish). Some scientists played the competitive career game and were, unlike men, criticised for it (Curie, Wu, Nyswander and Tu). We have seen the pragmatism of many of the women who navigated the contradictory demands on them by leveraging their relative degrees of social standing, or their connections with men, creating the opportunities they should have had. Today, we see how women compete unapologetically and more autonomously, taking up their place as equals in a man-made science.

Other women turn that science on its head. Many of the women in this book relied on a combination of conventional and undervalued, feminised skills usually excluded from descriptions of science: like intimacy with their research subjects (McClintock or Virginia Johnson), communication (Nightingale or McLaren), the work of editing and refining (Hypatia or Lovelace), building relationships (Hodgkin or Wong-Staal), caution (Kelsey) or restraint (Franklin) and the courage to resist division (McLaren or Hodgkin) or to adjust your perspective (Goodall). From races, machines, giants and frontiers, the women in this book suggest a new set of metaphors for the work science entails. Like motherhood. Or the caring hands of a scientist writing out her calculations. Or the intimacy of a relationship between humans, animals and the natural world. The metaphors we have for scientific discovery matter. It is no coincidence that the most consciously political scientists in this book combined storytelling with their technical work. They understood that invention would reach only as far as the collective human imagination would go. The women in this book offer another spark.

Sometimes, though, we get our prompters from the scientific findings themselves. From RNA to jumping genes; from the atom bomb to gene editing and the World Wide Web. As much as scientists since the Enlightenment have tried to break down our world into parts, modern science won't allow us to forget our interconnectedness.

Maybe, if I wrote this book again, I'd highlight the work of the mushroom biologists, who have turned their gazes from the cosmos to the depths of the Earth for new metaphors for progress. Less individualistic, less destructive and relentlessly alive.

As fungal filaments thread through the soil, plants and animals, living and dead, mushrooms show us the relationships we depend on.[1] They have forced scientists to revise the central metaphors that have connected science and capitalism in modernity. Evolution becomes 'involution', survival of the fittest an artefact of the past. The question is no longer how separate life forms came to triumph over others, but how life-sustaining *systems* emerge.[2]

The time is ripe for a new story of progress. In yet another era of human-made existential threats and possibilities, individuals won't help us. From the lives of singular scientists, we can turn to the ground they stand on. There, we will find a vision of a future-proof science. Its aim? Collective survival. As the renowned evolutionary biologist Lynn Margulis wrote in 1997 after spending her career studying microbes: 'Life did not take over the world by combat, but by networking.'[3]

That is the hopeful vision these twenty-one women leave behind. That is the spore in this book for the ever-emerging group of people and practices that we call science, and for every being they are connected with.

Acknowledgements

I always knew I wanted to write the story of the scientist at the heart of my PhD. I'd spent years with her in a very particular way, following her traces, reading her patterns. I discovered that I liked getting to know people in this way. I wanted to do more of it.

I am grateful to my first agent, Eli Keren, for signing me all those years ago, for continuing to believe in McLaren's story and in my capacity to write so many others.

Thank you to Sam Carter and Hannah Haseloff at Oneworld for the education in science writing you've given me, and for your thoughtful feedback over the months. I'm really proud of what we've achieved.

My supervisor, Professor Sarah Franklin, helped me name my own approach when it was an unknown seed in me. You gave me the confidence to do what I want to do endlessly – to describe people through their style of being in the world. With that, you gave me the gift of myself. A true teacher. Thank you.

The people at the British Library, especially Jonathan Pledge, supported me in my research, and opened up the archives in new ways to me. Anne McLaren's family gave me unprecedented access to additional letters and diaries. Thank you for trusting me with her legacy.

Elah Feder and Katie Hafner at the *Lost Women of Science* podcast and David Courtwright provided access to invaluable resources, Edward Watts spent time talking to me about Hypatia,

Richard Bates about Nightingale, and John Perlin generously shared his fantastic, unpublished manuscript on Eunice Foote with me.

The lovely people at The Chestnut Tree bookshop – you've made more of a difference than you know by providing such a warm, inspiring space for me to write in.

Thanks to my partner, Luong, for supporting me emotionally and in every other way through this project – for always supporting my writing, but putting me, as person, first. And to Huong, for lending me a desk, a cat, baking so many cakes and for being steadily in my corner during these transitional times.

Then, of course, my family back home, for loving me from afar through this process. Bruce, for his gentle kindness, and Mum, for your endless curiosity about my inner world – and for being the first inspiring scientist in my life.

Notes

Introduction: In the Shadows of Giants

1. Newton, Isaac, 'Isaac Newton to Robert Hooke', 5 February 1675, Historical Society of Pennsylvania.
2. Gage, Matilda Joslyn, *Woman as Inventor* (New York State Woman Suffrage Association, 1870), https://iiif.lib.harvard.edu/manifests/view/drs:2575141$3i; Lamm, Lisa, 'The Matilda Effect: How Women Are Becoming Invisible in Science', *Lost Women of Science* podcast, 26 October 2022, https://www.lostwomenofscience.org/news-events/the-matilda-effect-how-women-are-becoming-invisible-in-science.
3. Rossiter, Margaret W., 'The Matthew Matilda Effect in Science', *Social Studies of Science*, 23(2), 1993, pp. 325–41, *JSTOR*: https://www.jstor.org/stable/285482.
4. Ibid.
5. Knobloch-Westerwick, Silvia, Glynn, Carroll J., and Huge, Michael, 'The Matilda Effect in Science Communication: an Experiment on Gender Bias in Publication Quality Perceptions and Collaboration Interest', *Science Communication*, 35(5), 2013, pp. 603–25.
6. Doğan, İ., Baruh, L., Cemalcilar, Z., Kuru, O., Yıldırım, K. and Çarkoğlu, A., 'Biased Perceptions Against Female Scientists Affect Intentions to Get Vaccinated for COVID-19', *Public Understanding of Science*, 31(2), 2021, pp. 239–51, https://doi.org/10.1177/09636625211060472.

Chapter 1: Hypatia

1. Deakin, Michael A. B., 'Hypatia and Her Mathematics', *The American Mathematical Monthly* 101(3), 1994, pp. 234–43, doi:10.1080/00029890.1994.11996935.
2. Watts, Edward J., *Hypatia: The Life and Legend of an Ancient Philosopher* (Oxford: Oxford University Press, 2017), pp. 108–112.
3. Ibid., p. 7.
4. Ibid., p. 27.
5. Ibid., p. 26.
6. Ibid., p. 29.
7. *Epitome of the Ecclesiastical History of Philostorgius, Compiled by Photius, Patriarch of Constantinople*, trans. Edward Walford (London: Henry G. Bohn, 1855), accessed via https://www.tertullian.org/fathers/philostorgius.htm.
8. Watts, *Hypatia: The Life and Legend*, p. 26.
9. Kitagawa, Kate and Revell, Timothy, *The Secret Lives of Numbers* (London: Penguin, 2024), p. 48; Watts, *Hypatia: The Life and Legend*, p. 93.
10. Kitagawa and Revell, *The Secret Lives*, p. 48.
11. Watts, *Hypatia: The Life and Legend*, p. 94.
12. Ibid., p. 97.
13. Ibid., pp. 93–102; 102–6.
14. Ibid., p. 16.
15. Ibid., pp. 15–7.
16. Kitagawa and Revell, *The Secret Lives*, p. 49; Deakin, 'Hypatia and Her Mathematics', pp. 234–43.
17. Kitagawa and Revell, *The Secret Lives*, pp. 49–50.
18. Ibid., p. 51.

19. Whitfield, Bryan J., 'The Beauty of Reasoning: A Reexamination of Hypatia of Alexandria', *The Mathematics Educator*, 6(1), 1995, p. 18.
20. Deakin, 'Hypatia and Her Mathematics'.
21. Paraskevopoulos, Athanasios, 'Francois Viete and His Contribution to Mathematics', arXiv, 22 October 2022, https://doi.org/10.48550/arXiv.2210.12545.
22. Whitfield, 'The Beauty of Reasoning', pp. 18–9; Mahoney, M. S., *The Mathematical Career of Pierre de Fermat* (Princeton, N.J.: Princeton University Press, 1973).
23. Kitagawa and Revell, *The Secret Lives*, p. 51.
24. Rome, A., *Commentaires de Pappus et de Theon d'Alexandrie sur l'Almageste* (Vatican City: Biblioteca Vaticana, 1931–1943).
25. Kitagawa and Revell, *The Secret Lives*, p. 51; Deakin, 'Hypatia and Her Mathematics', p. 239.
26. Kitagawa and Revell, *The Secret Lives*, p. 51.
27. Watts, *Hypatia: The Life and Legend*, p. 31; Heath, T. L., *Euclid: The Thirteen Books of the Elements* (Cambridge: Cambridge University Press, 1990), pp. 54–8.
28. Watts, *Hypatia: The Life and Legend*, p. 26.
29. Ibid., p.3; Bernard, A., 'Theon of Alexandria and Hypatia,' *Cambridge History of Philosophy in Late Antiquity*, vol. 1, ed. L. Gerson (Cambridge: Cambridge University Press, 2010), pp. 417–36.
30. Watts, *Hypatia: The Life and Legend*, p.34; Bernard, 'Theon of Alexandria and Hypatia'.
31. Watts, *Hypatia: The Life and Legend*, p. 28; O'Meara, D., *Pythagoras Revived: Mathematics and Philosophy in Late Antiquity* (Oxford: Oxford University Press, 1989), pp. 97–9; Pépin, J., *Théologie Cosmique et Théologie Chrétienne* (Paris: Presses universitaires de France Vendôme, 1964), pp. 380–6.
32. Watts, *Hypatia: The Life and Legend*, pp. 34–5.
33. Ibid., p. 51.
34. Kitagawa and Revell, *The Secret Lives*, p. 52.
35. Socrates, *HE* 7.15, in Watts, *Hypatia: The Life and Legend*, p. 43.
36. Watts, *Hypatia: The Life and Legend*, pp. 49–50.
37. Ibid., p. 50.
38. Interview with Edward J. Watts, 14 June 2024.
39. Watts, *Hypatia: The Life and Legend*, p. 47.
40. Ibid.
41. Dzielska, *Hypatia of Alexandria*, trans. F. Lyra (Cambridge, MA: Harvard University Press, 1996), pp. 28–39.
42. Watts, *Hypatia: The Life and Legend*, p. 49.
43. Ibid., p. 50.
44. Ibid., p. 62; 78.
45. Synesius, 'Letter 124', Livius, https://www.livius.org/sources/content/synesius/synesius-letter-124.
46. Synesius, 'Letter 010', Livius, https://www.livius.org/sources/content/synesius/synesius-letter-010.
47. Watts, *Hypatia: The Life and Legend*, pp. 56–9.
48. Ibid., p. 79.
49. Ibid., pp. 82–3; 101–2.
50. Ibid.
51. Ibid., p. 105; for an example on signalling of clothing: Wilkinson, Kate, *Women and Modesty in Late Antiquity* (Cambridge: Cambridge University Press, 2015).
52. Watts, *Hypatia: The Life and Legend*, p. 105.
53. Socrates, *HE* 7.15.2, 3; Damascius, *The Life of Isisdore,* 43E.
54. Damascius, *The Life of Isisdore,* 43C; Harich-Schwartzbauer, *Hypatia,* pp. 268–70, in Watts, *Hypatia: The Life and Legend*, p. 75.
55. Watts, *Hypatia: The Life and Legend*, p.106.
56. Ibid.
57. Ibid., p. 113.
58. Ibid., p. 113.
59. Ibid., p. 114.
60. Ibid., p.115, citing Dio Chrysostom, *Or.* 32.35–6, and an analysis: Jones, C.P., *The Roman World of Dio Chrysostom* (Cambridge: Cambridge University Press, 1978), pp. 36–44.
61. Socrates, *HE* 7.15.
62. McPhee, David, 'Brenda Page Verdict: The Brutal Murder Mystery That Confounded Police for 44 Years', *Press and Journal*, https://www.pressandjournal.co.uk/fp/news/aberdeen-aberdeenshire/5414419/brenda-page-history; Smith, Helena, 'Crete "Struggling With Shame" After Rape and Murder of US Scientist', *Guardian*, 25 July 2019, https://www.theguardian.com/world/2019/jul/25/crete-struggling-shame-rape-murder-us-scientist.

63. Watts, *Hypatia: The Life and Legend*, p. 116.
64. Ibid.
65. Ibid., p. 117; Damascius, *The Life of Isadore,* 43H.
66. Watts, *Hypatia: The Life and Legend*, p. 118.
67. Ibid., p. 119.
68. Ibid.
69. Kitagawa and Revell, *The Secret Lives*, p. 54.
70. For example: Damascius, *Isid.* 106A-B.
71. Watts, *Hypatia: The Life and Legend*, pp. 122–34.
72. John Toland, *Tetradymus*, 1720, in Watts, *The Life and Legend*, p. 135.
73. 'Who Was Hypatia of Alexandria?', Constitutional Rights Foundation, 34(1), Fall 2018, https://teachdemocracy.org/wp-content/uploads/2025/01/Hypatia-of-Alexandria.pdf.
74. Watts, *The Life and Legend*, p. 140.
75. Whitfield, 'The Beauty of Reasoning'.
76. Watts, *The Life and Legend*, p.146.

Chapter 2: Margaret Cavendish

1. Fredrickson, Anne, 'First Lady', *Distillations Magazine*, Science History Institute Museum & Library, 16 April 2013, https://www.sciencehistory.org/stories/magazine/first-lady.
2. Wilkins, Emma, 'Margaret Cavendish and the Royal Society', *Notes and Records of the Royal Society of London*, 68(3), 2014, pp. 245–60, doi: 10.1098/rsnr.2014.0015; *The Diary of Samuel Pepys, A New and Complete Transcription Edited by Robert Latham and William Matthews* (London: G. Bell & Sons, 1970–83), vol. 8, pp. 243–4.
3. Cooke, Rachel, 'Pure Wit: The Revolutionary Life of Margaret Cavendish Review – English Eccentric and Talk of the Town', *Guardian*, 3 September 2023, https://www.theguardian.com/books/2023/sep/03/pure-wit-the-revolutionary-life-of-margaret-cavendish-francesca-peacock-review-english-eccentric-and-talk-of-the-town.
4. Lovascio, Domenico, 'Margaret Cavendish and Julius Caesar', *A Companion to the Cavendishes*, eds. Lisa Hopkins, Tom Rutter (York, UK: Arc Humanities Press, 2020), pp. 309–23, https://www.researchgate.net/publication/348748721_Chapter_18_MARGARET_CAVENDISH_AND_JULIUS_CAESAR/fulltext/600ec290a6fdccdcb87afe0b/Chapter-18-MARGARET-CAVENDISH-AND-JULIUS-CAESAR.pdf.
5. Fredrickson, 'First Lady'.
6. Ibid.
7. Peacock, Francesca, 'Her Blazing World', *Aeon*, ed. Sam Dresser, 28 May 2024, https://aeon.co/essays/the-contradictions-that-give-life-to margaret-cavendishs-story?s=03.
8. Peacock, 'Her Blazing World'.
9. Ibid.; Peacock, Francesca, *Pure Wit: The Revolutionary Life of Margaret Cavendish* (London: Bloomsbury, 2023), chapter 2.
10. Peacock, *Pure Wit*, pp. 35–6.
11. Ibid.; Peacock, 'Her Blazing World'.
12. Cavendish, William, *The Phanseys Of William Cavendish Marquis Of Newcastle Addressed To Margaret Lucas and Her Letters In Reply*, ed. Douglas Grant (London: Nonesuch Press, 1956); 'The Marriage Songe', p. 2, in Peacock, *Pure Wit*, p. 65.
13. Emre, Merve, 'Margaret Cavendish's "Mad" Imagination', *The New Yorker*, 29 January 2024, https://www.newyorker.com/magazine/2024/02/05/pure-wit-the-revolutionary-life-of-margaret-cavendish-francesca-peacock-book-review.
14. Peacock, 'Her Blazing World'.
15. Cavendish, Margaret, *Poems and Fancies* (London: Printed by T. R[oycroft]. for J. Martin, and J. Allestrye at the Bell in Saint Pauls Church Yard, 1653).
16. Emre, 'Margaret Cavendish's "Mad" Imagination'.
17. Ibid.
18. Marshall, Gwendolyn, 'Margaret Cavendish (1623 1673)', Internet Encyclopedia of Philosophy, https://iep.utm.edu/margaret-cavendish/; Simon, Ed, 'The Vitalist Multiverse of "Mad Madge" Cavendish', 3 Quarks Daily, 25 March 2024, https://3quarksdaily.com/3quarksdaily/2024/03/the-vitalist-multiverse-of-mad-madge-cavendish.html.
19. Cavendish, Margaret, *Philosophical Letters* (London, 1664) pp. 320–1.
20. Marshall, 'Margaret Cavendish (1623–1673)'.
21. Cavendish, Margaret, *Philosophical and Physical Opinions* (London: William Wilson, 1655); Marshall, 'Margaret Cavendish (1623–1673)'.
22. The Royal Society, 'History of the Royal Society', https://royalsociety.org/about-us/who-we-are/history/.

23. Boyle, Deborah, *The Well-Ordered Universe: The Philosophy of Margaret Cavendish* (Oxford: Oxford University Press, 2018); Rogers, John, *The Matter of Revolution: Science, Poetry, and Politics in the Age of Milton* (Ithaca, NY: Cornell University Press, 1996).
24. Crawford, Patricia, 'Women's Published Writings 1600–1700', *Women in English Society 1500–1800*, ed. Mary Prior (London: Methuen, 1985), pp. 211–82.
25. Peacock, 'Her Blazing World'.
26. Cited in Baker, Rupert, 'First Lady', The Royal Society, 8 March 2024, https://royalsociety.org/blog/2024/03/first-lady/.
27. Hintz, Carrie. *An Audience of One: Dorothy Osborne's Letters to Sir William Temple, 1652–1654* (Toronto: University of Toronto Press, 2005), https://doi.org/10.3138/9781442670778; Emre, Merve, 'Margaret Cavendish's "Mad" Imagination', https://www.newyorker.com/magazine/2024/02/05/pure-wit-the-revolutionary-life-of-margaret-cavendish-francesca-peacock-book-review; Peacock, 'Her Blazing World'.
28. Peacock, 'Her Blazing World'.
29. Lovascio, 'Margaret Cavendish and Julius Caesar'.
30. Bodleian Libraries, Oxford, *MS North, c.4* f. 146, cited in Peacock, *Pure Wit*, p. 228.
31. Cavendish, Margaret, *Observations upon Experimental Philosophy*, ed. Eileen O'Neil (Cambridge: Cambridge University Press, 2001). Originally published: London, A. Maxwell, 1668; Wilkins, 'Margaret Cavendish and the Royal Society'.
32. Mintz, S., 'The Duchess of Newcastle's Visit to the Royal Society', *The Journal of English and Germanic Philology.* 51, 1952, pp. 168–76, in Wilkins, 'Margaret Cavendish and the Royal Society'.
33. The Royal Society, 'History of the Royal Society'.
34. Cavendish, *Observations*, 1666.
35. Cavendish, *Further Observations upon Experimental Philosophy*, in *Observations* (Cambridge: Cambridge University Press, 2012) p. 209.
36. Hedesan, Georgiana, 'Hooke's Microscope, 1665', Cabinet, https://www.cabinet.ox.ac.uk/hookes-microscope-1665.
37. *The Diary of Samuel Pepys*, 21 January 1665; Gest, Howard, 'The Discovery of Microorganisms by Robert Hooke and Antoni van Leeuwenhoek, Fellows of the Royal Society', *Notes and Records of the Royal Society of London*, 58(2), 2004, p. 189.
38. Ford, Brian J., 'History of Optical Microscopes', Britannica, 16 January 2026, https://www.britannica.com/technology/microscope/History-of-optical-microscopes; 'Hooke's "*Micrographia*: or, Some physiological descriptions of minute bodies made by magnifying glasses"', The Diary of Samuel Pepys, 12 January 2026, https://www.pepysdiary.com/encyclopedia/8180/.
39. Cavendish, *Observations upon Experimental Philosophy*, sig. b1v; *Observations upon Experimental Philosophy*, p. 90 wrongly paginated as 86, in Wilkins, 'Margaret Cavendish and the Royal Society'.
40. Mintz, *op. cit.* (note 6), p. 176, in Wilkins, 'Margaret Cavendish and the Royal Society'.
41. Malcolm, Noel, *Aspects of Hobbes* (Oxford: Clarendon Press, 2002), p. 320; Wilkins, 'Margaret Cavendish and the Royal Society'.
42. Cavendish, *Observations upon Experimental Philosophy*, p. 8, in Wilkins, 'Margaret Cavendish and the Royal Society'.
43. Wilkins, 'Margaret Cavendish and the Royal Society'.
44. Ibid.
45. Cavendish, *Observations upon Experimental Philosophy*, p. 92.
46. R. Hooke, preface to *An Attempt for the Explication of the Phaenomena* ..., quoted in Jim Bennett, Michael Cooper, Michael Hunter and Lisa Jardine (eds) *London's Leonardo: The Life and Work of Robert Hooke* (Oxford: Oxford University Press, 2003), p. 85, in Wilkins, 'Margaret Cavendish and the Royal Society'.
47. Wilkins, 'Margaret Cavendish and the Royal Society'.
48. Cavendish, *Observations upon Experimental Philosophy*, 1666, p. 91 wrongly paginated as 87, in Wilkins, 'Margaret Cavendish and the Royal Society'.
49. Wilkins, 'Margaret Cavendish and the Royal Society'.
50. Ibid.
51. Ibid.
52. Peacock, 'Her Blazing World'.
53. Ibid.
54. Ibid.
55. Ibid.
56. Ibid.
57. Wilkins, 'Margaret Cavendish and the Royal Society'.

58. Hooke, R., *Micrographia*, sig. A1v; Glanvill, *Scepsis Scientifica*, sig. a3v (London, 1665). See also Hunter, Michael, *The Royal Society and its Fellows 1660–1700: The Morphology of an Early Scientific Institution* (British Society for the History of Science, Chalfont St Giles, 1982), esp. p. 9 and p. 25, in Wilkins, 'Margaret Cavendish and the Royal Society'.
59. Wilkins, 'Margaret Cavendish and the Royal Society'.
60. Emre, 'Margaret Cavendish's "Mad" Imagination'.
61. Whiting, Daniel, 'Margaret Cavendish, Duchess of Newcastle', Welbeck, 5 September 2025, https://www.welbeck.co.uk/margaret-cavendish-duchess-of-newcastle/.

Chapter 3: Ada Lovelace

1. January 5, 1841, she asked, 'What is Imagination?' [emphasis my own].
2. Lessenich, Rolf P., 'Byron, "The Sanity or Madness of Same-Sex Love"', *The Byron Journal*, 44, 2016, pp. 139–50; Miller, Laura, 'After the Bad Boy', *Slate*, 4 December 2018, https://slate.com/culture/2018/12/lord-byron-wife-daughter-biography-review.html.
3. Schiff, Stacy, 'Mad, Bad and Dangerous: The Legacy Left to Byron's Wife and Daughter', *New York Times*, 30 November 2018, https://www.nytimes.com/2018/11/30/books/review/miranda-seymour-in-byrons-wake-ada-lovelace-biography.html.
4. Markus, Julia, *Lady Byron and Her Daughters* (New York: W. W. Norton, 2015).
5. Tsiamis, C., Piperaki, E. T., Kalantzis, G., Poulakou, Rebelakou E., Tompros, N., Thalassinou, E., Spilipoulou, C., Tsakris, A., 'Lord Byron's Death: A Case of Late Malarial Relapse?', *Infez Med*, 23(3), 2015, pp. 288–95; Miller, 'After the Bad Boy'.
6. Essinger, James, *Ada's Algorithm: How Lord Byron's Daughter Ada Lovelace Launched the Digital Age Through the Poetry of Numbers* (London: Gibson Square, 2013).
7. Accessed via Craddock, Paul, 'The word "scientist" was coined in 1834, as a joke', Common Knowledge, 10 November 2023, https://paulcraddock.substack.com/p/the-word-scientist-was-coined-in.
8. Toole, Betty Alexandra, 'Poetical Science', *The Byron Journal*, 15, 1987, pp. 55–65; Morais, Betsy, 'Ada Lovelace, the First Tech Visionary', The New Yorker, 15 October 2013, https://www.newyorker.com/tech/annals-of-technology/ada-lovelace-the-first-tech-visionary.
9. Jollymore, Amy, 'Ada Lovelace, An Indirect And Reciprocal Influence', *Forbes*, 15 October 2013, https://www.forbes.com/sites/oreillymedia/2013/10/15/ada-lovelace-an-indirect-and-reciprocal-influence/.
10. Woolley, Benjamin, *The Bride of Science: Romance, Reason and Byron's Daughter* (London: Macmillan, 1999); 'Child of Poetry, Bride of Science', Guardian, 14 August 1999, https://www.theguardian.com/books/1999/aug/14/books.guardianreview.
11. Knox, Kevin C., *From Newton to Hawking: A History of Cambridge University's Lucasian Professors of Mathematics* (Cambridge: Cambridge University Press, 2003); Essinger, James, 'Charles Babbage and Ada Lovelace: The computer's most passionate partnership', The History Press, https://thehistorypress.co.uk/article/charles-babbage-and-ada-lovelace-the-computers-most-passionate-partnership/.
12. Hollings, C., Martin, U., and Rice, A., 'The Early Mathematical Education of Ada Lovelace', *BSHM Bulletin: Journal of the British Society for the History of Mathematics*, 32(3), 2017, pp. 221–34, https://doi.org/10.1080/17498430.2017.1325297.
13. De Morgan, Sophia Elizabeth, *Memoir of Augustus De Morgan* (London: Longmans, Green and co., 1882).
14. Cain Miller, Claire, 'Overlooked Obituaries: Ada Lovelace', *New York Times*, https://www.nytimes.com/interactive/2018/obituaries/overlooked-ada-lovelace.html.
15. Babbage, Charles, Letter from Babbage to Lovelace, 9 September 1843, via Wolfram, Stephen, 'Untangling the Tale of Ada Lovelace', 10 December 2015, https://writings.stephenwolfram.com/2015/12/untangling-the-tale-of-ada-lovelace/.
16. 'What and Where is the Black Country?', BBC [website], 15 March 2005, https://www.bbc.co.uk/blackcountry/content/articles/2005/03/15/where_is_the_black_country_feature.shtml, 'Ada Lovelace's Life and Impact on Scientific Computing', Britannica, https://www.britannica.com/video/Walter-Isaacson-life-impact-Ada-Lovelace/-218170; Wolfram, Stephen, 'Untangling the Tale of Ada Lovelace', 10 December 2015, https://writings.stephenwolfram.com/2015/12/untangling-the-tale-of-ada-lovelace/.
17. 'Programming Patterns: The Story of the Jacquard Loom', Science + Industry Museum, 25 June 2019, https://www.scienceandindustrymuseum.org.uk/objects-and-stories/jacquard-loom.
18. 'Charles Babbage's Difference Engines and the Science Museum', Science Museum, 18 July 2023, https://www.sciencemuseum.org.uk/objects-and-stories/charles-babbages-difference-engines-and-science-museum.

19. Swade, Doron interviewed on *In Our Time: Ada Lovelace*, BBC Radio, 6 March 2008, https://www.bbc.co.uk/sounds/play/b0092j0x.
20. Ibid.
21. Wolfram, Stephen, 'Untangling the Tale of Ada Lovelace'.
22. Ibid.
23. Peel, 1842, quoted in BBC documentary: *Calculating Ada: The Countess of Computing*, 2015.
24. Wolfram, Stephen, 'Untangling the Tale of Ada Lovelace'.
25. Morais, Betsy, 'Ada Lovelace, the First Tech Visionary', *The New Yorker*, 15 October 2013, https://www.newyorker.com/tech/annals-of-technology/ada-lovelace-the-first-tech-visionary.
26. Lovelace, Ada, Thursday morning, 27 July 1843, via https://blogs.bl.uk/science/2015/10/your-puzzle-mate-ada-lovelace-and-charles-babbage-1.html.
27. Morais, 'Ada Lovelace, the First Tech Visionary'.
28. Isaacson, Walter, *The Innovators: How a Group of Hackers, Geniuses, and Geeks Created the Digital Revolution* (London: Simon & Schuster UK, 2014).
29. Menabrea, L. F., 'Sketch of The Analytical Engine Invented by Charles Babbage', *Bibliothèque Universelle de Genève*, October 1842, No. 82. View full paper online at https://www.fourmilab.ch/babbage/sketch.html.
30. For example, Turing, A. M, 'Can Digital Computers Think?', BBC Third Programme, 15 May 1951.
31. Robertson, Alan, 'The Lovelace Test vs. The Turing Test: Do They Have Relevance to AI in 2025?', Medium, 4 February 2025, https://medium.com/@alanrobertson_85371/the-lovelace-test-vs-the-turing-test-do-they-have-relevance-to-ai-in-2025-a1e5bf2e98d4; Wakefield, Jane, 'Robots Face New Test of Creative Abilities', BBC News, 21 November 2014, https://www.bbc.co.uk/news/technology-30144069.
32. Swade, Doron, *In Our Time*; Swade, Doron, 'Turing, Lovelace, and Babbage', *The Turing Guide*, eds Jack Copeland et al. (Oxford: Oxford University Press, 2017) pp. 249–62, 506–8.
33. Babbage, Charles, *Passages from the Life of a Philosopher* (London: Longman, Green, Longman, Roberts, & Green, 1864).
34. For a record of the Lovelace–Babbage correspondence, see, Green, Christopher, D., 2000, Introduction to Ada Lovelace's Translation of, and Notes to, Luigi F. Menabrea's 'Sketch of the analytical engine invented by Charles Babbage, Esq.' (1842/1843), https://psychclassics.yorku.ca/Lovelace/intro.
35. Much of this correspondence is captured in the archives at New York Public Library and the British Library, London. For this account, see Wolfram, 'Untangling the Tale of Ada Lovelace', writings.stephenwolfram.com/2015/12/untangling-the-tale-of-ada-lovelace.
36. Ibid.
37. Woolley, *The Bride of Science*.
38. 'Child of Poetry, Bride of Science', *Guardian*, 14 August 1999, https://www.theguardian.com/books/1999/aug/14/books.guardianreview.
39. Ibid.; BBC documentary: *Calculating Ada: The Countess of Computing*, 2015.
40. Wolfram, 'Untangling the Tale of Ada Lovelace'.
41. Anon., 'Death of Lady Lovelace', *The Examiner* (4 December 1852), p. 771.
42. Letter to Charles Babbage on 5 July 1843.

Chapter 4: Eunice Foote

1. Kurland, Zoe, Hafner, Katie, Feder, Elah and the Lost Women of Science Initiative, 'The Woman Who Demonstrated the Greenhouse Effect', Scientific American, https://www.scientificamerican.com/article/the-woman-who-demonstrated-the-greenhouse-effect/.
2. Kohlstedt, Sally, quoted in Kurland et al., 'The Woman Who Demonstrated the Greenhouse Effect'.
3. 'Forgotten Founder of Climate Science: Eunice Newton Foote', BBC World Service, https://www.bbc.co.uk/sounds/play/w3ct38sx.
4. Willard, Emma, *A Plan for Improving Female Education*, 1819, http://college.cengage.com/history/ayers_primary_sources/improving_femal_education_willard.htm quoted in Perlin, unpublished manuscript.
5. Kurland et al., 'The Woman Who Demonstrated the Greenhouse Effect'.
6. Perlin, John, quoted in Brazil, Rachel, 'Eunice Foote: The Mother of Climate Change', *Chemistry World*. https://www.chemistryworld.com/culture/eunice-foote-the-mother-of-climate-change/4011315.article.
7. Interview with John Perlin, 3 June 2024.

8. Kurland et al., 'The Woman Who Demonstrated the Greenhouse Effect'.
9. Brazil, 'Eunice Foote: The Mother of Climate Change'.
10. Kurland et al., 'The Woman Who Demonstrated the Greenhouse Effect'.
11. Perlin, John, 'Following in the Footsteps of Eunice Foote, The Woman Who in 1856 Discovered Global Warming', 2023, unpublished manuscript, pp. 67–71.
12. Elizabeth Cady Stanton et al., eds, *History of Woman Suffrage*, 1881, vol. 1, p. 69, in Perlin, 'Following in the Footsteps of Eunice Foote'.
13. 'Eunice Foote & the Feminist Movement', Eunice Foote to USCSB [webpage], https://spotlight.library.ucsb.edu/starlight/eunice-foote-to-ucsb/feature/eunice-foote-the-feminist-movement.
14. Ibid.; Perlin, 'Following in the Footsteps of Eunice Foote'.
15. 'Eunice Foote & the Feminist Movement', Eunice Foote to USCSB.
16. 'Women's Rights Convention', *Seneca County Courier*, 21 July 1848, vol. 4, #20, in Perlin, 'Following in the Footsteps of Eunice Foote', p. 70.
17. Report of the Woman's Rights Convention, 19/20 July 1848, 'Declaration of Sentiments', pp. 10–11, in Perlin, 'Following in the Footsteps of Eunice Foote', p. 70.
18. Perlin, 'Following in the Footsteps of Eunice Foote', p. 70.
19. Kohlstedt, Sally, quoted in Kurland et al., 'The Woman Who Demonstrated the Greenhouse Effect'.
20. Perlin, 'Following in the Footsteps of Eunice Foote', p. 57.
21. Shapiro, Justin, Shapiro, Maura and Rein, Alison, 'Initial Conditions Episode 1: Eunice Foote: A Once Forgotten Climate Science Pioneer', AIP, https://www.aip.org/initialconditions/eunice-foote-once-forgotten-climate-science-pioneer.
22. Coulson, Thomas, *Joseph Henry: His Life and Work*, 1950, p. 194., in Perlin, 'Following in the Footsteps of Eunice Foote', p.200.
23. The Henderson Family Papers, RU 7075, Box 7, Folder 6, Smithsonian Institution Archives.
24. Henry, Joseph, *The Writings of Joseph Henry*, Volume II (Washington: The Smithsonian Institution, 1886) p.38.
25. Lincoln Phelps, Almira, *Familiar Lectures on Chemistry*, Preface, viii., 1838, in Perlin, 'Following in the Footsteps of Eunice Foote', p. 98.
26. Interview with John Perlin, 2 June 2024.
27. Perlin, John, *Let It Shine: The 6000-Year Story of Solar Energy*, 2022, p.74., in Perlin, 'Following in the Footsteps of Eunice Foote', p. 91.
28. Henry, Joseph, *Scientific Writings of Joseph Henry*, Volume II, 1886, p. 45., in Perlin, 'Following in the Footsteps of Eunice Foote', p. 92.
29. Henry, Joseph, *Scientific Writings of Joseph Henry*, Volume II, 1886, p. 46., in Perlin, 'Following in the Footsteps of Eunice Foote', p. 93.
30. Henry, Joseph, *Scientific Writings of Joseph Henry*, Volume II, 1886, p. 46., in Perlin, 'Following in the Footsteps of Eunice Foote', p. 94.
31. Fourier, Joseph, Memoire sur les Temperatures du Globe Terrestre et Espaces Planetaires, *Mem. Acad. r. Sci. Paris*, 2(7), 1827, p.587.
32. Foote, Eunice, 'Circumstances Affecting the Heat of the Sun's Rays': Art. XXXI, *The American Journal of Science and Arts,* 2nd series, v. XXII/no. LXVI, November 1856, pp. 382–3, https://ia800802.us.archive.org/4/items/mobot31753002152491/mobot31753002152491.pdf.
33. Perlin, unpublished, p. 97.
34. Ortiz, Joseph D. and Jackson, Roland, 'Understanding Eunice Foote's 1856 Experiments: Heat Absorption by Atmospheric Gases'. *Notes Rec.* 76, 2022, pp. 67–84, http://doi.org/10.1098/rsnr.2020.0031.
35. Perlin, 'Following in the Footsteps of Eunice Foote', p. 97.
36. Perlin, 'Following in the Footsteps of Eunice Foote', p. 98.
37. Foote, 'Circumstances Affecting the Heat of the Sun's Rays'.
38. Perlin, 'Following in the Footsteps of Eunice Foote', p. 99.
39. Foote, 'Circumstances Affecting the Heat of the Sun's Rays'.
40. Email correspondence, Sally Kohlstedt, 2024.
41. Translation by W. M. Connolley of Fourier 1827: MEMOIRE sur les temperatures du globe terrestre et des espaces planetaires, via http://www.wmconnolley.org.uk/sci/fourier_1827/fourier_1827.html.
42. 'Scientific Ladies: Experiments with Condensed Gases', 13 September 1856, *Scientific American*, vol. 12, p. 5, https://babel.hathitrust.org/cgi/pt?id=coo.31924080787660;view=1up;seq=9.

43. Helmholtz, Hermann, trans by John Tyndall, 'On the Interaction of Natural Forces' *Annual of Scientific Discovery or The Years of Facts in Science and Art for* 1858, p.139, in Perlin, 'Following in the Footsteps of Eunice Foote', p. 111.
44. Huddleston, Amara, 'Happy 200th Birthday to Eunice Foote, Hidden Climate Science Pioneer',Climate.gov,https://www.climate.gov/news-features/features/happy-200th-birthday-eunice-foote-hidden-climate-science-pioneer#:~:text=Henry%20then%20read%20a%20paper,159).
45. 'Scientific Ladies: Experiments with Condensed Gases', *Scientific American.*
46. Perlin, 'Following in the Footsteps of Eunice Foote', p. 112.
47. Kurland et al., 'The Woman Who Demonstrated the Greenhouse Effect'.
48. Interview with John Perlin, 3 June 2024.
49. Newton Leonard, Ermina, *Newton Genealogy*, Bernard Ammidown Leanord, De Pere, Wisconsin 1915, p. 720; see also *New York Tribune*, 26 August 1856, Volume XVI, Issue 4790, p. 7, in Perlin, 'Following in the Footsteps of Eunice Foote', p. 87.
50. Interview with John Perlin, 3 June 2024.
51. Wells, David A. (ed.), *Annual of Scientific Discovery* (Boston: Gould and Lincoln, 1857), p. 159, in Perlin, 'Following in the Footsteps of Eunice Foote'.
52. Foote, Elisha, 'ART. XXX.—On the Heat in the Sun's Rays', https://archive.org/details/mobot31753002152491/page/381/mode/2up?view=theater.
53. Perlin, 'Following in the Footsteps of Eunice Foote', p. 100.
54. Gregory Kohlstedt, Sally et al., *The Establishment of Science in America*, (New Brunswick, NJ: Rutgers University Press, 1999).
55. 'American Association for the Advancement of Science Session of 1856 Fourth Day', *New York Tribune*, 26 August 1856, p. 7, in Perlin, 'Following in the Footsteps of Eunice Foote'.
56. Quoted in Shapiro et al., 'Initial Conditions Episode 1: Eunice Foote'; Jackson, Roland, 'Eunice Foote, John Tyndall and a Question of Priority', *Notes and Records: The Royal Society Journal of the History of Science*, 74, no. 1, 2019, pp. 105–18.
57. Jackson, Roland, in Shapiro et al., 'Initial Conditions Episode 1: Eunice Foote'.
58. Tyndall, John, 'VII. Note on the Transmission of Radiant Heat Through Gaseous Bodies', *Proc. R. Soc. Lond.* 10, 1860, pp. 1037–39.
59. Brazil, 'Eunice Foote: The Mother of Climate Change'.
60. Jackson, Roland, quoted in Shapiro et al., 'Initial Conditions Episode 1: Eunice Foote'.
61. Kurland et al., 'The Woman Who Demonstrated the Greenhouse Effect'.
62. Perlin, 'Following in the Footsteps of Eunice Foote'; interview with John Perlin, 3 June 2024.
63. Perlin, in Brazil, 'Eunice Foote: The Mother of Climate Change'.
64. Shapiro et al., 'Initial Conditions Episode 1: Eunice Foote'.
65. Ibid.
66. Ibid.
67. Perlin, 'Following in the Footsteps of Eunice Foote', pp. 146–7.
68. The Henderson Family Papers 1868-1923. Box 6, Henderson Family General Correspondence 1868–1923. Folder 18 – Eunice Newton Foote, 1870–1871. Smithsonian Institution Archives.
69. Sorenson, R. P., 'Eunice Foote's Pioneering Research on CO_2 and Climate Warming: Search and Discovery Article # 70092, 2011', Search and Discovery, http://www.searchanddiscovery.com/documents/2011/70092sorenson/ndx_sorenson.pdf.
70. Schwartz, John, 'Overlooked No More: Eunice Foote, Climate Scientist Lost to History', *New York Times*, https://www.nytimes.com/2020/04/21/obituaries/eunice-foote-overlooked.html?_gl=1*85e0z5*_ga*MTkzNTg4MjUxNC4xNzE1NzY0NTIz*_ga_1CCM6YP0WF*MTcxNTc2NDUyMy4xLjAuMTcxNTc2NDUyMy42MC4wLjA.

Chapter 5: Florence Nightingale

1. DiGregorio, Sarah, 'The Problematic Myth of Florence Nightingale', 2023, https://lithub.com/the-problematic-myth-of-florence-nightingale/.
2. Correspondence with Dr Richard Bates, 18 February 2025.
3. Ibid.
4. Crawford, P., Greenwood, A., Bates, R., Memel, J., *Florence Nightingale at Home* (London: Palgrave Macmillan, 2020), p. 36.
5. Magnello, M. Eileen, 'Victorian Statistical Graphics and the Iconography of Florence Nightingale's Polar Area Graph', *BSHM Bulletin: Journal of the British Society for the History of Mathematics*, 27(1), 2012, pp. 13–37, in Crawford et al., *Florence Nightingale at Home*, p. 37.
6. Helmstadter, Carol, *Beyond Nightingale* (Manchester: Manchester University Press, 2020), p. 41.
7. Crawford et al., *Florence Nightingale at Home*, p. 37.

8. Crawford et al., *Florence Nightingale at Home*, p. 40.
9. Stark, Myra, 'Introduction' in *Florence Nightingale, Cassandra: An Essay* (New York: The Feminist Press, The City University of New York, 1979), p. 11–12.
10. Correspondence with Dr Richard Bates, 18 February 2025.
11. Bostridge, M., *Florence Nightingale: the Woman and Her Legend*, (London: Penguin, 2008), p. 51; Crawford et al., *Florence Nightingale at Home*, p. 87.
12. Nightingale, 'Lebenslauf', *CW* 1, 91, in Crawford et al., *Florence Nightingale at Home*, p. 86.
13. Helmstadter, Carol and Godden, Judith, *Nursing Before Nightingale*, (Abingdon, UK: Routledge, 2011), p. 5, 7 and 56.
14. Helmstadter and Godden, *Nursing Before Nightingale;* Correspondence with Dr Richard Bates, 18 February 2025.
15. 'The Art of Anaesthesia', Science Museum, London [webpage], https://www.sciencemuseum.org.uk/objects-and-stories/medicine/art-anaesthesia.
16. Helmstadter and Godden, *Nursing Before Nightingale.*
17. Fleming, Amy, 'Keep It Clean: The Surprising 130-Year History of Handwashing', *Guardian*, https://www.theguardian.com/world/2020/mar/18/keep-it-clean-the-surprising-130-year-history-of-handwashing#:~:text=Religious%20handwashing%20rituals%20have%20been,shut%20against%20such%20malevolent%20forces.
18. Helmstadter and Godden, *Nursing Before Nightingale;* Correspondence with Dr Richard Bates, 18 February 2025.
19. Correspondence with Dr Richard Bates, 18 February 2025.
20. Dickens, Charles, *Martin Chuzzlewit*, First Edition (London: Penguin, 2000), chapter 19.
21. Correspondence with Dr Richard Bates, 18 February 2025.
22. Strachey, Lytton, *Eminent Victorians* (New York: Chatto and Windus, 1918), p. 134.
23. Dolan, Josephine A., *History of Nursing*, Twelfth Edition (Philadelphia: W. B. Saunders, 1968), pp. 264–5; Ellis, Janice R and Hartley, Celia L., *Nursing in Today's World: Trends, Issues, and Management*, Ninth Edition (Philadelphia: Kluwer/Lippincott Williams & Wilkins, 2008), pp. 115–6; Kalisch, Philip A. and Kalisch, Beatrice J., *The Advance of American Nursing* (Philadelphia: J. B. Lippincott, 1995), pp. 79–84, in Grundmann, Christoffer H., 'The Calling of Nursing', http://thecresset.org/2010/Trinity/Grundmann_T10.html.
24. Miss Nightingale to Madame Mohl, in Cook, Sir Edward Tyas, *The Life of Florence Nightingale*, vol. 2 of 2 (London: Macmillan, 1913).
25. Dolan, *History of Nursing*, pp. 264–5; Ellis and Hartley, *Nursing in Today's World*, pp. 115–6; Kalisch and Kalisch, *The Advance of American Nursing*, pp. 79–84, in Grundmann, 'The Calling of Nursing'.
26. Poplin, Irene Schuessler, 'A Study of the Kaiserswerth Deaconess Institute's Nurse Training School in 1850–1851: Purposes and Curriculum', The University of Texas at Austin ProQuest Dissertations & Theses, 1988.
27. Ibid.
28. Donahue, Patricia, *Nursing, The Finest Art: An Illustrated History*, Second Edition. (St. Louis: C. V. Mosby, 1996), p. 200; Bostridge, *Florence Nightingale: The Woman and Her Legend*, pp.145–6, in Grundmann, 'The Calling of Nursing'.
29. 1851. The Institution of Kaiserswerth on the Rhine, for the practical training of deaconesses, under the direction of the Rev. Pastor Fliedner, embracing the support and care of a hospital, infant and industrial schools, and a female penitentiary. London, Colonial Ragged Training School, p. 3.
30. Farrell, Adrian, 'Florence Nightingale in the Crimean War', CanDoWisdom.com, https://candowisdom.com/leadership/florence-nightingale-crimea.
31. Ibid.
32. Woodham-Smith, Cecil, *Florence Nightingale: 1820–1910* (New York: McGraw-Hill Book Company, 1951), p. 74; Farrell, 'Florence Nightingale in the Crimean War'.
33. Lynn McDonald ed., *Florence Nightingale: An Introduction to her Life and Family* (Waterloo, Ontario, Canada: Wilfrid Laurier University Press, 2002), p. 327.
34. Correspondence with Dr Richard Bates, 18 February 2025.
35. Quoted in Helmstadter, *Beyond Nightingale*, p.43, from *Cook, The Life of Florence Nightingale.*
36. Helmstadter, *Beyond Nightingale*, p. 41.
37. Correspondence with Dr Richard Bates, 18 February 2025; Helmstadter, *Beyond Nightingale.*
38. Correspondence with Dr Richard Bates, 18 February 2025.
39. Helmstadter, *Beyond Nightingale*, p. 71.
40. Correspondence with Dr Richard Bates, 18 February 2025.
41. Helmstadter, *Beyond Nightingale*, pp. 73–4
42. Tunç U, 'Lessons from the Crimean War: How Hospitals were Transformed by Florence Nightingale and Others' *Infectious Diseases and Clinical Microbiology*, 2019; 2: 110–118.

43. Nightingale F., *Florence Nightingale: Measuring Hospital Care Outcomes: Excerpts From the Books Notes on Matters Affecting the Health, Efficiency, and Hospital Administration of the British Army Founded Chiefly on the Experience of the Late War, and Notes on Hospitals* (Oakbrook Terrace, Ill: Joint Commission on Accreditation of Healthcare Organizations, 1999), p. 41–5.
44. Fee, E, Garofalo, M. E., 'Florence Nightingale and the Crimean War', *Am J Public Health*, 100(9), 2010, p. 1591, doi: 10.2105/AJPH.2009.188607. Epub 2010 Jul 29. Erratum in: *Am J Public Health*, 101(5), 2011, p. 776.
45. Ibid.
46. Peszko, Magdalena,'Florence Nightingale the Angel of the Crimea', 2020. Archived at: https://web.archive.org/web/20200515075509/https://blogs.bl.uk/magnificent-maps/2020/05/florence-nightingale-the-angel-of-the-crimea.html.
47. Nightingale, F, [14 November 1854], 'Letter to Dr William Bowman' in *Florence Nightingale: Letters From the Crimea 1854–1856*, ed. S. M. Goldie (Manchester: Mandolin, 1997), pp. 36–9.
48. Aravind, M, Chung K. C., 'Evidence-Based Medicine and Hospital Reform: Tracing Origins Back to Florence Nightingale', *Plast Reconstr Surg*, 125(1), 2010, pp. 403–9, doi: 10.1097/PRS.0b013e3181c2bb89.
49. Richmond P. A., 'American Attitudes Toward the Germ Theory of Disease (1860–1880)', *J History Med Allied Sci*, 9, 1954, pp. 428–54.
50. Ibid.
51. Farrell, 'Florence Nightingale in the Crimean War'.
52. Hinton, Mike, *Victory Over Disease: Resolving the Medical Crisis in the Crimean War, 1854–1856* (Solihull: Helion, 2019): Bates, R., and Greenwood, A., 'Could Nightingale Get Cancelled? The Rise, Endurance, and Possible Fall of Florence Nightingale in British Historical Culture Since 1854', *Women's History Review*, 31(7), 2022, pp. 1080–1106, https://doi.org/10.1080/09612025.2022.2045110; Fee and Garofalo, 'Florence Nightingale and the Crimean War'.
53. Nightingale, F., [15 December 1854], 'Letter to Sidney Herbert' in *Florence Nightingale: Letters From the Crimea 1854–1856*, pp. 5–53.
54. Bates and Greenwood, 'Could Nightingale get cancelled?'
55. Hammer, Joshua, 'The defiance of Florence Nightingale', *Smithsonian magazine*, https://www.smithsonianmag.com/history/the-worlds-most-famous-nurse-florence-nightingale-180974155/#:~:text=The%20commission%20cleaned%20out%20latrines,42.7%20percent%20to%202.2%20percent.&text=Today%2C%20historians%20and%20public%20health,in%20the%20turnaround%20at%20Scutari.
56. Ibid.
57. Bostridge, *Florence Nightingale: the Woman and her Legend.*
58. Small, Hugh, 'Florence Nightingale's Statistical Diagrams', paper from *Stats & Lamps* Research Conference organised by the Florence Nightingale Museum at St. Thomas' Hospital, 18 March 1998, https://www.york.ac.uk/depts/maths/histstat/small.htm#:~:text=Like%20Minard's%2C%20Nightingale's%20most%20famous,happen%20if%20you%20reduced%20taxes.
59. Cohen, I. B., *The Triumph of Numbers: How Counting Shaped Modern Life* (New York; W. W. Norton, 2006). Chapter 9 is devoted to Florence Nightingale, p.133.
60. 'Florence Nightingale: The Pioneer Statistician', Science Museum, London [webpage], https://www.sciencemuseum.org.uk/objects-and-stories/florence-nightingale-pioneer-statistician.
61. Andrew, R. J., 'How Florence Nightingale Changed Data Visualization Forever', *Scientific American*, https://www.scientificamerican.com/article/how-florence-nightingale-changed-data-visualization-forever/.
62. McDonald, Lynn, 'Florence Nightingale: Social Theory and Methodology', Paper for the ASA Meetings, Toronto 1997, https://cwfn.uoguelph.ca/short-papers-excerpts/social-theory-and-methodology/#fn4059116644c083aacd0c61.
63. Nightingale, F., 'Note on the Supposed Protection Afforded Against Venereal Disease', London. Unpublished briefing note, 1862.
64. Miss Nightingale to Madame Mohl, in Tyas Cook, *The Life of Florence Nightingale*'.
65. Small, H., Chadwick, E., 'A Biographical Update', *J Med Biogr*, 30(2), 2022, pp. 118–124, doi: 10.1177/0967772020951083. Epub 2020 Aug 19; Correspondence with Dr Richard Bates, 18 February 2025.
66. Bates and Greenwood, 'Could Nightingale get cancelled?'
67. 'Florence Nightingale Times Obituary', Vauxhall History, accessed 15 December 2017, http://vauxhallhistory.org/florence-nightingale-times-obituary/.
68. Hammer, 'The defiance of Florence Nightingale'.
69. Bates and Greenwood, 'Could Nightingale get cancelled?'

70. Ibid.
71. McDonald, Lynn, 'The Nightingale System of Training', for a conference, 14 September 2010, Windsor, American Association for the History of Nursing and European Nursing History Group.
72. Ibid.
73. Undated notes on Netley Hospital, Add Mss 45826 f148.
74. McDonald, Lynn, 'Introduction', *Collected Works of Florence Nightingale*, (Waterloo, ON; Wilfrid Laurier University Press, 2009), vol. 12, pp. 10–1.
75. Ibid.
76. Correspondence with Dr Richard Bates, 18 February 2025.
77. Ibid.
78. McDonald, *Collected Works of Florence Nightingale*, pp. 10–1.
79. Crawford et al, *Florence Nightingale at Home*, p. 180.
80. Young, D. A. B., 'Florence Nightingale's Fever', *British Medical Journal*, 311, 1995, pp. 1697–700, as credited in Crawford et al., *Florence Nightingale at Home*, p. 179.
81. Bostridge, Mark, *Florence Nightingale: The Woman and Her Legend* (London: Penguin, 2008), p. 325, in Crawford et al, *Florence Nightingale at Home*, p. 184.
82. Nightingale to Sir John Lawrence, 26 September 1964, *CW* 9, 212.
83. Vallée, Gérard, 'Nightingale's Work on India', *CW* 9, 1–36, 27–8 in Crawford et al., *Florence Nightingale at Home*, p. 185; Crawford et al., *Florence Nightingale at Home*, p. 191–5.
84. Nightingale, Stational Reports, *CW* 9, 144. See also Nightingale, *Suggestions in Regard to Sanitary Works Required for Improving Indian Stations*, prepared by the Barrack and Hospital Improvement Commission, 15 July 1864, Blue Book (London: Eyre and Spottiswoode, 1864), *CW* 9, 302–730, 340–342, in Crawford et al., *Florence Nightingale at Home*, p. 195.
85. Kaiser-I-Hind, Bombay, 27 March 1887, translated in 'Miss Nightingale on Village Sanitation in India, Voice of India (April 1887)', p. 188–9, cited in Ichikawa, 'Writing as a Female National and Imperial Responsibility: Florence Nightingale's Scheme for Social and Cultural Reforms in England and India', *Victorian Literature and Culture*, 39, 2011, p. 103, in Crawford et al., *Florence Nightingale at Home*, p. 200.
86. Correspondence with Dr Richard Bates, 18 February 2025.
87. Penner, L., 'Engaging the Victorian Reading Public: Nightingale and the Madras Famine of 1876' in *Victorian Medicine and Social Reform. Nineteenth-Century Major Lives and Letters* (New York: Palgrave Macmillan, 2010), https://doi.org/10.1057/9780230106598_5.
88. Vallée, 'Nightingale's Work in India', *CW* 9, 10, in Crawford et al., *Florence Nightingale at Home*, p. 196.
89. McDonald, Lynn, 'Florence Nightingale: A Leading Anti-Racist', The Nightingale Society, https://nightingalesociety.com/papers/florence-nightingale-a-leading-anti-racist/.
90. Crawford et al., *Florence Nightingale at Home*, p. 196.
91. Ibid.
92. Nightingale, Florence, 'Rural Hygiene: Health Teaching in Towns and Villages', (London: Spottiswoode, 1894). Reprinted from a paper read at the Conference of Women Workers held in Leeds, 7–10 November 1893.
93. For example, Letter to Editor, *The Illustrated London News*, 7 July 1877, 22–3, *CW* 9, 761, in Crawford et al., *Florence Nightingale at Home*, p. 195.
94. Poovey, Mary, *Uneven Developments: The Ideological Work of Gender in Mid-Victorian England* (Chicago: University of Chicago Press, 1988), p. 191, in Crawford et al., *Florence Nightingale at Home*, p. 197.
95. Nightingale, F., 'Suggestions on a System of Nursing for Hospitals in India' (1865), *CW* 9, pp. 948–80.
96. Crawford et al., *Florence Nightingale at Home*, p. 197.
97. Bates and Greenwood, 'Could Nightingale get cancelled?'
98. Brooks, Grant and Nuku, Kerri, 'Why We Aren't Celebrating Florence's Birthday,' *Kai Tiaki Nursing New Zealand*, 26(3), 2020, p.34.

Chapter 6: Marie Curie

1. Harding, Warren G., 'Remarks on Presenting to Marie Curie a Gift of Radium From the American People' by Gerhard Peters and John T. Woolley, The American Presidency Project, https://www.presidency.ucsb.edu/node/359818.
2. Dry, Sarah, *Curie* (London: Haus Publishing, 2025).
3. Curie, Eve, *Madame Curie: A Biography* (Hachette, 1937), in Dry, *Curie*.
4. Ibid.
5. Dry, *Curie*.

6. Rockwell S., 'The Life and Legacy of Marie Curie', *Yale J Biol Med*, 76(4–6), 2003, p. 168.
7. Dry, *Curie*.
8. Ibid.
9. Ibid.
10. Ibid.
11. Quoted in Dry, *Curie*.
12. Hemmungs Wirtén, Eva, 'Making Marie Curie: Intellectual Property and Celebrity Culture in an Age of Information', University of Chicago, 2015, in Dry, *Curie*.
13. Coppes-Zantinga, Arty. R, and Coppes, Max J., 'Madame Marie Curie: A Giant Connecting Two Centuries', *AJR*, 171, 1998, p. 1453.
14. Dry, *Curie*.
15. Ibid.
16. Rockwell, 'The Life and Legacy of Marie Curie', p. 170.
17. Curie, *Madame Curie* in Dry, *Curie*.
18. Coppes-Zantinga and Coppes, 'Madame Marie Curie: A Giant Connecting Two Centuries', p. 1453.
19. Dry, *Curie*, p. 26 (Kindle edition).
20. Dry, *Curie*, p. 28 (Kindle edition).
21. Curie, *Pierre Curie*, p. 85, in Dry, *Curie*.
22. Dry, *Curie*.
23. Curie, *Madame Curie*, p. 119, in Dry, *Curie*.
24. Curie, *Pierre Curie*, p. 34, in Dry, *Curie*, p. 29 (Kindle edition)
25. Rockwell, 'The Life and Legacy of Marie Curie', p. 170.
26. Ibid.
27. Curie, *Madame Curie*, p. 136, in Dry, *Curie*, p. 35.
28. Des Jardins, Julie, 'Madame Curie's Passions', *Smithsonian magazine*, https://www.smithsonianmag.com/history/madame-curies-passion-74183598/.
29. Rockwell, 'The Life and Legacy of Marie Curie', p. 171.
30. Dry, *Curie*, p. 36.
31. Ibid., p. 37.
32. Des Jardins, 'Madame Curie's Passions'.
33. 'A Tale of Two Hands: The Story Behind the Two Famous Radiographs Captured by Wilhelm Roentgen', Cafe Roentgen, https://caferoentgen.com/2023/10/07/a-tale-of-two-hands-the-story-behind-the-two-famous-radiographs-captured-by-wilhelm-roentgen/; Macfarlane, Ross, 'X-Ray Anniversary' [Wellcome Library blog] https://wayback.archive-it.org/16107/20210313064104/http://blog.wellcomelibrary.org/2010/11/x-ray-anniversary/.
34. Dry, *Curie*, p. 43 (Kindle edition).
35. Grady, Denise, 'A Glow in the Dark, and a Lesson in Scientific Peril', *New York Times*, https://www.nytimes.com/1998/10/06/science/a-glow-in-the-dark-and-a-lesson-in-scientific-peril.html.
36. Rockwell, 'The Life and Legacy of Marie Curie', p. 172.
37. Dry, *Curie*, p. 47 (Kindle edition).
38. Coppes-Zantinga and Coppes, 'Madame Marie Curie: A Giant Connecting Two Centuries', p. 1454.
39. Dry, *Curie*, p. 47
40. Quinn, *Marie Curie: A Life*, 1995, p. 406, in Dry, *Curie*, p. 48 (Kindle edition).
41. Dry, *Curie*, p. 49 (Kindle edition).
42. Ibid.
43. Ibid.
44. Curie, *Madame Curie*, p. 169 in Dry, *Curie*, pp. 49–50 (Kindle edition).
45. Curie P., Curie M., Bémont, M. G., 'Sur une nouvelle substance fortement radio-active contenue dans la pechblende, *C R Acad Sci*, 127, 1898, pp. 1215–7.
46. Quinn S., *Marie Curie: A Life* (New York: Simon and Schuster; 1995).
47. Curie et al., 'Sur une nouvelle substance fortement radio-active contenue dans la pechblende'.
48. Ibid.
49. Dry, *Curie*, p. 52 (Kindle edition); Coppes-Zantinga and Coppes, 'Madame Marie Curie: A Giant Connecting Two Centuries', p. 1454; Curie, M., 'Les raies de Becquerel et le polonium', *Rev Gen Sci*, 10, 1899, pp. 41–50.
50. Coppes-Zantinga and Coppes, 'Madame Marie Curie: A Giant Connecting Two Centuries', p. 1454; Des Jardins, 'Madame Curie's Passions'.
51. Dry, *Curie*, pp. 53–4 (Kindle edition).

52. Dry, *Curie*, p. 53.
53. Stephanie Farrlow et al., eds, *The Science Book* (London: Penguin, 2025), p. 198.
54. Dry, *Curie*, p. 56 (Kindle edition).
55. Coppes-Zantinga and Coppes, 'Madame Marie Curie: A Giant Connecting Two Centuries', p. 1454.
56. Curie, *Pierre Curie*, p. 90, quoted in Dry, *Curie*, p. 59 (Kindle edition).
57. Dry, *Curie*, pp. 82–83 (Kindle edition).
58. Rose, Hilary, *Love, Power and Knowledge: Towards a Feminist Transformation of the Sciences* (Cambridge: Polity Press, 1994), p. 142.
59. 'Nobel Prize in Physics 1903', Nobel Prize, https://www.nobelprize.org/prizes/physics/1903/summary/#:~:text=share:%201/4-,Photo%20from%20the%20Nobel%20Foundation%20archive.,discovered%20by%20Professor%20Henri%20Becquerel%22.
60. 'Marie Curie – Biographical', Nobel Prize, https://www.nobelprize.org/prizes/physics/1903/marie-curie/biographical/#:~:text=Together%20with%20her%20husband%2C%20she,of%20her%20work%20in%20radioactivity.
61. Gunderman, Richard, 'The Curious History of the Novel Peace Prize', The Conversation, https://theconversation.com/the-curious-history-of-the-nobel-peace-prize-66609; Mirowski, Philip, 'The Neoliberal Ersatz Nobel Prize', in *Nine Lives of Neoliberalism*, eds Dieter Plehwe, Quinn Slobodian, Philip Mirowski (London: Verso, 2020), https://www.versobooks.com/books/3075-nine-lives-of-neoliberalism.
62. Quoted in Suddath, Claire, 'The Nobel Prize', https://time.com/archive/6909264/the-nobel-prize/.
63. Alfred Nobel's will, quoted in Friedman, Robert Marc, *The Politics of Excellence* (New York: Times Books, 2001), pp. 13–4.
64. Lichtman, M. A., 'Alfred Nobel and His Prizes: From Dynamite to DNA', *Rambam Maimonides Med J*, 8(3), 2017, e0035, doi: 10.5041/RMMJ.10311.
65. Novosad, P., Asher, S., Farquharson, C. and Iljazi, E., 'DP19551 Access to Opportunity in the Sciences: Evidence From the Nobel Laureates', CEPR Discussion Paper No. 19551 (Paris and London, CEPR Press, 2024), https://cepr.org/publications/dp19551.
66. In fact, many scientists today complain that the prize (which remains pretty faithful to Nobel's original vision) is no longer fit for purpose. The prize can still only be shared between a maximum of three recipients. Even more so than in the early twentieth century, science today relies on the sharing of knowledge, practices and materials across disciplines and nations. Science has always advanced through building on past discoveries, and recognising a single person, or even three, makes even less sense than it did in Curie's time.
67. Curie, *Pierre Curie*, pp. 103–4, in Dry, *Curie*, p. 65.
68. Curie, *Madame Curie*, pp. 205–6, in Dry, *Curie*, p. 65.
69. Dry, *Curie*, p. 66.
70. Lubenau, Joel O. and Pasquier, Jean-Luc, 'The Radioactive Remains of Pierre and Marie Curie', *The Invisible Light, Journal of The British Society of the History of Radiology*, 37, 2013, p. 12.
71. Lubenau and Pasquier, 'The Radioactive Remains of Pierre and Marie Curie', p. 13.
72. Curie, Marie, *Pierre Curie : With Autobiographical Notes*, trans. Charlotte and Vernon Kellogg (New York: Macmillan, 1923), pp. 191–2.
73. 'Marie Curie, Invisible Light, the Red Cross and WWI', British Red Cross, https://www.redcross.org.uk/stories/our-movement/our-history/marie-curie-invisible-light-the-red-cross-and-wwi.
74. Dry, *Curie*, pp. 71 75 (Kindle edition); Curie, E., *Madame Curie: A Biography by Eve Curie* (Garden City, NY: Doubleday, Doran and Company, Inc., Country Life Press, 1937); Jorgensen, Timothy J., 'Marie Curie and Her X-Ray Vehicles' Contribution to World War I Battlefield Medicine', https://theconversation.com/marie-curie-and-her-x-ray-vehicles-contribution-to-world-war-i-battlefield-medicine-83941.
75. Jorgensen, 'Marie Curie and Her X-Ray Vehicles' Contribution to World War I Battlefield Medicine'.
76. Ibid.
77. 'Frédéric and Irène Joliot-Curie', Britannica, https://www.britannica.com/biography/Frederic-and-Irene-Joliot-Curie.
78. Dry, *Curie*, p. 98 (Kindle edition).
79. Rockwell, 'The Life and Legacy of Marie Curie', p. 177; 'History of the Institut Curie' Institut Curie, https://mariecurie-usa.org/institut-curie-history/#:~:text=Institut%20Curie%20was%20known%20as%20%22Institut%20du%20Radium%22&text=When%20Marie%20Curie%20discovered%20radium,Construction%20was%20completed%20in%201914.

80. Skowronek, J., 'Current Status of Brachytherapy in Cancer Treatment – Short Overview', *J Contemp Brachytherapy*, 9(6), 2017, pp. 581–9, doi: 10.5114/jcb.2017.72607. Epub 2017 Dec 30.
81. Dry, *Curie*, p. 135 (Kindle edition).
82. Dry, *Curie*, p. 75 (Kindle edition).
83. Grady, 'A Glow in the Dark, and a Lesson in Scientific Peril'.
84. Marie Curie to Harlan S. Miner, 29 May 1925, cited in Quinn, *Marie Curie*, p. 413.
85. Irène Curie to Marie Curie, 5 August 1907, cited in Quinn, *Marie Curie*, p. 415 in Dry, *Curie*, p. 137 (Kindle edition).
86. Dry, *Curie*, p. 137 (Kindle edition).
87. Des Jardins, 'Madame Curie's Passions'.
88. Ibid.
89. Ibid.
90. Curie, *Madame Curie*, p. 386 in Dry, *Curie*, p. 135.

Chapter 7: Sabina Spielrein

1. Graf-Nold, Angela, 'The Zürich School of Psychiatry in Theory and Practice. Sabina Spielrein's Treatment at the Burghölzli Clinic in Zürich', *Journal of Analytical Psychology*, 46(1), 2001, p. 86.
2. Minder, Bernard 'A Document. Jung to Freud 1905: a Report on Sabina Spielrein', *Analytic Psychology*, 46(1), 2001, pp. 76–7; Lothane, Henry Zvi, *Tap Magazine*, 'The Untold Story of Sabina Spielrein', https://tapmagazine.org/all-articles/u9jmx46hm65ah3nutt46n6xuyjebnm.
3. Graf-Nold, 'The Zürich School of Psychiatry in Theory and Practice', pp. 73–104.
4. Quoted in Graf-Nold, 'The Zürich School of Psychiatry in Theory and Practice', p. 74.
5. Sells, Angela M., *Sabina Spielrein: The Woman and The Myth*, (Albany, NY: SUNY Press, 2017), p. 9; Jung 1904, p. 16, in Graf-Nold, 'The Zürich School of Psychiatry in Theory and Practice'.
6. Graf-Nold, 'The Zürich School of Psychiatry in Theory and Practice', p. 74.
7. Bigg, Marieke, *No Such Thing as Normal* (London: Profile Books, 2025), p. 13.
8. Álvaro González, L. C., 'Shadows at La Salpêtrière: Charcot and the Legend of the Wandering Jew', *Neurosciences and History*, 12(2), 2024, pp. 94–104.
9. Janssen, D. F., 'Krankung and Erkrankung: Sexual Trauma Before 1895', *Med Hist*, 63(4), 2019, pp. 411–34, doi:10.1017/mdh.2019.42; Library of Congress, 'Sigmund Freud: Conflict & Culture', https://www.loc.gov/exhibits/freud/; Cherry, Kentra, 'Anna O's Life and Impact on Psychology', Verywell Mind, https://www.verywellmind.com/who-was-anna-o-2795857#:~:text=At%20a%20Glance,with%20Breuer%20during%20her%20treatment.
10. Sells, *Sabina Spielrein*, p. 10.
11. Launer, John, 'Carl Jung's Relationship With Sabina Spielrein: A Reassessment', *International Journal of Jungian Studies*, 7(3), 2015, pp. 179–93.
12. Qtd. in Graf-Nold, 'The Zürich School of Psychiatry in Theory and Practice', p. 91.
13. Ibid.
14. Qtd. in Graf-Nold, 'The Zürich School of Psychiatry in Theory and Practice', p. 93.
15. Harris, A., '"Language Is There to Bewilder Itself and Others": Theoretical and Clinical Contributions of Sabina Spielrein' in *Sabina Spielrein and the Beginnings of Psychoanalysis: Image, Thought, and Language*, eds P. Cooper-White and F. B. Kelcourse (Abingdon, UK: Routledge/Taylor & Francis Group, 2019), pp. 151–94.
16. Graf-Nold, 'The Zürich School of Psychiatry in Theory and Practice', p. 98.
17. Graf-Nold, 'The Zürich School of Psychiatry in Theory and Practice', p. 85; Polonsky, Antony, 'The Position of The Jews in the Tsarist Empire, 1881–1905' in *The Jews in Poland and Russia, Vol. II* (Liverpool, UK: Liverpool University Press, 2010), pp. 3–39.
18. Wackenhut, I. and Willke, A., 'Sabina Spielrein, Survivor of Abuse and Psychoanalyst', Diss.Med. Fak. Med. Hochschule Hanover, 1995, p. 186, in Graf-Nold, 'The Zürich School of Psychiatry in Theory and Practice', p. 100.
19. Minder, B., 'A Document. Jung to Freud 1905', 2001, pp. 67–72, in Launer, 'Carl Jung's relationship with Sabina Spielrein'.
20. Launer, 'Carl Jung's Relationship With Sabina Spielrein', pp. 179–93.
21. Quoted in Launer, 'Carl Jung's Relationship With Sabina Spielrein'.
22. Carotenuto, A., ed., *Tagebuch einer heimlichen Symmetrie: Sabina Spielrein zwischen Jung und Freud*, (Freiburg: Kore, 1986).
23. Quoted in Launer, 'Carl Jung's Relationship With Sabina Spielrein'.
24. Quoted in Launer, McGuire, W., ed., *The Freud/Jung Letters* (London: Routledge & Kegan Paul, 1974), p. 235.

25. Carotenuto, *Tagebuch einer heimlichen Symmetrie*, p. 79, in Launer, 'Carl Jung's Relationship With Sabina Spielrein'.
26. Launer, 'Carl Jung's Relationship With Sabina Spielrein', pp. 179–93.
27. Graf-Nold, 'The Zürich School of Psychiatry in Theory and Practice', p. 80.
28. Spielrein, S., 'On the psychological content of a case of schizophrenia (in German)', *Jahrbuch für psychoanalytische und psychopathologische Forschungen* 3, 1911, pp. 329–400.
29. Spielrein, S. (1912). 'Destruction as Cause of Becoming', trans. S. K. Witt. *Psychoanalysis & Contemporary Thought* 18,1995, pp. 85–118; Spielrein, S., 'Die Destruktion als Ursache des Werdens', eds Eugen Bleuler and Sigmund Freud, *Jahrbuch Für Psychoanalytische* 4(1), 1912, p. 466.
30. Spielrein, 'Destruction as Cause of Becoming'.
31. Sells, *Sabina Spielrein*, p. 120.
32. Ibid.
33. Ibid.
34. Sells, *Sabina Spielrein*, p. 137.
35. Sells, *Sabina Spielrein*, pp. 111–27.
36. Spielrein, 'Die Destruktion als Ursache des Werdens,' p. 472, in Sells, *Sabina Spielrein*, p. 122.
37. Ibid.
38. Ibid.
39. Sells, *Sabina Spielrein*, pp. 122–3.
40. Sells, *Sabina Spielrein*, p. 27.
41. Freud, S., 'Beyond the Pleasure Principle', Loc 852, in Sells, *Sabina Spielrein*, p. 129.
42. Jung, C. G., 'The Letters of C. G. Jung to Sabina Spielrein,' trans. Barbara Wharton, *Journal of Analytical Psychology* 46(1), 2001, p.185., in Sells, *Sabina Spielrein*, pp. 130–131.
43. Ibid.
44. Ibid.
45. For example, Balsam, Rosemary, 'Women of the Wednesday Society: The Presentations of Drs. Hilferding, Spielrein, and Hug-Hellmuth', *American Imago* 60(3), 2003, p. 320; Sells, *Sabina Spielrein*, p. 157.
46. For example, Ovcharenko, Victor, 'Love, Psychoanalysis and Destruction', *Journal of Analytical Psychology* 44(3), 1999, p. 359, in Sells, *Sabina Spielrein*, pp. 25–6.
47. For example, most catastrophically, Carotenuto, Aldo, *A Secret Symmetry*, (New York: Pantheon, 1984; published in Italian in 1980); see also Launer, 'Carl Jung's Relationship With Sabina Spielrein'.
48. *I and Thou*, 1923, in Sells, *Sabina Spielrein*, p. 116.
49. Sells, *Sabina Spielrein*, p. 116.
50. Acton, William, *The Functions and Disorders of the Reproductive Organs* (Philadelphia: Lindsay and Blakiston, 1867), p.102, in Sells, *Sabina Spielrein*, p. 22.
51. Sells, *Sabina Spielrein*, p. 21.
52. Noth, Isabelle, '"Beyond Freud and Jung": Sabina Spielrein's Contribution to Child Psychoanalysis and Developmental Psychology', *Pastoral Psychol*, 64, 2015, pp. 279–86; Sells, *Sabina Spielrein*, p. 133, pp.179–180; Spielrein, S. (2002). Sabina Spielrein: Sämtliche Schriften, 2nd Ed. Mit einem Vorwort von Ludger Lütkehaus. Gießen: Psychosozial-Verlag/Edition Kore. (1st Ed. 1987.), pp. 199–200.
53. Rosen, Irving, 'Dr. Sabina Spielrein : Consequences of Feminism and Love', https://hekint.org/2020/12/29/dr-sabina-spielrein-consequences-of-feminism-and-love/.
54. Kojevnikov, Alexei and Rossiianov, Kirill, 'Abortions, Eugenics and Artificial Reproduction in the Soviet Union, 1920 1936', in Women, Gender, and Technosciences, 1900–*2020*, eds Grégory Dufaud and Isabelle Lémonon-Waxin (New York: Routledge, 2025).
55. Dolgova E. A., Streltsova E. A., '"Welcome to the Club": Position of Women in Soviet Science in the 1920s', *Sotsiologicheskie issledovaniya [Sociological Studies]*, no 2. 2019, pp. 97–107.
56. Miller, Martin, *Freud and the Bolsheviks: Psychoanalysis in Imperial Russia and the Soviet Union* (New Haven: Yale University Press, 1998), p. 60.
57. Sells, *Sabina Spielrein*, p. 33.
58. Etkind, Alexander, *Eros of the Impossible: The History of Psychoanalysis in Russia* (Boulder, CO: WestView, 1997), p. 5, 272 and 350, in Sells, *Sabina Spielrein*, p. 34.
59. Frankl, Viktor E., *Man's Search for Meaning*, (London: Rider, 1946; 2008 edition)
60. Sells, *Sabina Spielrein*, p. 37.
61. Sells, *Sabina Spielrein*, p. 34.
62. Ehrenburg, Ilya and Grossman, Vasily, *The Black Book: The Ruthless Murder of Jews by German-Fascist Invaders Throughout the Temporarily Occupied Regions of the Soviet Union and in the Death Camps of Poland During the War of 1941–1945* (Holocaust Library, 1980), p. 258 and 261, in Sells, *Sabina Spielrein*, p. 42.

Chapter 8: Barbara McClintock

1. Keller, Evelyn Fox, *A Feeling for the Organism.* (New York: W. H. Freeman and Company, 1983), p. 17.
2. Ibid., p. 20.
3. Ibid., p. 22.
4. McGrayne, Sharon Bertsch, *Nobel Prize Women in Science: Their Lives, Struggles, and Momentous Discoveries*, Second Edition, (Washington: Joseph Henry Press, 2001), p. 148.
5. Keller, *A Feeling for the Organism*, p. 23.
6. Ibid., p. 26.
7. Ibid., p. 27; McGrayne, *Nobel Prize Women in Science*, p. 149.
8. Keller, *A Feeling for the Organism*, pp. 27–8.
9. Ibid., p. 28.
10. McGrayne, *Nobel Prize Women in Science*, p. 149.
11. '150 Ways to Say Cornell: "...regardless of sex or color', Cornell University [webpage], https://rmc.library.cornell.edu/cornell150/exhibition/sexcolor/index.html#:~:text=The%20Cornell%20Women's%20News%20%2C%20Vol.&text=In%20May%20of%201925%20the,all%20faculty%2C%20students%20and%20staff.
12. Keller, *A Feeling for the Organism*, p. 28; Parkhurst, Genevieve, 'Does This Mean "Sex-War"?', *The Delineator*, March 1921.
13. Keller, *A Feeling for the Organism*, pp. 33–4.
14. McGrayne, *Nobel Prize Women in Science*, p. 150.
15. Account based on McClintock's recollections in Keller, *A Feeling for the Organism.*
16. Keller, *A Feeling for the Organism*, p. 31; McGrayne, *Nobel Prize Women in Science*, p. 150.
17. Keller, *A Feeling for the Organism*, p. 39.
18. Miko, Ilona, 'Gregor Mendel and the Principles of Inheritance', *Nature Education*, 1(1), 2008, p. 134, https://www.nature.com/scitable/topicpage/gregor-mendel-and-the-principles-of-inheritance-593/.
19. O'Connor, C. and Miko, I., 'Developing the chromosome theory', *Nature Education*, 1(1), 2008, p. 44; Hardy, Paul A., Zacharias, Helmut, 'Walther Flemming on Histology in Medicine 1878: A Newly Discovered Letter to His Father', *Annals of Anatomy – Anatomischer Anzeiger*, 191(2), 2009, pp. 171–85.
20. Brush, S. G., 'Nettie M. Stevens and the Discovery of Sex Determination by Chromosomes', *Isis*, 69(2), 1978, pp. 163–72, https://doi.org/10.1086/352001.
21. Morgan, T. H., 'Sex-Limited Inheritance in *Drosophila*', *Science* 32, 1910, pp. 120–2.
22. O'Connor and Miko, 'Developing the chromosome theory'.
23. Stewart, C., 'Nettie Stevens, The Matilda Project, https://www.thematildaproject.com/scientists/nettie-stevens; Morgan, T. H., 'The scientific work of Miss N. M. Stevens', *Science*, 36(928), 1972, pp. 468–70, https://doi.org/10.1126/science.36.928.468.
24. Keller, *A Feeling for the Organism*, p. 2.
25. Ibid., p. 39.
26. Ibid., p. 3.
27. Ibid., p. 44.
28. Ibid., pp. 44–5.
29. Ibid., p. 45.
30. Ibid., p. 51.
31. Ibid., p. 60; McClintock, B., 'The Order of the Genes C, Sh and Wx in Zea Mays with Reference to a Cytologically Known Point in the Chromosome', *Proc. Natl. Acad. Sci. U.S.A.*, 17(8), 1934, pp. 485–91, https://doi.org/10.1073/pnas.17.8.485.
32. Keller, *A Feeling for the Organism*, p. 63 and 73.
33. McGrayne, *Nobel Prize Women in Science*, p. 144
34. Ibid.
35. Keller, *A Feeling for the Organism*, pp. 107–112.
36. McGrayne, *Nobel Prize Women in Science*, p. 163
37. Keller, *A Feeling for the Organism*, p. 111–112.
38. McGrayne, *Nobel Prize Women in Science*, p. 164.
39. Qtd. in McGrayne, *Nobel Prize Women in Science*, p. 165.
40. Keller, *A Feeling for the Organism*, p. 112.
41. Ibid., p. 114
42. McClintock in Keller, *A Feeling for the Organism*, p. 115.
43. Keller, *A Feeling for the Organism*, p. 116.
44. McClintock in Keller, *A Feeling for the Organism*, p. 117.
45. Ibid.; McClintock in Keller, *A Feeling for the Organism*, p. 34–37.

46. Quoted in Holton, Gerald, *Thematic Origins of Scientific Thought* (London: Harvard University Press, 1988; revised edition), p. 378.
47. Keller, *A Feeling for the Organism*, p. 118.
48. Ibid., pp. 121–38; Pray, L. and Zhaurova, K., 'Barbara McClintock and the Discovery of Jumping Genes (Transposons)', *Nature Education*, 1(1), 2008, p. 169; Ravindran, S., 'Barbara McClintock and the Discovery of Jumping Genes', *Proc Natl Acad Sci U.S.A.*, 109(50), 2012, pp. 20198–9, doi: 10.1073/pnas.1219372109.
49. Keller, *A Feeling for the Organism*, pp. 121–38; Ravindran, 'Barbara McClintock and the Discovery of Jumping Genes'.
50. Keller, *A Feeling for the Organism*, p. 122 and pp. 92–8.
51. Ibid., p.130.
52. McClintock, Barbara, 'Chromosome Organisation and Gene Expression', *Cold Spring Harbor Symposia on Quantitative Biology* 16, 1951, p. 34.
53. Comfort, Nathaniel C., *The Tangled Field: Barbara McClintock's Search for the Patterns of Genetic Control* (Harvard University Press, 2001), p. 121; McGrayne, *Nobel Prize Women in Science*, p. 168.
54. McClintock in Keller, *A Feeling for the Organism*, p. 126.
55. McGrayne, *Nobel Prize Women in Science*, p. 168.
56. Ibid.
57. Ibid.
58. Keller, *A Feeling for the Organism*, p. 140.
59. Ibid., p. 142.
60. Comfort, *The Tangled Field*, p. 176.
61. In McGrayne, *Nobel Prize Women in Science*, p. 169.
62. Ibid.
63. In McGrayne, *Nobel Prize Women in Science*, p. 170.
64. Keller, *A Feeling for the Organism*, p. 189.
65. Ibid., p. 187.
66. Comfort, *The Tangled Field*, p. 2.
67. McGrayne, *Nobel Prize Women in Science*, p. 171; Doudna, J., 'By Studying Corn, Barbara McClintock Unlocked the Secrets of Life', *Smithsonian magazine*, https://www.smithsonianmag.com/smithsonian-institution/by-studying-corn-barbara-mcclintock-unlocked-secrets-life-180981555/
68. Pray and Zhaurova, 'Barbara McClintock and the Discovery of Jumping Genes (transposons)'.
69. For example, Keller, *A Feeling for the Organism.*
70. For example, Comfort, *The Tangled Field.*

Chapter 9: Dorothy Hodgkin

1. Ferry, Georgina, *Dorothy Hodgkin: A Life* (London: Granta,1998), p. 224.
2. Sharkey, Heather J., 'Mover and Shaker: Grace Mary Crowfoot, Intimate Conversations, and Sudanese History', *Égypte/Monde arabe* [Online], vol.14, 2016, doi: https://doi.org/10.4000/ema.3606.
3. Ibid.
4. Dodson, Eleanor, 'Dorothy Hodgkin (1910–1994): Crystallographer, Chemist, and Role Model' in *Women in European Academies: From Patronae Scientiarum to Path-Breakers*, eds Ute Frevert, Ernst Osterkamp and Günter Stock (Berlin, Boston: De Gruyter Akademie Forschung, 2021), pp. 307–20; Interview with Georgina Ferry, Judith Howard and Patricia Fara, *In Our Time: Dorothy Hodgkin*, BBC radio, https://www.bbc.co.uk/sounds/play/m0008wkk.
5. Dodson, 'Dorothy Hodgkin (1910–1994)'.
6. Ferry, *Dorothy Hodgkin: A Life*, p. 12.
7 Dodson, 'Dorothy Hodgkin (1910–1994)', p.309.
8. Quoted in Ferry, Georgina, 'Dorothy Hodgkin: on Proteins and Patterns', *The Lancet*, 384(9953), 2014, pp. 1496–7.
9. Ibid.
10. Dodson, 'Dorothy Hodgkin (1910–1994)', p.309.
11. 'Dorothy Hodgkin – Biographical', Nobel Prize, https://www.nobelprize.org/prizes/chemistry/1964/hodgkin/biographical/.
12. Ferry, 'Dorothy Hodgkin: on Proteins and Patterns'.
13. Ibid.

14. 'The Exceptional Life of Dorothy Crowfoot Hodgkin', BBC, https://www.bbc.co.uk/ideas/videos/the-exceptional-life-of-dorothy-crowfoot-hodgkin/p09x02lh.
15. Dodson, 'Dorothy Hodgkin (1910–1994)', p.310.
16. Yount, Lisa, *A to Z of Women in Science and Math*, (New York: Infobase Publishing, 2008), p. 128.
17. Ibid.
18. Rossmann, M., 'Full Biography of Dorothy Hodgkin', *Nat Struct Mol Biol*, 6, 1999, p. 412 (1999), https://doi.org/10.1038/8209; Dodson, 'Dorothy Hodgkin (1910–1994)', pp. 310–11; 'Dorothy Crowfoot Hodgkin. Nobel Prize in Chemistry 1964', Nobel Prize, https://www.nobelprize.org/womenwhochangedscience/stories/dorothy-hodgkin; Donald, Athene, 'Dorothy Hodgkin and the Year of Crystallography', *Guardian*, https://www.theguardian.com/science/occams-corner/2014/jan/14/dorothy-hodgkin-year-of-crystallography.
19. Dodson, 'Dorothy Hodgkin (1910–1994)', p. 311.
20. See also: Bigg, M., *Anne McLaren and the Human Embryo Research Debates in Britain, 1982–1990: Sociological Biography in the Analysis of Public Debate* [Apollo – University of Cambridge Repository, 2019], https://doi.org/10.17863/CAM.57928.
21. Bernal, John Desmond and Crowfoot, Dorothy, 'X-Ray Photographs of Crystalline Pepsin', *Nature*, 133, 1934, pp. 794–5, in Dodson, 'Dorothy Hodgkin (1910–1994)', p. 311.
22. 'Our History', Cavendish Laboratory, Department of Physics, University of Cambridge [webpage], https://www.phy.cam.ac.uk/about/our-history/.
23. Gonçalves-Maia, R., 'From X-rays to Biomolecular Structure: D. Hodgkin, R. Franklin and A. Yonath', *Rev. Virtual Quim*, 4(6), 2012, pp. 818–83.
24. Yount, *A to Z of Women in Science and Math*.
25. Ibid.
26. Donald, 'Dorothy Hodgkin and the Year of Crystallography'.
27. Yount, *A to Z of Women in Science and Math*.
28. Kelly, Patrick, 'Penicillin: From Fleming to the Pharmacy', https://www.youtube.com/watch?v=mhXmkDapHWg.
29. Fleming, A., 'On the Antibacterial Action of Cultures of a Penicillium, with Special Reference to Their Use in the Isolation of *B. influenzae.*', *Br J Exp Pathol*, (10)3, 1929, pp. 226–36.
30. Rosen, William, *Miracle Cure* (New York: Viking Press, 2018).
31. Hutchings, M. I., Truman, A. W., Wilkinson, B., 'Antibiotics: Past, Present and Future', *Curr Opin Microbiol*, 51, 2019, pp. 72–80, doi: 10.1016/j.mib.2019.10.008. Epub 2019 Nov 13; Davies, Sally and Sugden, Rebecca, 'What if antibiotics stopped working?', The King's Fund, https://www.kingsfund.org.uk/insight-and-analysis/long-reads/nhs-if-antibiotics-stopped-working.
32. Gaynes, R., 'The Discovery of Penicillin—New Insights After More Than 75 Years of Clinical Use', *Emerg Infect Dis*, 23(5), 2017, pp. 849–53, doi: 10.3201/eid2305.161556; Bud, R., 'Upheaval in the Moral Economy of Science? Patenting, Teamwork and the World War II Experience of Penicillin', *History and Technology*, 24(2), 2008, pp. 173–90, https://doi.org/10.1080/07341510701810955; American Chemical Society International Historic Chemical Landmarks, 'Discovery and Development of Penicillin', http://www.acs.org/content/acs/en/education/whatischemistry/landmarks/flemingpenicillin.html.
33. Yount, *A to Z of Women in Science and Math*.
34. Crowfoot, Dorothy, Bunn, Charles W., Rogers-Low, Barbara W. and Turner-Jones, A., 'The X-Ray Crystallographic Investigation of the Structure of Penicillin' in *The Chemistry of Penicillin*, eds H. T. Clarke, J. R. Johnson and R. Robinson (Princeton: Princeton University Press, 1949), pp. 310–367.
35. Yount, *A to Z of Women in Science and Math*.
36. Ibid.
37. Ibid.
38. Ibid.
39. Hodgkin, Dorothy Crowfoot, Kamper, Jennifer, Lindsey, June, MacKay, Maureen, Pickworth, Jenny, Robertson, John Harry, Shoemaker, Clara Brink, White, Johnson G., Prosen, R.J. and Trueblood, Kenneth Nyitray. 'The Structure of Vitamin B12. I. An Outline of the Crystallographic Investigation of Vitamin B12', *Proc R Soc Lond A*, 242, 1957, pp. 228–63; Hodgkin, Dorothy Crowfoot, Pickworth, Jenny, Robertson, John Harry, Prosen, R. J., Sparks, Robert A. and Trueblood, Kenneth Nyitray, 'The Structure of Vitamin B12. II. The Crystal Structure of a Hexacarboxylic Acid Obtained by the Degradation of Vitamin B', *Proc R Soc Lond A*, 251, 1959, pp. 306–52.
40. Yount, *A to Z of Women in Science and Math*.
41. Quoted in Yount, *A to Z of Women in Science and Math*, p. 128; Crowfoot, Dorothy et al., 'The X-ray Crystallographic Investigation of the Structure of Penicillin' in *The Chemistry of Penicillin*; Dodson, 'Dorothy Hodgkin (1910–1994)', p. 313.

42. Ferry, 'Dorothy Hodgkin: on Proteins and Patterns'.
43. Yount, *A to Z of Women in Science and Math.*
44. Ibid., p. 129.
45. Ferry, 'Dorothy Hodgkin: on Proteins and Patterns'.
46. Donald, 'Dorothy Hodgkin and the Year of Crystallography'.
47. Ferry, Georgina, *Dorothy Crowfoot Hodgkin: Patterns, Proteins and Peace: A Life in Science*, (London: Bloomsbury, 2020); McWilliams, Brittany, 'Crystal-Clear Penicillin, 1945', TS-Digest, The Scientist, https://www.the-scientist.com/crystal-clear-penicillin-1945-71026.
48. Ferry, *Dorothy Hodgkin: A Life*, p. 167
49. Donald, 'Dorothy Hodgkin and the Year of Crystallography'.
50. Perutz, M. F., address delivered at a memorial service in the University Church, Oxford, on 4 March 1995. Accessed at https://www.crystallography.org.uk/old-bca-website/obits/dh.html.
51. Glusker, Jenny P., 'Obituary: Dorothy Crowfoot Hodgkin (1910–1994)', *Protein Science* 3, 1994, pp. 2465–9.
52. Ferry, *Dorothy Hodgkin: A Life*, p. 224.
53. McWilliams, 'Crystal-Clear Penicillin, 1945'.
54. Ibid.
55. Gamble, J., 'Politics: When Hodgkin Met Thatcher', *Nature*, 514, 2014, p. 304, https://doi.org/10.1038/514304a.
56. Donald, 'Dorothy Hodgkin and the Year of Crystallography'.
57. 'Thatcher and Hodgkin: How Chemistry Overcame Politics', BBC News, https://www.bbc.co.uk/news/uk-politics-28801302.
58. Dodson, 'Dorothy Hodgkin (1910–1994)', p. 315; 'Thatcher and Hodgkin : How Chemistry Overcame Politics', BBC News.
59. Dorothy Crowfoot Hodgkin. Nobel Prize in Chemistry 1964', Nobel Prize.
60. McGrayne, S. B., *Nobel Prize Women in Science* (New York: Carol Publishing Group, 1993) in Glusker, 'Obituary: Dorothy Crowfoot Hodgkin (1910–1994)'.
61. Marshall, Katherine, 'Sleight of Hand', The Royal Society blog, https://royalsociety.org/blog/2021/09/sleight-of-hand/; Glusker, 'Obituary: Dorothy Crowfoot Hodgkin (1910–1994)'.
62. Ferry, Georgina, '"I Think With My Hands" – Scientist Dorothy Crowfoot Hodgkin', Event at the Leeds Philosophical and Literary Society, https://www.leedsphilandlit.org.uk/shop/i-think-with-my-hands-scientist-dorothy-crowfoot-hodgkin/.

Chapter 10: Chien-Shiung Wu

1. Yuan, Jada, 'Discovering Dr. Wu', *Washington Post*, https://www.washingtonpost.com/lifestyle/2021/12/13/chien-shiung-wu-biography-physics-grandmother/.
2. Hammond, Richard, *Chien-Shiung Wu: Pioneering Nuclear Physicist* (New York: Infobase Publishing, 2010), p.17.
3. Yuan, 'Discovering Dr. Wu'; Weinstock, Maia, 'Chien-Shiung Wu: Courageous Hero of Physics', https://findingada.com/shop/a-passion-for-science-stories-of-discovery-and-invention/chien-shiung-wu-courageous-hero-of-physics/.
4. Weinstock, 'Chien-Shiung Wu: Courageous Hero of Physics'.
5. Yuan, 'Discovering Dr. Wu'.
6. Weinstock, 'Chien-Shiung Wu: Courageous Hero of Physics'; Broyles, Michael L., 'Cyclotrons', EBSCO Knowledge Advantage, EBSCO, https://www.ebsco.com/research-starters/science/cyclotrons; '2. The Cyclotron', Bancroft Library, UC Berkeley, https://web.archive.org/web/20120527183442/http://bancroft.berkeley.edu/Exhibits/physics/bigscience02.html.
7. Yuan, 'Discovering Dr. Wu'.
8. For example, Blakemore, Erin, 'How Chien-Shiung Changed the Laws of Physics', *National Geographic*, https://www.nationalgeographic.com/premium/article/chien-shiung-wu-chinese-american-woman-physics-manhattan-project; McGrayne, Sharon Bertsch, *Nobel Prize Women in Science: Their Lives, Struggles, and Momentous Discoveries*. Second Edition, (Washington: Joseph Henry Press, 2001); Yuan, 'Discovering Dr. Wu'.
9. Blakemore, 'How Chien-Shiung Changed the Laws of Physics'; McGrayne, *Nobel Prize Women in Science*; Tsjeng, Zing, *Forgotten Women: The Scientists* (London: Hachette, 2018).
10. Quoted in McGrayne, *Nobel Prize Women in Science*, p. 260.

11. McGrayne, *Nobel Prize Women in Science*, p. 263.
12. Ibid.
13. Meitner, L., Frisch, O.R., 'Disintegration of Uranium by Neutrons: a New Type of Nuclear Reaction', *Nature*, 143,1939, pp. 239–40, https://doi.org/10.1038/143239a0; Frisch, O., 'Physical Evidence for the Division of Heavy Nuclei Under Neutron Bombardment', *Nature*, 143, 1939, p. 276, https://doi.org/10.1038/143276a0.
14. Nomination Archive: Lise Meitner, Nobel Prize, https://www.nobelprize.org/nomination/archive/show_people.php?id=6097.
15. Weinstock, 'Chien-Shiung Wu: Courageous Hero of Physics'.
16. Yuan, 'Discovering Dr. Wu'.
17. McGrayne, *Nobel Prize Women in Science*, p. 262.
18. Ibid.
19. Yuan, 'Discovering Dr. Wu'.
20. McGrayne, *Nobel Prize Women in Science*, p. 264.
21. Quoted in McGrayne, *Nobel Prize Women in Science*, p. 264.
22. McGrayne, *Nobel Prize Women in Science*, p. 263.
23. Yuan, 'Discovering Dr. Wu'; McGrayne, *Nobel Prize Women in Science*, p. 263.
24. Yuan, 'Discovering Dr. Wu'; McGrayne, p. 263
25. McGrayne, *Nobel Prize Women in Science*, p. 265.
26. Ibid.
27. Blakemore, 'How Chien-Shiung Changed the Laws of Physics'.
28. Yuan, 'Discovering Dr. Wu'; Chiang, Tsai-Chien, *Madame Wu Chien-Shiung: The First Lady of Physics Research* (Singapore: World Scientific, 2014), p. 98.
29. '1998 Research Corporation Award' interview in *New York Post*, 22 January 1959, in Chiang, *Madame Wu Chien-Shiung*, p. 98.
30. Chiang, *Madame Wu Chien-Shiung*, p. 98.
31. Ibid.
32. 'When State Politics Meets Gender Politics: Chien-Shiung Wu and the Manhattan Project', Lady Science, https://www.ladyscience.com/where-state-politics-meets-gender-politics-chienshiung-wu-and-the-manhattan-project/no40.
33. Ibid.; McGrayne, *Nobel Prize Women in Science*, p. 266.
34. Chiang, *Madame Wu Chien-Shiung*, p. 98.
35. McGrayne, *Nobel Prize Women in Science*, p. 272.
36. Yuan, 'Discovering Dr. Wu'.
37. McGrayne, *Nobel Prize Women in Science*, p. 273.
38. Quoted in Lederman, Leon M., Teresi, Dick, *The God Particle: If the Universe Is the Answer, What Is the Question?* (Boston, MA: Houghton Mifflin Harcourt, 1993).
39. McGrayne, *Nobel Prize Women in Science*, p. 273.
40. Yuan, 'Discovering Dr. Wu'.
41. McGrayne, *Nobel Prize Women in Science*, p. 273.
42. Ibid.
43. Yuan, 'Discovering Dr. Wu'.
44. Ibid.
45. McGrayne, *Nobel Prize Women in Science*, p. 273.
46. Chiang, Tsai-Chien, 'Wu Chien-Shiung: A Brief Biography', *AIP Conf Proc*, 1697(1), 2015, 040004. https://doi.org/10.1063/1.4937640; Chiang, *Madame Wu Chien-Shiung*.
47. Yuan, 'Discovering Dr. Wu'.
48. Ibid.
49. McGrayne, *Nobel Prize Women in Science*, p. 276.
50. Yuan, 'Discovering Dr. Wu'.
51. Ibid.
52. Ibid.
53. McGrayne, *Nobel Prize Women in Science*, p. 276.

Chapter 11: Frances Oldham Kelsey

1. *Dr. Frances Kelsey, The Devil in the Details: Chapter One* podcast, Lost Women of Science Initiative, https://www.lostwomenofscience.org/podcast-episodes/the-devil-in-the-details-chapter-one.
2. Ibid.

3. Oldham Kelsey, Frances, 'Autobiographical Reflections' (U.S. Food and Drug Administration, 1991–1993), p. 3.
4. Ibid.
5. Warsh, Cheryl, *Frances Oldham Kelsey, the FDA, and the Battle Against Thalidomide* (Oxford: Oxford University Press, 2024).
6. Kelsey, 'Autobiographical Reflections', p. 21; *Dr. Frances Kelsey, The Devil in the Details: Chapter One* podcast, Lost Women of Science Initiative.
7. *Dr. Frances Kelsey, The Devil in the Details: Chapter One* podcast, Lost Women of Science Initiative.
8. Lam C., Patel P., 'Food, Drug, and Cosmetic Act' [Updated 2023 Jul 31] (Treasure Island (FL): StatPearls [Internet]), https://www.ncbi.nlm.nih.gov/sites/books/NBK585046/; *Dr. Frances Kelsey, The Devil in the Details: Chapter One* podcast, Lost Women of Science Initiative; Kelsey, 'Autobiographical Reflections', p. 97.
9. Shanks, Dennis G., 'Antimalarial Drug Supply Issues During the Second World War', JMVH, 32(3), 2024, pp.34–7, https://jmvh.org/article/antimalarial-drug-supply-issues-during-world-war-ii/.
10. *Dr. Frances Kelsey, The Devil in the Details: Chapter Three* podcast, Lost Women of Science Initiative, https://www.lostwomenofscience.org/podcast-episodes/the-devil-in-the-details-chapter-three; Kelsey, 'Autobiographical Reflections', pp. 28–9.
11. Christine Kelsey (daughter) recalls in *Dr. Frances Kelsey, The Devil in the Details: Chapter One* podcast, Lost Women of Science Initiative.
12. Kelsey, 'Autobiographical Reflections'.
13. Ibid., p.31.
14. Ibid., p.29.
15. Ibid., p.32.
16. *Dr. Frances Kelsey, The Devil in the Details: Chapter One* podcast, Lost Women of Science Initiative.
17. Ibid.
18. Kelsey, 'Autobiographical Reflections', p. 47.
19. Ibid., p.33.
20. U.S. FDA, 'About FDA: Mary Engle Pennington: The 'Cold Chain' of Food Safety', https://www.fda.gov/about-fda/fda-history-exhibits/mary-engle-pennington-cold-chain-food-safety, in Warsh, *Frances Oldham Kelsey*.
21. See: Young, James Harvey, *Pure Food: Securing the Federal Food and Drugs Act of 1906* (Princeton, NJ: Princeton University Press, 1989).
22. Warsh, *Frances Oldham Kelsey*, pp. 15–7.
23. Ibid., p. 17.
24. Ibid., p. 44.
25. *Dr. Frances Kelsey, The Devil in the Details: Chapter One* podcast, Lost Women of Science Initiative.
26. Kelsey, 'Autobiographical Reflections', p. 54.
27. Ibid.
28. *Dr. Frances Kelsey, The Devil in the Details: Chapter One* podcast, Lost Women of Science Initiative.
29. Ibid.
30. Kelsey, 'Autobiographical Reflections', p. 45; *Dr. Frances Kelsey, The Devil in the Details: Chapter One* podcast, Lost Women of Science Initiative.
31. Kelsey, 'Autobiographical Reflections', p.47.
32. Ibid., p.49.
33. Ibid., p.51
34. Ibid., p. 52; *Dr. Frances Kelsey, The Devil in the Details: Chapter One* podcast, Lost Women of Science Initiative.
35. Ibid.
36. Ibid.; *Dr. Frances Kelsey, The Devil in the Details: Chapter One* podcast, Lost Women of Science Initiative.
37. Kelsey, 'Autobiographical Reflections', p. 63 and 73; *Dr. Frances Kelsey, The Devil in the Details: Chapter One* podcast, Lost Women of Science Initiative.
38. *Dr. Frances Kelsey, The Devil in the Details: Chapter Two* podcast, Lost Women of Science Initiative, https://www.lostwomenofscience.org/podcast-episodes/the-devil-in-the-details-chapter-two; Vanderbes, Jennifer, *Wonder Drug: The Hidden Victims of America's Secret Thalidomide Scandal* (London: Harper Collins, 2023), p. 65.

39. *Dr. Frances Kelsey, The Devil in the Details: Chapter Two* podcast, Lost Women of Science Initiative.
40. Florence, A. L., 'Is Thalidomide to Blame?' *Br Med J*, 2, 1960, p. 1954, doi:10.1136/bmj.2.5217.1954.
41. *Dr. Frances Kelsey, The Devil in the Details: Chapter Two* podcast, Lost Women of Science Initiative.
42. Kelsey, 'Autobiographical Reflections', p.55.
43. *Dr. Frances Kelsey, The Devil in the Details: Chapter Two* podcast, Lost Women of Science Initiative.
44. Ibid.; Kelsey, 'Autobiographical Reflections', p.57.
45. Ibid., p.59.
46. *Dr. Frances Kelsey, The Devil in the Details: Chapter Two* podcast, Lost Women of Science Initiative.
47. *Dr. Frances Kelsey, The Devil in the Details: Chapter Three* podcast, Lost Women of Science Initiative, https://www.lostwomenofscience.org/podcast-episodes/the-devil-in-the-details-chapter-three.
48. *Dr. Frances Kelsey, The Devil in the Details: Chapter Three* podcast, Lost Women of Science Initiative; Kelsey, 'Autobiographical Reflections', p. 62.
49. Ibid., p. 63.
50. *Dr. Frances Kelsey, The Devil in the Details: Chapter Three* podcast, Lost Women of Science Initiative.
51. *Dr. Frances Kelsey, The Devil in the Details: Chapter One* podcast, Lost Women of Science Initiative.
52. *Dr. Frances Kelsey, The Devil in the Details: Chapter Three* podcast, Lost Women of Science Initiative; Nulsen, R. O., 'Trial of Thalidomide in Insomnia Associated With the Third Trimester', *Am J Obstet Gynecol*, 81, 1961, pp. 1245–8, doi: 10.1016/s0002-9378(15)33360-3.
53. *Dr. Frances Kelsey, The Devil in the Details: Chapter Three* podcast, Lost Women of Science Initiative.
54. *Dr. Frances Kelsey, The Devil in the Details: Chapter Four* podcast, Lost Women of Science Initiative, https://www.lostwomenofscience.org/podcast-episodes/the-devil-in-the-details-chapter-four.
55. Ibid.; Kelsey, 'Autobiographical Reflections', p.67.
56. Ibid., p. 69.
57. *Dr. Frances Kelsey, The Devil in the Details: Chapter Four* podcast, Lost Women of Science Initiative.; Kelsey, 'Autobiographical Reflections', p. 69.
58. Ibid., p. 70.
59. *Dr. Frances Kelsey, The Devil in the Details: Chapter Four* podcast, Lost Women of Science Initiative.
60. Warsh, *Frances Oldham Kelsey*.
61. Mintz, Morten, '"Heroine" of FDA Keeps Bad Drug Off Market', *Washington Post*. Access via https://s3.amazonaws.com/s3.documentcloud.org/documents/2206185/archives-heroine-of-fda-keeps-bad-drug-off-the.pdf.
62. President John F. Kennedy, News Conference 40, 1 August 1962. State Department Auditorium, Washington, DC accessed at https://www.jfklibrary.org/archives/other-resources/john-f-kennedy-press-conferences/news-conference-40#:~:text=THE%20PRESIDENT%3A%20Good%20afternoon.,harmful%20or%20worthless%20drug%20products.
63. *Dr. Frances Kelsey, The Devil in the Details: Chapter Five* podcast, Lost Women of Science Initiative, https://www.lostwomenofscience.org/podcast-episodes/the-devil-in-the-details-chapter-five.
64. 'Frances Oldham Kelsey: Medical reviewer famous for averting a public health tragedy', FDA, https://www.fda.gov/about-fda/fda-history-exhibits/frances-oldham-kelsey-medical-reviewer-famous-averting-public-health-tragedy.
65. *Dr. Frances Kelsey, The Devil in the Details: Chapter Five* podcast, Lost Women of Science Initiative; Kelsey, 'Autobiographical Reflections', p. 78.
66. Kendall, Laura, *My Prosthetic Life: A Memoir*, independently published, 2022, quoted in *Dr. Frances Kelsey, The Devil in the Details: Chapter Five* podcast, Lost Women of Science Initiative.
67. Kelsey, 'Autobiographical Reflections', p. 58.
68. *Dr. Frances Kelsey, The Devil in the Details: Chapter Five* podcast, Lost Women of Science Initiative; USA Thalidomide Survivors [website] https://usthalidomide.org/; 'The True Story of Thalidomide in the US', USA Thalidomide Survivors, https://usts-013021.wpmudev.host/our-story-thalidomide-babies-us/.

Chapter 12: Marie Nyswander

1. *The Man Who Declared War on Drugs* podcast, On the Media, WNYC, https://www.wnyc.org/story/man-who-declared-war-drugs/.

2. Trickey, Erick, 'Inside the Story of America's 19th-Century Opiate Addiction', *Smithsonian magazine*, https://www.smithsonianmag.com/history/inside-story-americas-19th-century-opiate-addiction-180967673/.
3. 'History of Heroin', UN Office on Drugs and Crime, https://www.unodc.org/unodc/en/data-and-analysis/bulletin/bulletin_1953-01-01_2_page004.html.
4. Interview with David Courtwright, 5 June 2024.
5. *Honolulu Star-Bulletin*, in sapiens.org/culture/mead-freeman/
6. McBroom, Pat, 'World Pioneer in Public Health Education Dies in Berkeley at the Age of 104', News Archive, UC Berkeley, https://newsarchive.berkeley.edu/news/media/releases/98legacy/12-21-1998.html.
7. *The Doctor & The Fix: Episode 1* podcast, Lost Women of Science Initiative, https://www.lostwomenofscience.org/podcast-episodes/season-4-episode-1-the-accidental-calling; Telephone interview: David Courtwright with Dorothy Bird Nyswander, 17 February 1992 (courtesy of DC).
8. Biographical information has been assimilated largely from David Courtwright's 1981 Oral History interview, as well as the *Lost Women of Science* podcast, with special thanks to Dr Courtwright and the podcast producers for providing additional primary source material; *The Doctor & The Fix: Episode 1* podcast, Lost Women of Science Initiative.
9. Campbell, Nancy D., *Discovering Addiction: The Science and Politics of Substance Abuse Research.* (Michigan: University of Michigan Press, 2007), JSTOR, http://www.jstor.org/stable/j.ctvnjbdtz.
10. *The Doctor & The Fix: Episode 1* podcast, Lost Women of Science Initiative.
11. Campbell, *Discovering Addiction.*
12. *The Doctor & The Fix: Episode 1* podcast, Lost Women of Science Initiative; Campbell, *Discovering Addiction.*
13. Courtwright, David, *Dark Paradise: A History of Opiate Addiction in America* (Harvard University Press, 2001), in Glenn, Melody, *Mother of Methadone*, (Boston, MA: Beacon Press, 2025).
14. Glenn, Melody, *Mother of Methadone*, (Boston, MA: Beacon Press, 2025).
15. Hari, Johann, Chasing the Scream: The Search for the Truth About Addiction (London: Bloomsbury, 2015), in Glenn, *Mother of Methadone.*
16. Hari, Johann, *Chasing the Scream: The Search for the Truth About Addiction* (London: Bloomsbury, 2015).
17. Ibid.
18. Oral History Interview, 1981.
19. Ibid., p.14.
20. Center for Racial and Disability Justice, 'The Long Shadow of Institutionalization: How History Warns Against Trump and RFK Jr.'s 'Wellness Farms', Medium, https://nlawcrdj.medium.com/the-long-shadow-of-institutionalization-how-history-warns-against-trump-and-rfk-jr-s-4fb495ffaf3d; Campbell, Nancy D., Olsen, J. P. and Walden, Luke, *The Narcotic Farm: The Rise and Fall of America's First Prison for Drug Addicts* (Lexington, KY; The University Press of Kentucky, 2021).
21. According to one account in *The Doctor & The Fix: Episode 1* podcast, Lost Women of Science Initiative.
22. Oral History Interview, 1981, p. 47.
23. Ibid., Nyswander.
24. Oral History Interview, 1981.
25. Ibid., p.21.
26. *The Doctor & The Fix: Episode 2* podcast, Lost Women of Science Initiative, https://www.lostwomenofscience.org/podcast-episodes/season-4-episode-2-the-pleasure-center.
27. Nyswander, M., 'Withdrawal Treatment of Drug Addiction', *N Engl J Med*, 242(4), 1950, pp.128–30.
28. Oral History Interview, 1981, p. 23.
29. Interview with David Courtwright, 5 June, 2024; Nyswander Robinson, Marie, *The Power of Sexual Surrender* (New York: Doubleday & Company, 1959), Preface.
30. Nyswander, *The Power of Sexual Surrender.*
31. Ibid.
32. Oral History Interview, 1981, p. 23
33. Ibid.
34. 'History of Heroin', UN Office on Drugs and Crime.
35. Foreign Policy Association, Pamphlet No. 24. Series of 1923–1924, 4.
36. Courtwright, David, 'When Gotham Was Heroin's Capital', *Vital*, https://www.vitalcitynyc.org/articles/history-of-opioids-in-new-york-city.

37. *The Doctor & The Fix: Episode 2* podcast, Lost Women of Science Initiative; also see: Zimmering, Paul, Toolan, James, Safrin, Renate, Wortis, S. Bernard, 'Heroin Addiction in Adolescent Boys', *The Journal of Nervous and Mental Disease* 114(1), 1951, pp. 19–34.
38. *The Doctor & The Fix: Episode 2* podcast, Lost Women of Science Initiative.
39. 'The Treatment of Patients-II', *The New Yorker*, 26 June 1965, p. 32, quoted in *The Doctor & The Fix: Episode 2* podcast, Lost Women of Science Initiative.
40. Ibid.
41. *The Doctor & The Fix: Episode 2* podcast, Lost Women of Science Initiative.
42. Dufton, Emily, in *The Doctor & The Fix: Episode 2* podcast, Lost Women of Science Initiative.
43. Dufton, Emily, quoted in *The Doctor & The Fix: Episode 4* podcast, Lost Women of Science Initiative, https://www.lostwomenofscience.org/podcast-episodes/season-4-episode-4-the-morality-of-molecules.
44. Hari, *Chasing the Scream*, in Glenn, *Mother of Methadone.*
45. Hari, *Chasing the Scream.*
46. Ibid, p. 18.
47. Ibid.
48. M. Nyswander interview in Glenn, *Mother of Methadone.*
49. *The Doctor & The Fix: Episode 3* podcast, Lost Women of Science Initiative, https://www.lostwomenofscience.org/podcast-episodes/season-4-episode-3-the-prepared-mind.
50. Ibid.
51. Ibid.
52. Ibid.
53. Ibid.
54. *The Doctor & The Fix: Episode 2* podcast, Lost Women of Science Initiative.
55. Mary Jeanne Kreek, 2017, Oral History interview, in *The Doctor & The Fix: Episode 3* podcast, Lost Women of Science Initiative ; Interview with Mary Jeanne Kreek, The Rockefeller University, https://www.rockefeller.edu/about/history/oral-history-project/interview-mary-jeanne-kreek/.
56. Interview with Mary Jeanne Kreek, The Rockefeller University.
57. *The Doctor & The Fix: Episode 3* podcast, Lost Women of Science Initiative; also see, *New York Times Magazine*, 15 October 1967.
58. Oral History Interview, 1981, p. 36.
59. *The Doctor & The Fix: Episode 3* podcast, Lost Women of Science Initiative.
60. *The Doctor & The Fix: Episode 4* podcast, Lost Women of Science Initiative.
61. Dole, V. P., Nyswander, M., 'A Medical Treatment for Diacetylmorphine (Heroin) Addiction. A Clinical Trial With Methadone Hydrochloride', *JAMA*, 193, 1965, pp. 646–50, doi: 10.1001/jama.1965.03090080008002.
62. Hentoff, Nat, 'The Treatment of Patients–I', *The New Yorker,* 19 June 1965, p. 102.
63. 'Givers, Doers, Changers,' *Vogue*, May 1968, vol. 151, p. 279.
64. Courtwright, D. T., 'The Prepared Mind: Marie Nyswander, Methadone Maintenance, and the Metabolic Theory of Addiction', *Addiction*, 92(3), 1997, pp. 257–65.
65. *The Doctor & The Fix: Episode 3* podcast, Lost Women of Science Initiative.
66. Courtwright, 'The Prepared Mind'.
67. *The Doctor & The Fix: Episode 4* podcast, Lost Women of Science Initiative.
68. Hentoff, Nat, 'The Treatment of Patients-II', *The New Yorker*, https://www.newyorker.com/magazine/1965/07/03/the-treatment-of-patients.
69. Hentoff, Nat, 'The Treatment of Patients-II', in *The Doctor & The Fix: Episode 3* podcast, Lost Women of Science Initiative.
70. *The Doctor & The Fix: Episode 4* podcast, Lost Women of Science Initiative; *The Doctor & The Fix: Episode 1* podcast, Lost Women of Science Initiative.
71. Quoted in Glenn, *Mother of Methadone.*
72. *The Doctor & The Fix: Episode 1* podcast, Lost Women of Science Initiative.
73. *The Doctor & The Fix: Episode 4* podcast, Lost Women of Science Initiative; Dufton, Emily *Grass Roots: The Rise and Fall and Rise of Marijuana in America* (New York: Basic Books, 2017).
74. Oral History Interview, 1981, p.39.
75. Nixon, Richard, 'Remarks About an Intensified Program for Drug Abuse Prevention and Control', 17 June 1971, The American Presidency Project, https://www.presidency.ucsb.edu/documents/remarks-about-intensified-program-for-drug-abuse-prevention-and-control.
76. *The Doctor & The Fix: Episode 4* podcast, Lost Women of Science Initiative.
77. Baum, Don, 'Legalize It All', *Harper's Magazine*, https://harpers.org/archive/2016/04/legalize-it-all/.
78. *The Doctor & The Fix: Episode 4* podcast, Lost Women of Science Initiative.

79. Ibid.
80. Warsh, Cheryl, *Frances Oldham Kelsey, the FDA, and the Battle Against Thalidomide* (Oxford: Oxford University Press, 2024).
81. *The Doctor & The Fix: Episode 4* podcast, Lost Women of Science Initiative; Lingayat, Isha et al., 'Chapter 25 – Ethical Defilements in Clinical Research and Potential Roles of Institutional Review Board' in *Advances in Pharmaceutical Product Development and Research, Pharmacokinetics and Toxicokinetic Considerations. Volume II*, ed. Rakesh Kumar Tekade (Academic Press, 2022), pp. 727–50.
82. Oral History Interview, 1981.

Chapter 13: Rosalind Franklin

1. Rose, Hilary, 'In the Shadow of Men', *Guardian*, https://www.theguardian.com/books/2002/jun/15/featuresreviews.guardianreview15.
2. Maddox, Brenda, *Rosalind Franklin: The Dark Lady of DNA* (New York: Harper Perennial, 2002), p. 82.
3. Elkin, Lynne Osman, 'Rosalind Franklin and the Double Helix', *Physics Today*, 56 (3), 2003, p. 42–8.
4. *The Science Show: No Acknowledgement, No Thanks for Rosalind Franklin Who Made the Discovery of DNA Structure Possible* podcast, ABC Radio National, https://www.abc.net.au/listen/programs/scienceshow/no-acknowledgement,-no-thanks-for-rosalind-franklin-who-made-th/13004604.
5. Colin Franklin, in *The Science Show: No Acknowledgement* podcast, ABC Radio National.
6. Ibid.
7. *The Science Show: No Acknowledgement* podcast, ABC Radio National.
8. Maddox, *Rosalind Franklin*, p. 55.
9. Ibid., p. 83.
10. Ibid., p. 79.
11. Sayre, Anne, 'Collection of the Microbiological Society Archives at the University of Maryland, Baltimore County', Elkin, 'Rosalind Franklin and the Double Helix'.
12. Quoted in Maddox, *Rosalind Franklin*, p. 89.
13. 'Once Worthless Things That Have Suddenly Become of Value', *Popular Science Monthly*, December 1918, p.80, https://books.google.com.au/books?id=EikDAAAAMBAJ&pg=PA80&redir_esc=y#v=onepage&q&f=false; Resnick, Brian, 'What America Looked Like: Collecting Peach Pits for WWI Gas Mask', *The Atlantic*, https://www.theatlantic.com/national/archive/2012/02/what-america-looked-like-collecting-peach-pits-for-wwi-gas-masks/252294/.
14. Maddox, *Rosalind Franklin*, p. 9; 'The Holes in Coal: Research at BCURA and in Paris, 1942-1951', Profiles in Science: Rosalind Franklin, NIH National Library of Medicine, https://profiles.nlm.nih.gov/spotlight/kr/feature/coal.
15. Maddox, *Rosalind Franklin*, p. 79.
16. Ibid.
17. Ibid., p. 82.
18. Glynn, Jenifer, 'Remembering My Sister Rosalind Franklin', *The Lancet*, 379(9821), 2012, pp. 1094–5.
19. Maddox, *Rosalind Franklin*, p. 100.
20. Ibid; Rose, Hilary, *Love, Power, Knowledge: Towards a Feminist Transformation of the Sciences*, (Cambridge, UK: Polity Press, 1994), p. 151.
21. McGrayne, Sharon Bertsch, *Nobel Prize Women in Science: Their Lives, Struggles, and Momentous Discoveries*, Second Edition, (Washington: Joseph Henry Press, 2001), '13: Rosalind Elsie Franklin'.
22. Dahm, Ralf, 'Friedrich Miescher and the discovery of DNA', *Developmental Biology*, 278(2), 2005, pp. 274–88.
23. Maddox, *Rosalind Franklin*, pp. 122–6.
24. Ibid., p. 123.
25. Ibid., pp. 122–6.
26. Astbury, W. and Bell, F., 'X-Ray Study of Thymonucleic Acid', *Nature*, 141, 1938, pp. 747–8.
27. Maddox, *Rosalind Franklin*, pp. 122–6; Avery, Oswald T., MacLeod, Colin M., McCarty, Maclyn, 'Studies on the Chemical Nature of the Substance Inducing Transformation of Pneumococcal Types : Induction of Transformation by a Desoxyribonucleic Acid Fraction Isolated From Pneumococcus Type III', *J Exp Med*, 79(2), 1944, pp. 137–58.
28. McGrayne, *Nobel Prize Women in Science*, '13: Rosalind Elsie Franklin'.

29. Rose, *Love, Power, Knowledge*, p. 151.
30. *The Science Show: No Acknowledgement* podcast, ABC Radio National.
31. Maddox, *Rosalind Franklin*, p. 178.
32. McGrayne, *Nobel Prize Women in Science*, '13: Rosalind Elsie Franklin'.
33. Ibid.
34. Ibid.
35. Baldwin, Melinda, *Making Nature: The History of a Scientific Journal* (Chicago and London: University of Chicago Press, 2015), p. 152.
36. Crick, Francis, *What Mad Pursuit* (New York: Basic Books, 1988), p. 58–9, in Baldwin, *Making Nature*, p.152.
37. Tony Baker, cited in *The Science Show: No Acknowledgement* podcast, ABC Radio National.
38. Maddox, p. 189.
39. Watson, J. D., *The Double Helix* (London: Orion, 2012; first published 1968), p. 98.
40. Baldwin, *Making Nature*, p. 153.
41. *The Science Show: No Acknowledgement* podcast, ABC Radio National.
42. Ibid.
43. Private letter, 7 March 1953, quoted in Maddox, *Rosalind Franklin*, p. 13.
44. Maddox, *Rosalind Franklin*, p. 188.
45. Ibid., p. 196.
46. Quoted in Maddox, *Rosalind Franklin*, p. 174.
47. Rosalind Franklin in a letter to Ellis Franklin, ca. summer 1940, in 'Biographical Overview', Profiles in Science: Rosalind Franklin, NIH National Library of Medicine, https://profiles.nlm.nih.gov/spotlight/kr/feature/biographical.
48. Maddox, *Rosalind Franklin*, p. 195
49. Rose, Hilary, 'In the Shadow of Men'.
50. Watson, J. D., *The Double Helix*.
51. Maddox, *Rosalind Franklin*.
52. Baldwin, *Making Nature*, p. 153.
53. Rose, *Love, Power, Knowledge*, p. 150.
54. Ibid.
55. John Maddox interview on *Nova*, PBS, 'The Secret of Photo 51'; Maddox, John, 'Valediction from an Old Hand', *Nature*, 378, 1995, pp. 521–2, in Baldwin, *Making Nature*, p. 168.
56. Rose, *Love, Power, Knowledge*, p. 16.
57. Glynn, 'Remembering my sister Rosalind Franklin'.
58. Sansom, Clare, 'Rosalind Franklin. A Life in Close-Up', *BBK*, December 2020, https://www.cryst.bbk.ac.uk/dept/franklin.pdf.
59. Glynn, 'Remembering My Sister Rosalind Franklin'.

Chapter 14: Katherine Johnson

1. Shetterly, Margot Lee, *Hidden Figures* (New York: William Morrow Paperbacks, 2016), p. 250.
2. Ibid., p. 245.
3. Fountain, Nigel, 'Katherine Johnson Obituary', *Guardian*, https://www.theguardian.com/science/2020/feb/24/katherine-johnson-obituary.
4. Shetterly, *Hidden Figures*, p. 71; Fox, Margalit, 'Katherine Johnson Dies at 101: Mathematician Broke Barriers at NASA', *New York Times*, https://www.nytimes.com/2020/02/24/science/katherine-johnson-dead.html.
5. Mrs Johnson told The Associated Press in 1999, in Fox, 'Katherine Johnson Dies at 101'.
6. Ibid.
7. Mullin, Rick, 'The Rise of Environmental Justice', *Chemical & Engineering News*. Retrieved 10 September 2020.
8. Fox, 'Katherine Johnson Dies at 101'.
9. Shetterly, *Hidden Figures*, p. 71.
10. Ibid., p.72.
11. Ibid., p. 73.
12. Ibid.
13. Quoted in Shetterly, *Hidden Figures*, p. 74.
14. Shetterly, *Hidden Figures*, p. 74; Fox, 'Katherine Johnson Dies at 101'.
15. Ibid.
16. Shetterly, *Hidden Figures*, pp. 73–4.
17. Fox, 'Katherine Johnson Dies at 101'.

18. Shetterly, *Hidden Figures*, p. 75.
19. Ibid.
20. Ibid., p. 76.
21. Shetterly, *Hidden Figures*.
22. Shetterly, *Hidden Figures*, p. 76
23. Fox, 'Katherine Johnson Dies at 101'.
24. Ibid.
25. Ibid.
26. Shetterly, *Hidden Figures*, p. 229.
27. Richmond, Emily, 'The Forgotten School in Brown v. Board of Education', *The Atlantic*, 16 May 2014.
28. Shetterly, *Hidden Figures*, p. 121
29. Ibid, p. 122
30. Ibid.
31. Ibid.
32. Johnson, Katherine, *My Remarkable Journey: A Memoir* (New York: HarperCollins, 2021).
33. Shetterly, *Hidden Figures*, p. 124.
34. Ibid., p.129.
35. Ibid., p. 123.
36. Ibid., p.122.
37. Ibid., p. 129.
38. Fox, 'Katherine Johnson Dies at 101'.
39. Shetterly, *Hidden Figures*.
40. *Daily Press*, Newport News; *New York Times*, in Shetterly, *Hidden Figures*, p. 151.
41. Shetterly, *Hidden Figures*, p. 98.
42. Ibid., p. 99; *Norfolk Journal and Guide*, 1950.
43. Quoted in Shetterly, *Hidden Figures*, p. 152.
44. Shetterly, *Hidden Figures*, p. 152.
45. Ibid., p. 188.
46. Fox, 'Katherine Johnson Dies at 101'; Shetterly, *Hidden Figures*, p. 189.
47. Shetterly, *Hidden Figures*, p. 190.
48. Ibid.
49. Quoted in Shetterly, *Hidden Figures*, p. 191.
50. Skopinski, T. H. and Johnson, Katherine G., 'Determination of Azimuth Angle at Burnout for Placing a Satellite Over a Selected Earth Position', NASA Technical Note 19980227091, NASA NTRS, https://ntrs.nasa.gov/citations/19980227091.
51. Shetterly, *Hidden Figures*, p. 192.
52. Fountain, Nigel, 'Katherine Johnson Obituary'.
53. Shetterly, *Hidden Figures*.
54. Shetterly, *Hidden Figures*, p. 233.
55. Ibid.
56. Ibid., p. 234.
57. Ibid., p. 237.
58. Ibid.
59. Ibid., p. 244.
60. Ibid., p. 245.
61. Ibid., pp. 240–1.
62. Ibid.
63. Ibid., p. 246.
64. Ibid., p. 241.
65. Fountain, Nigel, 'Katherine Johnson Obituary'.
66. Deiss, Heather, 'Katherine Johnson: A Lifetime of STEM', NASA, https://www.nasa.gov/learning-resources/katherine-johnson-a-lifetime-of-stem/#:~:text=Johnson%20worked%20at%20the%20agency,three%20NASA%20Special%20Achievement%20Awards.
67. Shetterly, *Hidden Figures*, p. 243.

Chapter 15: Anne McLaren

1. This chapter is based on the research included in my PhD dissertation: Bigg, M. (2019), *Anne McLaren and the Human Embryo Research Debates in Britain, 1982–1990: Sociological Biography in the Analysis of Public Debate*, [Apollo – University of Cambridge Repository] https://doi.org/10.17863/CAM.57928, and the oral history interview: Anne McLaren, British Library

Mammalian Developmental Biology 2007 by Sarah Franklin and Martin H. Johnson, unless referenced otherwise.

2. Aberconway, Christabel, *A Wiser Woman: A Book of Memories* (Hutchinson of London, 1966), p. 65.
3. Anne McLaren Archives, British Library, London.
4. Anne McLaren interview with Max Blythe, 3 July 1998, Part 1. The Royal College of Physicians and Oxford Brookes University Medical Sciences Video Archive MSVA 188.
5. Weismann, August, *The Germ Plasm: A Theory of Heredity* (New York: Charles Scribner's Sons, 1893), Preface, p. x.
6. Anne McLaren interview with Max Blythe, 3 July 1998, Part 1.
7. Ibid.
8. Anne's letter to her mother, n.d. ?1948/9, unarchived.
9. Ibid.
10. Ibid.
11. Anne McLaren interview with Max Blythe, 3 July 1998, Part 1.
12. McLaren, Anne, 'Protective Effect of Aged Vaccines on a Neurotropic Virus Infection in Mice' *Nature*, 172, 1953, p. 38.
13. Ann Clarke and Martin H. Johnson FRS, 'Anne Laura Dorinthea McLaren DBE. 26 April 1927–7 July 2007', *Biogr Mems Fell R Soc*, 74, 2023, pp. 309–37.
14. Cohen, J., Trounson, A., Dawson, K., Jones, H., Hazekamp, J., Nygren, K. G., Hamberger, L., 'The Early Days of IVF Outside the UK', *Hum Reprod Update*, 11(5), 2005, pp. 439–59, doi: 10.1093/humupd/dmi016. Epub 2005 May 27; Pinteric, Aubrey, 'The Idea of Ectogenesis in Daedalus; or Science and The Future (1924), by John Burdon Sanderson Haldane', Embryo Project Encyclopedia, ASU, https://embryo.asu.edu/pages/idea-ectogenesis-daedalus-or-science-and-future-1924-john-burdon-sanderson-haldane.
15. Interview with Hilary Rose, 12 September 2018, in Bigg, *Anne McLaren and the Human Embryo Research Debates in Britain.*
16. McLaren, in fact, worked in the lab of Conrad Hal Waddington in the Institute of Animal Genetics in Edinburgh for fifteen years, between 1959 and 1974, and he coined the term as early as 1942. Waddington was a similarly polymathic and committedly interdisciplinary scientist. McLaren said of his lab that it was 'the most wonderful scientific atmosphere I have come across anywhere'.
17. McLaren, in WHO Keynote, Geneva, 1989, p. 14, British Library Archives.
18. Murray, Andrew, 'Obituary Anne McLaren and Donald Michie', *Morning Star*. Wednesday, 11 July 2007, p. 8.

Chapter 16: Jane Goodall

1. For example, Greene, Meg, *Jane Goodall: A Biography* (New York: Prometheus Books, 2008), p. 1.
2. Ibid., p. 4.
3. Ibid., p.9.; Goodall, Jane, *My Life With the Chimpanzees* (New York: Simon and Schuster, 1996).
4. Dale, Steve, 'An Interview with Jane Goodall: What I Learned From Dogs', Studio One Networks, https://thedogdaily.com/you_dog/moments/archive/goodall_interview/.
5. Goodall, *My Life*, p. 20.
6. *On Being with Krista Tippett* podcast, 'Jane Goodall, In Memoriam – What It Means to Be Human', 3 October 2025 (2020 interview), https://onbeing.org/programs/jane-goodall-what-it-means-to-be-human/.
7. Ibid.
8. Goodall, *My Life*, p. 16.
9. Lassieur, Alison, 'When I Was a Kid: Childhood Experiences of Famous People', *National Geographic World* (September 1999), p. 11.
10. *On Being with Krista Tippett* podcast, 'Jane Goodall, In Memoriam'.
11. Ibid.
12. Goodall, *My Life*, p. 24.
13. Ibid., p. 25.
14. Greene, *Jane Goodall: A Biography*, p. 15.
15. Goodall, Jane with Berman, Philip, *Reason for Hope: A Spiritual Journey* (New York: Warner Books, 1999), p. 40.

16. Goodall, *My Life*, p. 30.
17. Greene, *Jane Goodall: A Biography*, p. 19.
18. Ibid.
19. Jane Goodall, *My Life with the Chimpanzees.*
20. Greene, *Jane Goodall: A Biography*, p. 28.
21. Correspondence, Sy Montgomery, April 2025.
22. Jane Goodall, *My Life with the Chimpanzees.*
23. Greene, *Jane Goodall: A Biography*, pp. 25–6.
24. Dr. Jane Goodall for the Leakey oral history project for the Leakey Foundation and the UC Berkeley Bancroft Library.
25. Lee, Luaine, 'Return to Gombe' with Jane Goodall on Animal Planet, *Knight Ridder/Tribune News Service*, 23 February 2004.
26. Segal, Daniel A., 'Witnessing Chimpanzee–Human Closeness: Jane Goodall at Gombe and Since', *Anthropological Quarterly*, 90(4), 2017, p. 1196.
27. Ibid., p. 1196–7; Dr. Jane Goodall for the Leakey oral history project for the Leakey Foundation and the UC Berkeley Bancroft Library.
28. Dr. Jane Goodall for the Leakey oral history project for the Leakey Foundation and the UC Berkeley Bancroft Library.
29. Ibid.
30. *On Being with Krista Tippett* podcast, 'Jane Goodall, In Memoriam'.
31. Ibid.
32. Ibid.
33. For example, Goodall and Berman, *Reason for Hope*, p. 67.
34. Segal, 'Witnessing Chimpanzee–Human Closeness', p. 1207.
35. Ibid., p. 1206.
36. Heyes, C. M., 'Social Learning in Animals: Categories and Mechanisms', *Biol Rev* 69, 1994, pp. 207–31.
37. Darwin, Charles, *On the Origin of Species by Means of Natural Selection, or the Preservation of Favoured Races in the Struggle for Life*, First Edition. (London: John Murray, 1859).
38. Whiten, Andrew, 'Cultural Evolution in Animals', *Annual Review of Ecology, Evolution, and Systematics*, vol. 50, 2019, pp. 27–48.
39. Segal, 'Witnessing Chimpanzee–Human Closeness', p. 1197.
40. Correspondence with Sy Montgomery, April 2025.
41. Goodall, Jane, 'Being a Woman Was Crucial to My Success in a Male-Dominated Field', *Time*, https://time.com/5192249/jane-goodall-sexism-gender-equality-documentary/.
42. Goodall, Jane, 'My Life Among Wild Chimpanzees', *National Geographic*, https://www.nationalgeographic.com/pdf/jane-goodall-1963.pdf.
43. Greene, *Jane Goodall: A Biography*, p. 66; 83–4.
44. Karbo, Karen, 'Jane Goodall, How a Woman Redefined Mankind', *National Geographic*, https://www.nationalgeographic.com/culture/article/jane-goodall-book-excerpt-praise-difficult-women.
45. *On Being with Krista Tippett* podcast, 'Jane Goodall, In Memoriam'.
46. Ibid.
47. Alem, S., Perry, C. J., Zhu, X., Loukola, O. J., Ingraham, T. et al. 'Associative Mechanisms Allow for Social Learning and Cultural Transmission of String Pulling in an Insect', *PLOS Biol*, 14, 2016, e1002564.
48. Attenborough, David, 'Wild Crows Inhabiting the City Use It to Their Advantage', BBC Wildlife, https://www.youtube.com/watch?v=BGPGknpq3e0&t=2s.
49. Jane Goodall Institute, 'Strategy: Sustainable Livelihoods', https://janegoodall.org/our-work/our-approach/sustainable-livelihoods/; 'Strategy: Healthy Habitats', https://janegoodall.org/our-work/our-approach/healthy-habitats/; Jane Goodall Institute (Canada), 'Our Work: Our Community-Led Approach', https://janegoodall.ca/what-we-do/our-approach/#:~:text=Four%20pillars.,and%20encourage%20us%20every%20day.; 'Jane Goodall and Gombe National Park', Google Art and Culture, https://artsandculture.google.com/story/jane-goodall-and-gombe-national-park/bQWBn_PEN9Mqrg?hl=en.
50. 'Dr. Jane Goodall's Final Message To The World', *Famous Last Words*, Netflix, https://www.youtube.com/watch?v=lfLKHY52ERc; 'How Jane Goodall's Death Initiated NetFlix's Newest Show', *New York Times*, https://www.nytimes.com/2025/10/03/arts/television/jane-goodall-netflix-famous-last-words.html.

Chapter 17: Virginia Johnson

1. Maier, Thomas, *Masters of Sex* (New York: Basic Books, 2009), pp. 220–3.
2. Ibid.
3. Ibid.
4. Ibid., p. 23.
5. Ibid., p.7.
6. Ibid., p. 9.
7. Interviewed in Maier, *Masters of Sex*, p. 22.
8. Maier, *Masters of Sex*, p. 24.
9. Ibid., p. 13.
10. Ibid., p. 15.
11. Ibid., p. 16.
12. Ibid., p. 17.
13. Ibid., p. 21.
14. Ibid.
15. Ibid., p. 32.
16. Temkin, E., 'Contraceptive Equity: the Birth Control Center of the International Workers Order', *Am J Public Health*, 97(10), 2007, pp. 1737–45, doi: 10.2105/AJPH.2006.098145. Epub 2007 Aug 29. Erratum in: *Am J Public Health*, 97(12), 2007, p. 2122.
17. Maier, *Masters of Sex*.
18. Ibid., p. 23.
19. Ibid., pp. 25–8.
20. Ibid., p. 28.
21. Ibid., p. 111.
22. Ibid, p. 83.
23. Ibid., p. 111.
24. Lejárraga, Juan, 'Virginia E. Johnson, 1925–2013', Medium, https://medium.com/sexology-101/virginia-e-johnson-1925-2013-sexologist-ccdbf347031b.
25. Maier, *Masters of Sex*, p. 44.
26. Ibid., p. 103.
27. Ibid., p. 104.
28. Ibid.
29. Ibid.
30. Ibid., p. 105.
31. Ibid., pp. 105–7.
32. Ibid., p. 94.
33. Hayman, Suzie, 'Virginia Johnson Obituary', *Guardian*, https://www.theguardian.com/lifeandstyle/2013/jul/28/virginia-johnson.
34. Bullough, Vern, *Science in the Bedroom: A History of Sex Research* (New York: HarperCollins, 1994).
35. Hayman, 'Virginia Johnson Obituary'.
36. Morrow, Ross, *Sex Research and Sex Therapy. A Sociological Analysis of Masters and Johnson*, (Abingdon, UK: Routledge, 2008), p. 3.
37. Hayman, 'Virginia Johnson Obituary'.
38. Hayman, 'Virginia Johnson Obituary'.
39. Maier, *Masters of Sex*, p. 185.
40. Ibid., p. 185.
41. Ibid., p. 186.
42. Escoffier, Jeffrey, 'The Sexual Revolution, 1960-1980', 2015, http://www.glbtqarchive.com/ssh/sexual_revolution_S.pdf
43. Maier, *Masters of Sex*, p. 186.
44. Ehrenreich, Barbara, Hess, Elizabeth, Jacobs, Gloria, *Re-Making Love: The Feminization of Sex* (New York: Anchor Books, 1986); Maier, *Masters of Sex*.
45. Maier, *Masters of Sex*, p. 186.
46. Ibid., p. 261.
47. Ibid.
48. Ibid.
49. Ibid., p. 262.
50. Ibid.
51. Ibid., p. 185.
52. Bigg, Marieke, *This Won't Hurt* (London: Hodder & Stoughton, 2023), p. 77; Angel, Katherine, *Tomorrow Sex Will Be Good Again* (London: Verso, 2021)

53. Hayman, 'Virginia Johnson Obituary'.
54. Maier, *Masters of Sex*, pp. 128–9.
55. Ibid., p. 237.
56. Ibid., p. 346.
57. Ibid., p. 202 and 369.

Chapter 18: Youyou Tu

1. Tu, Youyou, 'Biographical', Nobel Prize, https://www.nobelprize.org/prizes/medicine/2015/tu/biographical/.
2. Neill, U. S., 'From Branch to Bedside: Youyou Tu is Awarded the 2011 Lasker~DeBakey Clinical Medical Research Award for Discovering Artemisinin as a Treatment for Malaria', *J Clin Invest.*, 121(10), 2011, pp. 3768–73, doi: 10.1172/jci60887.
3. Tu, Youyou, 'Biographical', Nobel Prize.
4. Ibid.
5. Hatton, Celia, 'Nobel Prize Winner Tu Youyou Helped by Ancient Chinese Remedy', BBC News, https://www.bbc.co.uk/news/blogs-china-blog-34451386.
6. Tu, Youyou, 'Biographical', Nobel Prize.
7. Mahdi, J. G., Mahdi, A. J., Mahdi, A. J., Bowen, I. D., 'The Historical Analysis of Aspirin Discovery, Its Relation to the Willow Tree and Antiproliferative and Anticancer Potential', *Cell Prolif*, 39(2), 2006, pp. 147–55, doi: 10.1111/j.1365-2184.2006.00377.x.
8. Hsu, E., 'Reflections on the "Discovery" of the Antimalarial Qinghao', *Br J Clin Pharmacol*, 61(6), 2006, pp. 666–70; Li, Ruo-lan, Zhang, Qing, Liu, Jia, He, Li-ying, Huang, Qin-wan, Peng, Wei, Wu, Chun-jie, 'Processing Methods and Mechanisms for Alkaloid-Rich Chinese Herbal Medicines: A Review', *Journal of Integrative Medicine*, 19(2), 2021, pp. 89–103.
9. Tu, Youyou, 'Biographical', Nobel Prize.
10. McNeil Jr, Donald G., 'For Intrigue, Malaria Drug Gets the Prize', *New York Times*, https://www.nytimes.com/2012/01/17/health/for-intrigue-malaria-drug-artemisinin-gets-the-prize.html.
11. Perlez, Jane, 'Answering an Appeal by Mao Led Tu Youyou, a Chinese Scientist, to a Nobel Prize', *New York Times*, https://www.nytimes.com/2015/10/07/world/asia/tu-youyou-chinese-scientist-nobel-prize.html.
12. Hao, Cindy, 'Lasker Award Rekindles Debate Over Artemisinin's Discovery', https://www.science.org/content/article/lasker-award-rekindles-debate-over-artemisinins-discovery.
13. Zhang, J. F., Zhou, K. D., Zhou, Y. Q., Fu, L. S., Wang, H. S., Song, S. Y., *A Detailed Chronological Record of Project 523 and the Discovery and Development of Qinghaosu (Artemisinin)* (Guangzhou, China: Yangcheng Evening News Publishing Company, 2006).
14. Perlez, 'Answering an Appeal by Mao Led Tu Youyou, a Chinese Scientist, to a Nobel Prize'.
15. Tu, Youyou, 'Biographical', Nobel Prize.
16. Gao, Helen, 'How Did Women Fare in China's Communist Revolution?', *New York Times*, https://www.nytimes.com/2017/09/25/opinion/women-china-communist-revolution.html.
17. Hatton, 'Nobel Prize Winner Tu Youyou Helped by Ancient Chinese Remedy'.
18. Lowe, Derek, 'Chloroquine, Past and Present', *Science*, https://www.science.org/content/blog-post/chloroquine-past-and-present.
19. Burns, W. R., 'East Meets West: How China Almost Cured Malaria', *Endeavour*, 32(3), 2008, pp. 101–6.
20. *Saving Lives, Buying Time: Economics of Malaria Drugs in an Age of Resistance*, eds K. J. Arrow, C. Panosian, H. Gelband (Institute of Medicine (US) Committee on the Economics of Antimalarial Drugs; Washington DC: National Academies Press (US), 2004).
21. Lowe, 'Chloroquine, Past and Present'.
22. *Saving Lives, Buying Time: Economics of Malaria Drugs in an Age of Resistance*, eds K. J. Arrow, C. Panosian, H. Gelband.
23. Andersag, H., Breitner, S., Jung, H., March 1941. Quinoline Compound and Process of Making the Same. US Patent 2,233,970.
24. Coatney, G. R., 'Pitfalls in a Discovery: The Chronicle of Chloroquine', *Am. J Trop Med Hyg*, 12, 1963, pp. 121–8.
25. Chamoli, R. O., Painuly, A. M., Bahuguna, Gaurav, 'The Story of the Antimalarial Drug Artemisinin: Fiftieth Anniversary of Its Discovery', *Science and Culture*, 91(5–6), 2025, pp. 288–92.
26. Wellems, T. E., Plowe, C., 'Chloroquine-Resistant Malaria', *J Infect Dis*, 184, 2001, pp. 770–6; Buxton, P., 'The Use of the New Insecticide DDT in Relation to the Problems of Tropical Medicine', *Trans R Soc Trop Med Hyg*, 38, 1945, pp. 367–400.

27. Perlez, 'Answering an Appeal by Mao Led Tu Youyou, a Chinese Scientist, to a Nobel Prize'.
28. Neill, 'From Branch to Bedside'.
29. Ibid.
30. Hsu, 'Reflections on the "Discovery"', in Neill, 'From Branch to Bedside'.
31. McNeil, 'For Intrigue, Malaria Drug Gets the Prize'.
32. Neill, 'From Branch to Bedside'.
33. Ibid.
34. Neill, 'From Branch to Bedside'.
35. Hsu, 'Reflections on the "Discovery"'.
36. Su, X. Z., Miller, L. H., 'The Discovery of Artemisinin and the Nobel Prize in Physiology or Medicine', *Sci China Life Sci*, 58(11), 2015, pp. 1175–9, doi: 10.1007/s11427-015-4948-7.
37. Neill, 'From Branch to Bedside'.
38. Mckenna, Phil, 'Nobel Prize Goes to Modest Woman Who Beat Malaria for China', *New Scientist*, https://www.newscientist.com/article/mg21228382-000-the-modest-woman-who-beat-malaria-for-china/.
39. In, Neill, 'From Branch to Bedside'.
40. Ibid.
41. Cui, L., Su, X. Z., 'Discovery, Mechanisms of Action and Combination Therapy of Artemisinin', *Expert Rev Anti Infect Ther*, 7(8), 2009, pp. 999–1013, doi: 10.1586/eri.09.68.
42. Ibid.
43. Neill, 'From Branch to Bedside'.
44. Ibid. (Tu finally published the results in collaboration with the Institute of Biophysics, Chinese Academy of Sciences in: Collaboration Research Group for Qinghaosu, 'A New Sesquiterpene Lactone — Qinghaosu [in Chinese]. Kexue Tongbao. 3, 1977, p. 142.)
45. [No authors listed], 'Antimalaria studies on Qinghaosu', *Chin Med J (Engl)*, 92(12), 1979, pp. 811–6.
46. Vezina, Kenrick, 'Tu Youyou', EBSCO Knowledge Advantage, EBSCO, https://www.ebsco.com/research-starters/womens-studies-and-feminism/tu-youyou.
47. Mckenna, 'Nobel Prize Goes to Modest Woman Who Beat Malaria for China'.
48. Miller, L. H., Su, X., 'Artemisinin: Discovery From the Chinese Herbal Garden', *Cell*. 146(6), 2011, pp. 855–8, doi: 10.1016/j.cell.2011.08.024. Epub 2011 Sep 9; Correspondence with Miller and Su, August 2024.
49. Ibid.
50. Miller and Su, 'Artemisinin: Discovery from the Chinese Herbal Garden'; Correspondence with Miller and Su, August, 2024; McNeil, 'For Intrigue, Malaria Drug Gets the Prize'.
51. Jiang, J. B., Li, G. Q., Guo, X. B., Kong, Y. C., Arnold, K., 'Antimalarial Activity of Mefloquine and Qinghaosu', *Lancet*, 320(8293), 1982, pp. 285–8.
52. Neill, 'From Branch to Bedside'.
53. Correspondence with Nicholas White, April 2025.
54. Correspondence with Miller and Su, August 2024.
55. Altman, Lawrence K., 'Nobel Prize Awarded to 3 Scientists for Parasite Fighting Therapies', *New York Times*, https://www.nytimes.com/2015/10/06/science/william-c-campbell-satoshi-omura-youyou-tu-nobel-prize-physiology-medicine.html.
56. Bhatt, S., Weiss, D. J., Cameron, E. et al., 'The Effect of Malaria Control on *Plasmodium falciparum* in Africa Between 2000 and 2015', *Nature*, 526, 2015, pp. 207–11, doi: 10.1038/nature15535.
57. World Health Organization, 'Traditional Medicine Has a Long History of Contributing to Conventional Medicine and Continues to Hold Promise', https://www.who.int/news-room/feature-stories/detail/traditional-medicine-has-a-long-history-of-contributing-to-conventional-medicine-and-continues-to-hold-promise.

Chapter 19: Flossie Wong-Staal

1. Accessed via 'HIV/AIDS Timeline', New York City AIDS Memorial, https://www.nycaidsmemorial.org/timeline.
2. 'HIV/AIDS Timeline', New York City AIDS Memorial.
3. Poiesz, B. J., Ruscetti, F. W., Gazdar, A. F., Bunn, P. A., Minna, J. D., Gallo, R. C., 'Detection and Isolation of Type C Retrovirus Particles From Fresh and Cultured Lymphocytes of a Patient With Cutaneous T-Cell Lymphoma', *Proc Natl Acad Sci U.S.A.*, 77, 1980, pp. 7415–9, doi: 10.1073/pnas.77.12.7415.

4. Husband, Jeffrey McKelvy, quoted in Flam, Faye, 'Flossie Wong-Staal, Who Unlocked the Mystery of H.I.V., Dies at 73', *New York Times*, https://www.nytimes.com/2020/07/17/science/flossie-wong-staal-who-unlocked-mystery-of-hiv-dies-at-73.html#:~:text=%E2%80%9CShe%20said%2C%20'I%20don,a%20Flossie%2C'%E2%80%9D%20Mr.
5. Oral History (OH) interview with Flossie Wong-Staal, PhD, about the National Institutes of Health's response to AIDS, National Institutes of Health (NIH), 10 December 1997, with Victoria Harden, PhD, Director, Office of NIH History, and Caroline Hannaway, PhD, Historical Contractor, NIH, p. 1.
6. Ibid.
7. Ibid., p. 2.
8. Ibid., p. 3.
9. Alberts, B., Johnson, A., Lewis, J. et al. *Molecular Biology of the Cell*, Fourth edition (New York: Garland Science; 2002).
10. 'History of Genetic Engineering and the Rise of Genome Editing Tools', Synthego, https://www.synthego.com/learn/genome-engineering-history; Pai-Dhungat, Jayant, 'Restriction Enzymes–Molecular Scissors', *JAPI*, 73(6), 2025, p. 98, https://www.japi.org/article/japi-73-6-98; 'Fact Sheet: Human Genome Project', NIH Research Institute, https://www.genome.gov/about-genomics/educational-resources/fact-sheets/human-genome-project.
11. OH. p. 5.
12. Ibid., p. 7.
13. Martin, Caroline, 'Dr. Flossie Wong-Staal: Cracking the Code', A World of Women in STEM, https://www.wowstem.org/post/flossie-wong-staal.
14. OH, p. 7.
15. Ibid., p. 9.
16. Flam, 'Flossie Wong-Staal, Who Unlocked the Mystery of H.I.V., Dies at 73'.
17. OH, p. 13.
18. 'Flossie Wong-Staal – Discovery of Human Retroviral Transactivator Genes', Cold Spring Harbor Laboratory meeting, HIV/AIDS Research: Its History and Future, Genentech Center for the History of Molecular Biology and Biotechnology at Cold Spring Harbor Laboratory Archives. Meeting 13–16 October 2016, accessed at https://www.youtube.com/watch?v=fyVPIxAqg1c&ab_channel=WS.
19. OH, p. 13.
20. Ibid., p. 10.
21. Flam, 'Flossie Wong-Staal, Who Unlocked the Mystery of H.I.V., Dies at 73'.
22. OH, pp. 16–7.
23. Ibid., p.17.
24. Ibid., p. 18.
25. Ibid., p. 55.
26. Ibid., p. 21.
27. Gallo, Robert, *Virus Hunting: Cancer, AIDS, and the Human Retrovirus: A Story of Scientific Discovery* (New York: Basic Books, 1991).
28. 'Pat Buchanan's Greatest Hits', *Washington Post*, https://www.washingtonpost.com/archive/politics/1987/02/04/pat-buchanans-greatest-hits/416e2224-f7cd-4271-8c9d-0712712df6f3/.
29. 'AIDS and the Holocaust', *New York Times*, https://www.nytimes.com/1988/10/23/books/l-aids-and-the-holocaust-156488.html, in King, Jack, 'The Drama That Raged Against Reagan's America', BBC Culture, https://www.bbc.com/culture/article/20201019-the-drama-that-raged-against-reagans-america.
30. 'The Denver Principles (1983)', https://data.unaids.org/pub/externaldocument/2007/gipa1983denverprinciples_en.pdf
31. 'The Changing Response to AIDS', Council on Foreign Relations, https://education.cfr.org/learn/timeline/changing-response-aids; 'Snapshots of an Epidemic: An HIV/AIDS Timeline', amfAR, https://www.amfar.org/about-hiv-aids/snapshots-of-an-epidemic-hiv-aids/; King, 'The Drama That Raged'; 'HIV/AIDS Timeline', New York City AIDS Memorial.
32. Cannon, Lou, 'Ronald Reagan: Domestic Affairs', UVA Miller Center, https://millercenter.org/president/reagan/domestic-affairs.
33. King, 'The Drama That Raged'.
34. Cannon, Lou, *President Reagan: The Role of a Lifetime* (New York: PublicAffairs, 1991), p. 9.
35. 'Flossie Wong-Staal – Discovery of Human Retroviral Transactivator Genes', Cold Spring Harbor Laboratory meeting, HIV/AIDS Research: Its History and Future.
36. Shaw, G. M., Hahn, B. H., Arya, S. K., Groopman, J. E., Gallo, R. C., Wong-Staal, F., 'Molecular Characterization of Human T-Cell Leukemia (Lymphotropic) Virus Type III in

the Acquired Immune Deficiency Syndrome', *Science*, 226(4679), 1984, pp. 1165–71, doi: 10.1126/science.6095449.

37. Alizon, M., Sonigo, P., Barré-Sinoussi, F., Chermann, J. C., Tiollais, P., Montagnier, L, Wain-Hobson, S., 'Molecular Cloning of Lymphadenopathy-Associated Virus', *Nature*, 312(5996), 1984, pp. 757–60. doi: 10.1038/312757a0.
38. Ratner, L., Haseltine, W., Patarca, R. et al., 'Complete Nucleotide Sequence of the AIDS virus, HTLV-III'. *Nature*, 313, 1985, pp. 277–84, https://doi.org/10.1038/313277a0.
39. Wain-Hobson, S., Sonigo, P., Danos, O., Cole, S., Alizon, M., 'Nucleotide Sequence of the AIDS Virus, LAV'. *Cell*, 40(1), 1985, pp. 9–17, doi: 10.1016/0092-8674(85)90303-4.
40. 'Flossie Wong-Staal – Discovery of Human Retroviral Transactivator Genes', Cold Spring Harbor Laboratory meeting, HIV/AIDS Research: Its History and Future.
41. Ramakrishnan, Venki, *Gene Machine: The Race to Decipher the Secrets of the Ribosome* (London: Oneworld, 2018), pp. 98–9.
42. OH, p. 35.
43. Ibid., p. 35.
44. Ibid., p. 42.
45. Ibid., p. 27.
46. Ibid., p. 28; Franchini, Genoveffa, 'Flossie Wong-Staal (1946–2020)', *Science* 369, 2020, p. 1308, doi:10.1126/science.abe4095.
47. OH, p. 21.
48. 'The First AIDS Drugs', NIH National Cancer Institute, Center for Cancer Research, https://ccr.cancer.gov/news/landmarks/article/first-aids-drugs
49. Flam, 'Flossie Wong-Staal, Who Unlocked the Mystery of H.I.V., Dies at 73'.
50. 'The First AIDS Drugs', NIH National Cancer Institute.
51. Franchini, 'Flossie Wong-Staal (1946–2020)'; Flam, 'Flossie Wong-Staal, Who Unlocked the Mystery of H.I.V., Dies at 73'.
52. OH, p. 29.
53. Ibid.
54. Ibid., p. 30.
55. 'The Changing Response to AIDS', Council on Foreign Relations.
56. Gallo, Robert C. et al., 'Frequent Detection and Isolation of Cytopathic Retroviruses (HTLV-III) From Patients With AIDS and at Risk for AIDS', *Science*, 224, 1984, pp. 500–3.
57. Barré-Sinoussi, F., Chermann, J. C., Rey, F., Nugeyre, M. T., Chamaret, S., Gruest, J., Dauguet, C., Axler-Blin, C., Vézinet-Brun, F., Rouzioux, C., Rozenbaum, W., Montagnier, L., 'Isolation of a T-Lymphotropic Retrovirus From a Patient at Risk for Acquired Immune Deficiency Syndrome (AIDS)', *Science*, 220(4599), 1983, pp. 868–71, doi: 10.1126/science.6189183.
58. OH, p. 33.
59. Flam, 'Flossie Wong-Staal, Who Unlocked the Mystery of H.I.V., Dies at 73'.
60. Coghlan, Andy, 'Was Robert Gallo Robbed of the Nobel Prize?', *New Scientist*, https://www.newscientist.com/article/dn14881-was-robert-gallo-robbed-of-the-nobel-prize/#:~:text=Although%20this%20was%20settled%20out,Nature%20on%202%20April%201987.
61. Commentary, 'The Chronology of AIDS Research', *Nature*, 326, 1987, pp. 435–6, https://doi.org/10.1038/326435a0; Coghlan, Andy, 'Was Robert Gallo Robbed of the Nobel Prize?'
62. Crewdson, John, *Science Fictions: A Scientific Mystery, a Massive Cover-Up, and the Dark Legacy of Robert Gallo* (New York: Little, Brown and Company, 2002); Cohen, Jon and Enserink, Martin, 'HIV, HPV Researchers Honored, But One Scientist Is Left Out', *Science*, 322, 2008, pp. 174–5, doi:10.1126/science.322.5899.174.
63. Cohen and Enserink, 'HIV, HPV Researchers Honored, But One Scientist Is Left Out'.
64. Flam, 'Flossie Wong-Staal, Who Unlocked the Mystery of H.I.V., Dies at 73'.
65. Ibid.
66. OH, p. 43.
67. 'In Memoriam: Flossie Wong-Staal, Ph.D', NIH National Cancer Institute, https://ccr.cancer.gov/news/article/in-memoriam-flossie-wong-staal-phd.

Chapter 20: Jennifer Doudna

1. Jarvis, Lisa M., 'A Day With Jennifer Doudna: Trying to Keep Up With One of the World's Most Sought-After Scientists', Chemical & Engineering News, https://cen.acs.org/biological-chemistry/gene-editing/A-day-with-Jennifer-Doudna-Trying-to-keep-up-with-one-of-the-world-most-sought-after-scientists/98/i9.

2. 'Jennifer Doudna on the Brave New World Being Ushered in by Gene Editing', *The New Yorker*, https://www.newyorker.com/books/book-currents/jennifer-doudna-on-the-brave-new-world-being-ushered-in-by-gene-editing.
3. Doudna, Jennifer A., 'Biographical', Nobel Prize, https://www.nobelprize.org/prizes/chemistry/2020/doudna/biographical/.
4. Ibid.
5. Ibid.
6. Devlin, Hannah, 'Jennifer Doudna: "I Have to Be True to Who I Am as a Scientist"', *Guardian*, https://www.theguardian.com/science/2017/jul/02/jennifer-doudna-crispr-i-have-to-be-true-to-who-i-am-as-a-scientist-interview-crack-in-creation.
7. Doudna, 'Biographical', Nobel Prize.
8. Pollack, Andrew, 'Jennifer Doudna, a Pioneer Who Helped Simplify Gene Editing', *New York Times*, https://www.nytimes.com/2015/05/12/science/jennifer-doudna-crispr-cas9-genetic-engineering.html.
9. Doudna, 'Biographical', Nobel Prize.
10. Ibid.
11. Ibid.
12. Varshavsky, Alexander, 'Discovering the RNA Double Helix and Hybridization', *Cell*, 127(7), 2006, pp. 1295–7.
13. Quoted in Isaacson, Walter, *The Code Breaker* (New York: Simon & Schuster, 2022).
14. Doudna, 'Biographical', Nobel Prize.
15. Ibid.
16. Isaacson, *The Code Breaker*, pp. 70–1.
17. Brenner, S., Jacob, F., Meselson, M., 'An Unstable Intermediate Carrying Information From Genes to Ribosomes for Protein Synthesis', *Nature*, 190, 1961, pp. 576–81; Gros, F., Hiatt, H., Gilbert, W. et al., 'Unstable Ribonucleic Acid Revealed by Pulse Labelling of *Escherichia coli*', *Nature*, 190, 1961, pp. 581–5; Jacob, F., Monod, J., 'Genetic Regulatory Mechanisms in the Synthesis of Proteins', *J Mol Biol*, 3, 1961, pp. 318–56; Pollack, 'Jennifer Doudna, a Pioneer Who Helped Simplify Gene Editing'.
18. Bernhardt, H. S., 'The RNA World Hypothesis: The Worst Theory of the Early Evolution of Life (Except for All the Others)(a)', *Biol Direct*, 7, 2012, p. 23, doi: 10.1186/1745-6150-7-23; Joyce, G. F. and Szostak, J. W., 'Protocells and RNA Self-Replication', *Cold Spring Harb Perspect Biol*, 10(9), 2018, a034801, doi: 10.1101/cshperspect.a034801.
19. Quoted in Isaacson, *The Code Breaker.* Author's interviews with Jack Szostak and Jennifer Doudna; Doudna, Jennifer, 'Towards the Design of an RNA Replicase', PhD thesis (Harvard University, May 1989).
20. Franklin, Sarah, *Biological Relatives: IVF, Stem Cells, and the Future of Kinship* (Durham: Duke University Press, 2013).
21. Quoted in Isaacson, *The Code Breaker*. Author's interviews with Richard Lifton, Jennifer Doudna, Jack Szostak; Greengard Prize citation for Jennifer Doudna, 2 October 2018; Doudna, Jennifer and Szostak, Jack, 'RNA-Catalysed Synthesis of Complementary-Strand RNA', *Nature*, 339, 1989, pp. 519–22; Doudna, J., Couture, S. and Szostak, J., 'A Multisubunit Ribozyme That Is a Catalyst of and Template for Complementary Strand RNA Synthesis', *Science*, 251, 1991, pp. 1605–8; Doudna, J., Usman, N., and Szostak, J., 'Ribozyme-Catalyzed Primer Extension by Trinucleotides', *Biochemistry*, 32, 1993, pp. 2111–5.
22. Marino, M., 'Biography of Jennifer A. Doudna', *Proc Natl Acad Sci U.S.A.*, 101(49), 2004, pp. 16987–9, doi: 10.1073/pnas.0408147101. Epub 2004 Dec 1.
23. Doudna and Szostak, 'RNA-Catalysed Synthesis of Complementary-Strand RNA'.
24. Doudna, Jennifer and Sternberg, Samuel, *A Crack in Creation: Gene Editing and the Unthinkable Power to Control Evolution* (Boston, MA: Houghton Mifflin Harcourt, 2017), p. 45; Isaacson, *The Code Breaker.*
25. Doudna, 'Biographical', Nobel Prize.
26. Isaacson, *The Code Breaker.*
27. Ibid, p. 80.
28. Cate, J. H., Cech, T. R., Doudna, J. A. et al., 'Crystal Structure of a Group I Ribozyme Domain: Principles of RNA Packing', *Science*, 273, 1996, pp. 1678–85.
29. Doudna, 'Biographical', Nobel Prize; Pollack, 'Jennifer Doudna, a Pioneer Who Helped Simplify Gene Editing'.
30. Franklin, *Biological Relatives.*
31. Doudna, 'Biographical', Nobel Prize.
32. In Isaacson, *The Code Breaker.* Author's interviews with Jillian Banfield and Jennifer Doudna; Doudna and Sternberg, *A Crack in Creation*, p. 39; 'Deep Surface Biospheres', Banfield Lab page, Berkeley University website.

33. Doudna, 'Biographical', Nobel Prize; Isaacson, *The Code Breaker*, p. 103.
34. Doudna, 'Biographical', Nobel Prize.
35. Isaacson, *The Code Breaker*, p. 111.
36. Isaacson, *The Code Breaker*. Author's interview with Luciano Marraffini.
37. In Isaacson, *The Code Breaker*. Author's interview with Erik Sontheimer.
38. Doudna, 'Biographical', Nobel Prize.
39. Isaacson, *The Code Breaker*, p. 213.
40. Ibid.
41. Ibid., p. 148.
42. Ibid., p. 149.
43. Deltcheva, Elitza, Chylinski, Krzysztof, Charpentier, Emmanuelle et al., 'CRISPR RNA Maturation by Trans-Encoded Small RNA and Host Factor RNase III', *Nature*, 471, 2011, pp. 602–7; Isaacson, *The Code Breaker*.
44. Isaacson, *The Code Breaker*, p. 149.
45. Quoted in Isaacson, *The Code Breaker*.
46. Ibid., p. 152.
47. Doudna, 'Biographical', Nobel Prize.
48. Jinek, Martin et al., 'A Programmable Dual-RNA–Guided DNA Endonuclease in Adaptive Bacterial Immunity'. *Science*, 337, 2012, pp. 816–21, doi:10.1126/science.1225829.
49. In Isaacson, *The Code Breaker*. Author's interviews with Jennifer Doudna and Martin Jinek.
50. Cohen, Jon, 'How the Battle Lines Over CRISPR Were Drawn', *Science*, https://www.science.org/content/article/how-battle-lines-over-crispr-were-drawn.
51. Ibid.
52. Regalado, Antonio, 'Two Nobel Prize Winners Want to Cancel Their Own CRISPR Patents in Europe', MIT Technology Review, https://www.technologyreview.com/2024/09/25/1104475/nobel-prize-winners-cancel-crispr-patents-europe/.
53. Doudna, 'Biographical', Nobel Prize.
54. Doudna and Sternberg, *A Crack in Creation*, Prologue.
55. Cohen, Jon, 'The Untold Story of the "Circle of Trust" Behind the World's First Gene-Edited Babies', *Science*, https://www.science.org/content/article/untold-story-circle-trust-behind-world-s-first-gene-edited-babies.
56. Cohen, Jon, 'After Last Week's Shock, Scientists Scramble to Prevent More Gene-Edited Babies', *Science*, https://www.science.org/content/article/after-last-weeks-shock-scientists-scramble-prevent-more-gene-edited-babies.
57. Krämer, Katrina, 'How crispr went from niche to Nobel', *Chemistry World*, https://www.chemistryworld.com/features/how-crispr-went-from-niche-to-nobel/4012604.article.

Chapter 21: Timnit Gebru

1. Vinge, Vernor, 'The Coming Technological Singularity: How to Survive in the Post-Human Era', VISION-21 Symposium, March 30–1, 1993, https://edoras.sdsu.edu/~vinge/misc/singularity.html; Mucci, Tim, 'The History of AI', IBM, https://www.ibm.com/think/topics/history-of-artificial-intelligence.
2. 'Timnit Gebru: The Computer Scientist Fighting for a Fairer World', History of Data Science, https://www.historyofdatascience.com/timnit-gebru-the-computer-scientist-fighting-for-a-fairer-world/; Noble, Safiya, 'The 100 Most Influential People of 2022: Timnit Gebru', *Time*, https://time.com/collection/100-most-influential-people-2022/6177822/timnit-gebru/.
3. Noble, 'The 100 Most Influential People of 2022: Timnit Gebru'.
4. Perrigo, Billy, 'Why Timnit Gebru Isn't Waiting for Big Tech to Fix AI's Problems', *Time*, https://time.com/6132399/timnit-gebru-ai-google/.
5. '2023 Great Immigrants: Timnit Gebru, Founder and Executive Director, Distributed AI Research Institute', Carnegie Corporation of New York, https://www.carnegie.org/awards/honoree/timnit-gebru/; Gebru, Timnit, 'A Message From Dr. Timnit Gebru', *The Radical AI Podcast*, https://www.youtube.com/watch?v=-JYFpv63ELE&ab_channel=TheRadicalAIPodcast.
6. Harris, John, 'There Was All Sorts of Toxic Behaviour': Timnit Gebru on Her Sacking by Google, AI's Dangers and Big Tech's Biases, *Guardian*, https://www.theguardian.com/lifeandstyle/2023/may/22/there-was-all-sorts-of-toxic-behaviour-timnit-gebru-on-her-sacking-by-google-ais-dangers-and-big-techs-biases.
7. Perrigo, 'Why Timnit Gebru Isn't Waiting for Big Tech to Fix AI's Problems'

8. Harris, 'There Was All Sorts of Toxic Behaviour'.
9. Perrigo, 'Why Timnit Gebru Isn't Waiting for Big Tech to Fix AI's Problems'.
10. Harris, 'There Was All Sorts of Toxic Behaviour'.
11. Ibid.
12. Black in AI [website], https://www.blackinai.org/.
13. '2023 Great Immigrants: Timnit Gebru', Carnegie Corporation of New York.
14. Chicago Quantum Exchange, Device Physics [webpage], https://chicagoquantum.org/research-areas/device-physics.
15. 'Timnit Gebru: The Computer Scientist Fighting for a Fairer World', History of Data Science.
16. Gebru, Timnit 'An Interview with . . . Timnit Gebru, Founder and Executive Director, the Distributed AI Research Institute', National Academy of Engineering, https://www.nae.edu/317962/An-Interview-with-Timnit-Gebru-founder-and-executive-director-the-Distributed-AI-Research-Institute.
17. Microsoft Azure, What Is Computer Vision? [webpage], https://azure.microsoft.com/en-au/resources/cloud-computing-dictionary/what-is-computer-vision#:~:text=Computer%20vision%20allows%20computers%20to%20interpret%20and,as%20healthcare%2C%20manufacturing%2C%20security%2C%20and%20autonomous%20systems; Hyperscience, What Is Computer Vision? [webpage], https://www.hyperscience.ai/resource/computer-vision/.
18. Harris, 'There Was All Sorts of Toxic Behaviour'.
19. Gebru, T., Krause, J., Wang, Y., Chen, D., Deng, J., Aiden, E. L. and Fei-Fei, L., 'Using Deep Learning and Google Street View to Estimate the Demographic Makeup of Neighborhoods Across the United States', *Proc Natl Acad Sci U.S.A.*, 114(50), 2017, pp. 13108–13, https://doi.org/10.1073/pnas.1700035114.
20. 'Timnit Gebru: The Computer Scientist Fighting for a Fairer World', History of Data Science.
21. Gebru, T. et al., 'Datasheets for Datasets', Proceedings of the 5th Workshop on Fairness, Accountability, and Transparency in Machine Learning, Stockholm, Sweden, 2018. Accessed at: https://www.microsoft.com/en-us/research/uploads/prod/2019/01/1803.09010.pdf.
22. Angwin, Julia, Larson, Jeff, Mattu, Surya and Kirchner, Lauren, 'Machine Bias', ProPublica, https://www.propublica.org/article/machine-bias-risk-assessments-in-criminal-sentencing.
23. Ibid.
24. Ibid.
25. Ibid.
26. Harcourt, Bernard E., 'Risk as a Proxy for Race: The Dangers of Risk Assessment', 27 *Fed Sent'g Rep* 237 (2015), https://scholarship.law.columbia.edu/faculty_scholarship/2564; Angwin et al., 'Machine Bias'.
27. Bergman, Dave and Stryker, Cole, 'What Is Backpropagation?, IBM, https://www.ibm.com/think/topics/backpropagation; Mucci, 'The History of AI'.
28. Perrigo, 'Why Timnit Gebru Isn't Waiting for Big Tech to Fix AI's Problems'.
29. Pennington, Hilary and Gebru, Timnit, 'Making AI Systems More Just With Hilary Pennington and Dr. Timnit Gebru', Ford Foundation', 2023, Accessed at: https://www.fordfoundation.org/news-and-stories/videos/on-what-matters-social-justice-leaders-share-their-big-plans-for-a-more-just-world/making-ai-systems-more-just-with-hilary-pennington-and-dr-timnit-gebru/.
30. Mucci, 'The History of AI'.
31. Else, Holly, 'Nature's 10: Timnit Gebru', *Nature*, 15 December 2021, https://www.nature.com/immersive/d41586-021-03621-0/index.html#section-wEfZ0Jxqu4; Buolamwini, Joy, Gebru, Timnit, 'Gender Shades: Intersectional Accuracy Disparities in Commercial Gender Classification', Proceedings of the 1st Conference on Fairness, Accountability and Transparency, *PMLR* 81, 2018, pp. 77–91; 'Dr. Timnit Gebru: Translating 'Gender Shades' into Corporate Governance', Klover AI, https://www.klover.ai/dr-timnit-gebru-translating-gender-shades-into-corporate-governance/; Gershgorn, Dave, 'How a 2018 Research Paper Led Amazon, Microsoft, and IBM to Curb Their Facial Recognition Programs', OneZero, https://onezero.medium.com/how-a-2018-research-paper-led-to-amazon-and-ibm-curbing-their-facial-recognition-programs-db9d6cb8a420.
32. 'Dr. Timnit Gebru', Klover AI; 'How a 2018 Research Paper', OneZero.
33. Lohr, Steve, 'Facial Recognition Is Accurate if You're a White Guy', *New York Times*, https://www.nytimes.com/2018/02/09/technology/facial-recognition-race-artificial-intelligence.html.

34. Snow, Jacob, 'Amazon's Face Recognition Falsely Matched 28 Members of Congress With Mugshots', ACLU, https://www.aclu.org/news/privacy-technology/amazons-face-recognition-falsely-matched-28; 'How a 2018 Research Paper', OneZero.
35. Hern, Alex, 'BM Quits Facial-Recognition Market Over Police Racial-Profiling Concerns', *Guardian*, https://www.theguardian.com/technology/2020/jun/09/ibm-quits-facial-recognition-market-over-law-enforcement-concerns; 'Dr. Timnit Gebru', Klover AI; 'How a 2018 Research Paper', OneZero.
36. Greene, Jay, 'Microsoft Won't Sell Police Its Facial-Recognition Technology, Following Similar Moves by Amazon and IBM', *Washington Post*, https://www.washingtonpost.com/technology/2020/06/11/microsoft-facial-recognition/; Paul, Kari, 'Amazon to Ban Police Use of Facial Recognition Software for a Year', *Guardian*, https://www.theguardian.com/technology/2020/jun/10/amazon-rekognition-software-police-black-lives-matter; Jakubowska, Ella and Naranjo, Diego, 'Ban Biometric Mass Surveillance', EDRi, https://edri.org/wp-content/uploads/2020/05/Paper-Ban-Biometric-Mass-Surveillance.pdf; 'Dr. Timnit Gebru', Klover AI.
37. Else, 'Nature's 10: Timnit Gebru'.
38. Ibid.
39. Harris, 'There Was All Sorts of Toxic Behaviour'.
40. Else, 'Nature's 10: Timnit Gebru'; Harris, 'There Was All Sorts of Toxic Behaviour'.
41. Bender, Emily M., Gebru, Timnit, McMillan-Major, Angelina and Mitchell, Margaret, 'On the Dangers of Stochastic Parrots: Can Language Models Be Too Big? 🦜' in *Proceedings of the 2021 ACM Conference on Fairness, Accountability, and Transparency (FAccT '21)*, pp. 610–23 (Association for Computing Machinery, New York, NY, USA), https://doi.org/10.1145/3442188.3445922.
42. Hao, Karen, 'We Read the Paper That Forced Timnit Gebru Out of Google. Here's What It Says', MIT Technology Review, https://www.technologyreview.com/2020/12/04/1013294/google-ai-ethics-research-paper-forced-out-timnit-gebru/.
43. Hao, 'We Read the Paper'.
44. Ibid.
45. Harris, 'There Was All Sorts of Toxic Behaviour'.
46. Ibid.; Behal, Anuj, '"In the End, You Feel Blank": India's Female Workers Watching Hours of Abusive Content to Train AI', https://www.theguardian.com/global-development/2026/feb/05/in-the-end-you-feel-blank-indias-female-workers-watching-hours-of-abusive-content-to-train-ai.
47. Gebru, Timnit 'An Interview with . . . Timnit Gebru'.
48. Harris, 'There Was All Sorts of Toxic Behaviour'.
49. Gebru, Timnit 'An Interview with . . . Timnit Gebru'.
50. Tsanni, A., 'How Satellite Images and AI Could Help Fight Spatial Apartheid in South Africa', MIT Technology Review, 19 January 2024, https://www.technologyreview.com/2024/01/19/1086837/satellite-images-ai-spatial-apartheid-south-africa/.
51. Ibid.
52. Gebru, Timnit 'An Interview with . . . Timnit Gebru'.
53. Ibid.
54. Ibid.
55. Ibid.
56. Distributed AI Research Institute, Meet DAIR [webpage], https://www.dair-institute.org/team/.
57. Gebru, Timnit 'An Interview with . . . Timnit Gebru'.

Conclusion

1. Kerridge, Richard, 'Entangled Life by Merlin Sheldrake review – a brilliant 'door opener' book', *Guardian*, https://www.theguardian.com/books/2020/aug/27/entangled-life-by-merlin-sheldrake-review-a-brilliant-door-opener-book.
2. Ibid.
3. Margulis, Lynn, *Microcosmos: Four Billion Years of Microbial Evolution*, (Oakland, CA: University of California Press, 1997).